Handbook of Natural Energy Resources

Handbook of Natural Energy Resources

Editor: David McCartney

R CALLISTO REFERENCE

www.callistoreference.com

Callisto Reference,
118-35 Queens Blvd., Suite 400,
Forest Hills, NY 11375, USA

Visit us on the World Wide Web at:
www.callistoreference.com

ISBN: 978-1-63239-955-7 (Hardback)

Cataloging-in-Publication Data

Handbook of natural energy resources / edited by David McCartney.
      p. cm.
Includes bibliographical references and index.
ISBN 978-1-63239-955-7
1. Power resources. 2. Renewable energy sources.
3. Renewable natural resources. I. McCartney, David.
TJ163.2 .H36 2018
333.79--dc23

Table of Contents

Preface..VII

Chapter 1   **Influence of Solid Concentration on the Flow Characteristics and Settling Rate of Coal-Water Slurries**..1
Kenechukwu Emmanuel Ugwu, Anthony Chibuzo Ofomatah, Samson Ifeanyi Eze

Chapter 2   **Correlation among Vitrinite Reflectance Ro%, Pyrolysis Parameters, and Atomic H/C ratio: Implications for Evaluating Petroleum Potential of Coal and Carbonaceous Materials**..5
Hsien-Tsung Lee, Li-Chung Sun

Chapter 3   **Comparative Analysis of the Fuel Properties of Ethylester Biodiesels from *Cyperus esculentus*, *Sesamum indicum* and *Colocynthus vulgaris* Seed Oils**..........................21
Anekwe Ozioma Juliana, Ajiwe Vincent Ishmael Egbulefu

Chapter 4   **Saccharification of *Ulva lactuca* via *Pseudoalteromonas piscicida* for Biofuel Production**..26
El-Naggar M. M., Abdul-Raouf U. M., Ibrahim H. A. H., El-Sayed W. M. M.

Chapter 5   **A Variable Step Size MPPT Method for Stand-Alone PV Energy Systems**..........................34
Tahar Tafticht, Yamina Azzoug

Chapter 6   **Dye-Sensitized Solar Cells using Natural Dyes Extracted from Roselle (*Hibiscus Sabdariffa*) Flowers and Pawpaw (*Carica Papaya*) Leaves as Sensitizers**..........................39
Danladi Eli, Muhammad Ahmad, Idodo Maxwell, Danladi Ezra, Aungwa Francis, Sunday Sarki

Chapter 7   **Jordanian Oil Shales: Variability, Processing Technologies, and Utilization Options**..44
Hani Muhaisen Alnawafleh, Feras Younis Fraige, Laila Abdullah Al-khatib, Mohammad Khaleel Dweirj

Chapter 8   **Transesterification of Palm Oil to Biodiesel and Optimization of Production Conditions i.e. Methanol, Sodium Hydroxide and Temperature**..........................48
Shaila Siddiqua, Abdullah Al Mamun, Sheikh Md. Enayetul Babar

Chapter 9   **Renewable Energy: A Solution to Hazardous Emissions**..........................55
Ahmed Bilal Awan

Chapter 10   **Foraging Patterns of Birds in Resource Partitioning in Tropical Mixed Dry Deciduous Forest, India**..61
Nirmala Thivyanathan

Chapter 11   **Preparation of Charcoal Pellets from Eucalyptus Wood with Different Binders**..........................75
Alejandro Amaya, Mariana Corengia, Andrés Cuña, Jorge De Vivo, Andrés Sarachik, Nestor Tancredi

Chapter 12   **Pricing for Natural Gas**...**81**
Valentyna Novosad

Chapter 13   **Effect of the Shape Surface of Absorber Plate on Performance of Built-in-Storage Solar Water Heater**...**85**
Omer Khalil Ahmad, Ahmed Hassan Ahmed, Obiad Majeed Ali

Chapter 14   **Natural Gas Pricing in Eastern Europe**...**93**
Valentyna Novosad

Chapter 15   **Water-Pumping using Powered Solar System - More than an Environmentally Alternative: The Case of Toshka, Egypt**...**97**
Ahmed G. Abo-Khalil, Sameh S. Ahmed

Chapter 16   **Volume Models for Single Trees in Tropical Rainforests in Tanzania**.....................................**104**
Abel Malyango Masota, Eliakimu Zahabu, Rogers Ernest Malimbwi, Ole Martin Bollandsås, Tron Haakon Eid

Chapter 17   **Effect of Water Depth and Temperature on the Productivity of a Double Slope Solar Still**..**115**
T. A. Babalola, A. O. Boyo, R. O. Kesinro

Chapter 18   **Enhancing Biomass Energy Efficiency in Rural Households of Ethiopia**.................................**119**
Dagninet Amare, Asmamaw Endeblhatu, Awole Muhabaw

Chapter 19   **Optical Modeling of Thin-Film Amorphous Silicon Solar Cells Deposited on Nano-Textured Glass Substrates**...**126**
Mohammad Ismail Hossain, Wayesh Qarony

Chapter 20   **A Comparison between Statistical Analysis and Grey Model Analysis on Assessed Parameters of Petroleum Potential from Organic Materials**...**132**
Hsien-Tsung Lee

Chapter 21   **Energy Pricing for Households in Europe**...**154**
Novosad Valentyna, Kolosova Viktoria

Chapter 22   **Feasibility and Estimation of Technical Potential and Calculation of Payback Period of Roof-Top Solar PV System in the City of Majmaah, Province of Riyadh, K.S.A**..**158**
Ahmed-Bilal Awan, Taher Shaftichi, Ahmed G. Abu-Khalil

Chapter 23   **Photoelectrochemical Cell based on Natural Pigments and ZnO Nanoparticles**..................**165**
Getachew Yirga, Sisay Tadesse, Teketel Yohannes

Chapter 24   **Biofuel Energy for Mitigation of Climate Change in Ethiopia**...**175**
Abreham Berta, Belay Zerga

Permissions

List of Contributors

Index

Preface

Every book is a source of knowledge and this one is no exception. The idea that led to the conceptualization of this book was the fact that the world is advancing rapidly; which makes it crucial to document the progress in every field. I am aware that a lot of data is already available, yet, there is a lot more to learn. Hence, I accepted the responsibility of editing this book and contributing my knowledge to the community.

Energy that is produced from natural sources is known as sustainable energy. The aim of producing sustainable energy is to meet the demands of our present and future generations without adversely affecting the ecological balance of our planet. Some common forms of natural energy are tidal power, bioenergy, wind energy and geothermal energy. This book strives to provide a fair idea about this discipline and to help develop a better understanding of the latest advances within this field. For all those who are interested in natural energy resources, this book can prove to be an essential guide. The readers would gain knowledge that would broaden their perspective about the subject.

While editing this book, I had multiple visions for it. Then I finally narrowed down to make every chapter a sole standing text explaining a particular topic, so that they can be used independently. However, the umbrella subject sinews them into a common theme. This makes the book a unique platform of knowledge.

I would like to give the major credit of this book to the experts from every corner of the world, who took the time to share their expertise with us. Also, I owe the completion of this book to the never-ending support of my family, who supported me throughout the project.

Editor

Influence of solid concentration on the flow characteristics and settling rate of coal-water slurries

Kenechukwu Emmanuel Ugwu[*]**, Anthony Chibuzo Ofomatah, Samson Ifeanyi Eze**

National Centre for Energy Research and Development, University of Nigeria, Nsukka, Nigeria

Email address

kenecis@yahoo.co.uk(K. E. Ugwu), ofomatony@yahoo.co.uk(A. C. Ofomatah), ezeifeanyi4@yahoo.com(S. I. Eze)

Abstract: Coal–water slurries were prepared with a sub-bituminous coal at varying coal and water ratios using an anionic liquid soap as a surfactant. The slurry properties, viscosity and stability were determined. The rheological properties of the slurries were investigated to ascertain the characteristics of the slurries. The results showed that increasing coal (solid) concentration increased the density of the slurries. Higher density gave rise to more viscous slurries. The behaviour of the slurries changed from Newtonian to non-Newtonian at higher solid concentrations. The slurries with lower solid concentration settled faster than the higher solid concentration slurries.

Keywords: Coal, Concentration, Flow, Viscosity, Slurries, Settling Rate

1. Introduction

Coal is a heterogeneous material, consisting of two–phase systems, the organic and the inorganic constituents [1]. It is a primary fuel, which over the years, was in use in its solid form to generate electricity and heat, both domestically and industrially. It has also been applied in metal and chemical industries. The usage of coal in its solid form limited its applications and availability in comparison to liquid fuels. Production of liquid fuels from coal has been successful through direct liquefaction via hydrogenation and by coal gasification as applied in the Fischer-Tropsch process [2]. Simpler and cheaper methods of coal liquefaction in solvents including oil, organic solvents and water have been developed and applied in countries including USA, Russia and China [3, 4].

Coal-water slurry is the fuel formed from the mixture of coal and water. Additives are usually included to improve stability, homogeneity and dispersability of the mixture [5]. The primary purpose of coal-liquid mixtures is to make solid coal behave as an essentially liquid fuel that can be transported, stored, and burned in a manner similar to heavy fuel oil [6]. Coal-water slurry has been successfully tested and operated in various boilers for power generation and in chemical and metallurgical industries [7].

Important slurry characteristics are stability, pumping, atomizability and combustion characteristics [8]. Numerous studies have been made on the rheological properties of coal-water slurries in an effort to obtain acceptable fluidity while maintaining sufficient stability against sedimentation of the coal particles [9]. Studies on the effects of particle size distribution and packing characteristics [10] as well as effects of the use of additives on the preparation of coal-water slurries have been undertaken [5]. The use of coal in water as a fuel is possible because coal contains carbon with high energy content depending on the type of coal. Coals of different classifications have been used to prepare slurry fuels [11]. The viscosity of a fluid is an important rheological property that affects the pumpability of a fuel and also the atomization of the fuel in the combustion chamber [12, 13].

The main objective of this investigation is to research on the slurrying, rheology(flow characteristics), and stability properties of a Nigerian coal. The rheology was investigated from the study of the viscosity of the slurries under different conditions. The stability was determined from the settling rate experiment.

2. Experimental

The coal used in this investigation was a sub-bituminous coal from Okobo Enjema in Kogi State, Nigeria and was provided by Zuma-828 Coal Limited, a coal mining company developing the Zuma coal mine. The coal was used as received, without any modification. The coal pieces were crushed, followed by grinding and screening through 250 micron sieve. The undersize coal was collected and packaged with plastic bags to prevent oxidation.

Proximate analyses were done to characterize the coal sample according to ASTM standards [14].

2.1. Physical and Chemical Properties Determination

The density of the slurry was determined with a density bottle. The flash point was measured with a Pemsky Martin semi-automatic multi flash closed cup flash point tester (Made in Japan). The viscosity was measured using a NDJ–5s Digital Viscometer (SearchTech Instruments, England).

2.2. Coal Slurries Preparation

The coal particles that passed through the mesh size sieve 60 (250 microns) were used to prepare the coal slurries with deionized water and an anionic liquid soap (surfactant). The coal was dried in the oven at 105°C for 2 hours and then cooled in a dessicator. The necessary amount of deionized water and liquid soap were added into a 250mL volume beaker. The water and liquid soap mixture were put in a 500 mL plastic jar of the mixer and agitated for 2 minutes, followed by the addition of the pulverized coal. The coal/water/liquid soap mixture was agitated for 10 minutes in the mixer at 1200 rpm to ensure the homogenization of the slurries. Coal slurries with solid concentrations of 30%, 40%, 50%, and 55% (v/v) were prepared. The total volume of the coal-water slurry in the jar was kept constant at 250mL in all the experiments to standardize the mixing. The amount and type of liquid soap was kept constant at 1% on the basis of the weight of dry coal. At the end of the mixing, the measurements were taken.

2.3. Property Measurements

Rheological behaviour was investigated with a rotational viscometer. The viscosities of the different solid content slurries were measured at rotational speed of 6 rpm, 12 rpm, 30 rpm and 60 rpm at room temperature (27°C) to understand whether the mixtures were Newtonian or Non-Newtonian; shear-thinning or shear-thickening. These were done with appropriate spindles. Also, at constant shear rate of 30 rpm, the viscosities of the different solid content slurries were measured and results were taken at 30 seconds intervals. These data showed the character of fluid. Another experiment was conducted measuring the viscosities of the 40% coal concentration slurry as the temperature was varied and the Viscometer was in a circulating water bath using Spindle No. 3 at 30 rpm.

The stability characteristic of coal slurries was determined by a settling test in a 50-mL graduated cylinder in which slurry of particular concentration is poured into the cylinder and allowed to settle under gravity after agitation. The height of the supernatant liquid was determined over a period of 6 days.

This study was conducted at the laboratory of the National Center for Energy Research and development, University of Nigeria, Nsukka between October, 2012 and April, 2013.

3. Results and Discussion

The values presented are averages of several readings per sample and of triplicate samples.

Table 1 show the results of the proximate analyses for the coal which was from a single source. It was a sub-sub-bituminous coal as the fixed carbon content was 46.55%.

Table 2 gives the result of some properties of the slurries. The density increased as the concentration of coal in the slurries increased. Similarly, the viscosities increased with increase in density for room temperature measurements conducted at 30 rpm.

Table 1. *Proximate Analyses of Okobo, Nigeria coal sample.*

Proximate Analysis (As Received)	
Moisture (wt %)	8.17
Ash (wt %)	9.35
Volatile Matter (wt %)	35.93
Fixed Carbon (wt %)	46.55
Total	100.00

Table 2. *Summary of some Coal Water Fuel (CWF) Properties*

CWF	Density	Flash Point Viscosity	(%Coal)
30%wt	0.4766	No Flash	29.84
40%wt	0.5344	No Flash	200.7
50%wt	0.6985	No Flash	724.6
55%wt	0.6213	No Flash	914.8

3.1. Rheological Properties of the Slurries

Viscosity is the most important rheological characteristics of Coal-water slurries [15]. The viscosity was evaluated as a function of shear rate, time and temperature, respectively. The rheological characteristics of the slurries were determined from the view of whether they were Newtonian or non-Newtonian fluid. Figure 1, which was a plot viscosity vs. shear rate, confirmed the character of fluid at coal concentrations over 30% as non-Newtonian. The viscosity of the fluids changed as the shear rate was varied. This behaviour may be attributed to the higher particle interactions at low shear rates. It is possible that higher shear rates will break the interactions and closeness thereby reducing the viscosity [16]. At 30% coal concentration and using 30 rpm and 60 rpm, the fluid was Newtonian.

Figure 2 showed that shear stress decreased with shear rate at above 40%wt coal. This is called pseudoplastic or shear-thinning behaviour. The behaviour of the slurry at 30% coal concentration and below could be like a Bingham fluid. The slurry characteristics changed as the coal concentration was over 40% wt.

From Figure 3, which was a plot of viscosity with time at constant shear rate of 30 rpm, the slurries exhibited a thixotropic behaviour with 40%wt and 50%wt at 30 rpm. The fluid at 55%wt exhibited rheopectic behaviour after 200s. At 30% coal concentration, the viscosity of the fluid remained unchanged over time.

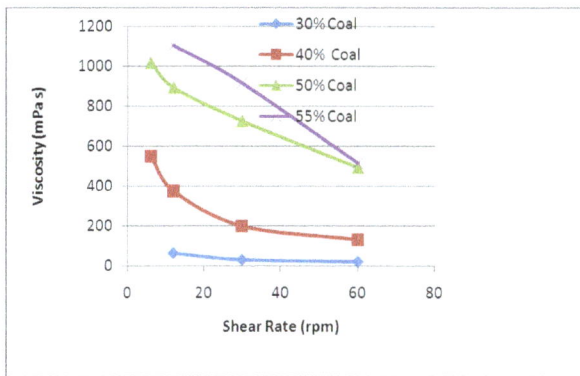

Figure 1. Viscosity as a function of shear rate for coal slurries at room temperature (27 °C).

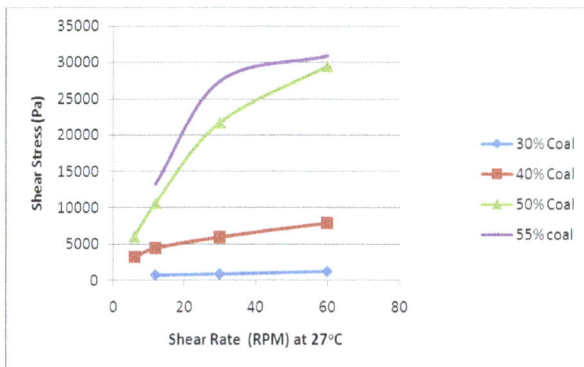

Figure 2. Shear Stress Vs Shear Rate

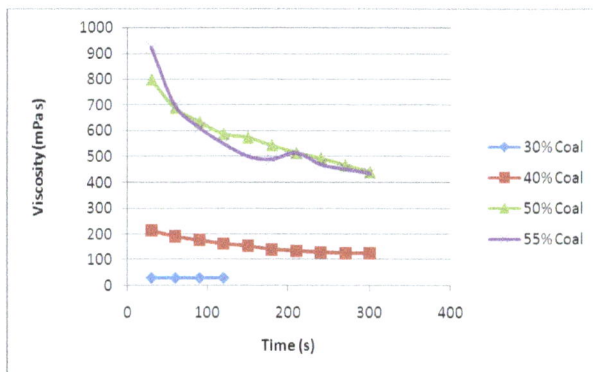

Figure 3. Plot of Viscosity vs Time at 30 RPM

Figure 4 is the viscosity-temperature relationship for 40% w/w coal-water slurry. In this figure, the viscosity of the fluid decreased as temperature increased. When the temperature is less than 333 K, the viscosity slowly decreased with increase in temperature. Above 333 K, there is a sharp decrease in viscosity, indicating flocculation.

The plot of the Interface Height Vs Time is given in Figure 5. As the storage time increased, the coal particles began to settle. The 30% coal concentration settled fastest, followed by the 40% coal concentration and finally the 50% coal concentration was the last to settle. The 55% coal concentration was a thick gel and the experiment could not

be conducted with it at room temperature. This behaviour may be accounted for by the inter-particle distance which is supposedly further with the lower concentration slurry, thereby allowing a faster motion of each particle since they move unhindered. The implication is that the 30% coal concentration slurry is easier to pour followed by the 40% coal concentration slurry and then the 50% coal slurry.

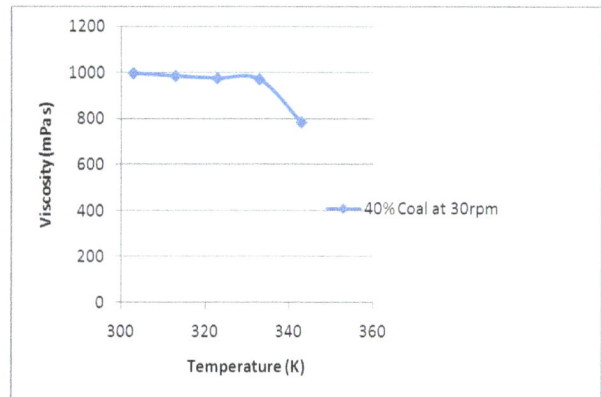

Figure 4. Viscosity temperature relationship for coal-water slurry (40% w/w)

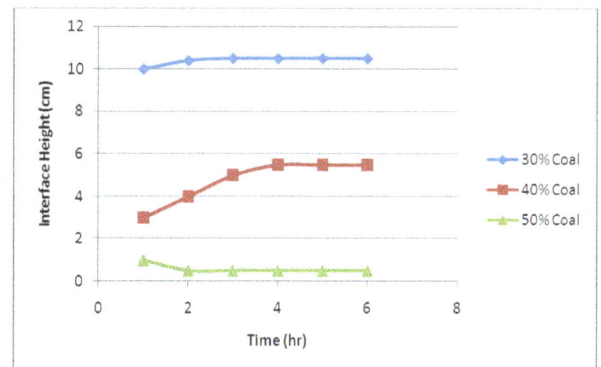

Figure 5. Interface Height Vs Time

These findings on the flow characteristics and settling rate of the coal-water slurries prepared from the studied coal sample provided vital information that would be useful for designing of transport system (pumping and pipeline), storage and the utilization of coal-water slurries as an energy source [15]. The stability of coal-water slurry is important in the pumping and storage of the fuel [17].

Slurries with less than 40% coal concentration were less viscous and more stable. These slurries may be more suitable for transportation via pipelines. The viscosity level of slurry affects the application for the slurry. Other factors like particle size distribution and additives are important to obtain slurries to meet specific requirements [18, 19].

4. Conclusion

Coal slurries were found to exhibit different flow characteristics, from Newtonian at low solid coal concentrations to shear-thinning at higher solid concentrations. The behaviour of the coal slurries at 40%, 50%

and 55% coal concentrations showed that the viscosities changed with time while it did not change at 30% coal concentration. The viscosities decreased with increase in temperature. The slurry with 30% coal concentration settled faster than the slurries with higher coal concentration. This result is important in the designing of pumps for transportation of coal slurries and also for the use of coal slurries in combustion systems. The result will be useful in making decisions on the storage of coal slurries.

Acknowledgement

The authors hereby acknowledge the NCERD, University of Nigeria, Nsukka, for supporting us through this project.

References

[1] H. H. Schobert, K.D. Bartle and L.J. Lynch, Coal Science, Symposium Series 461 , Washington, DC: American Chemical Society, 1991.

[2] A. Steynberg. and M. Dry (eds), Fischer-Tropsch Technology, Studies in Surface Science and Catalysis, Philadelphia: Elsevier, 2004.

[3] Z. Aktas and E.T. Woodburn, "Effect of addition of surface active agent on the viscosity of a high concentration slurry of a low rank British coal in water", Fuel Processing Technology, vol. 62, pp.1-15, 2000.

[4] G. O. Davies (eds), Coal Liquid Mixtures, Second European Conference on Coal Liquid Mixtures , London: The Institution of Chemical Engineers symposium series; No. 95, 1985.

[5] J. Natoli, R.C. Mahar and R. Barret and B.R. Bosein." Polyacrylate thickeners for coal water slurries: Slurry formulation, stability and rheology", in: Coal Liquid Mixtures, G.O.Davies (eds.), Second European Conference on Coal Liquid Mixtures, London: The Institution of Chemical Engineers symposium series; No. 95, 1985, pp. 17-36.

[6] A.P. Burdukov, V.I. Popov, V.G. Tomilov and V.D. Fedosenko, "The rheodynamics and combustion of coal-water mixtures," Fuel, vol.81, pp.927-933, 2002.

[7] K. Thambimuthu, Developments in Coal-Liquid Mixtures, London: IEA Coal Research, 1994.

[8] V. Kalpesh. And D. Shyam, "Review of Charcoal-Diesel slurry: An Alternative fuel for Compression Ignition Engine", International Journal of Advanced Engineering Research and Studies, vol. 1(111), pp.143-147, 2012.

[9] G. Atesok, F. Boylu, A.A. Sirkeci and H. Dincer, "The effect of coal properties on the viscosity of coal-water slurries", Fuel, vol. 81, pp.1855-1858, 2002.

[10] P. Buranasrisak and M.H. Narasingha, "Effects of Particle Size Distribution and Packing Characteristics on the Preparation of Highly-Loaded Coal-Water Slurry," International Journal of Chemical Engineering and Applications, vol. 3(1), pp. 31-35, 2012.

[11] Q.D. Nguyen, C. Logos and T. Semmler, "Rheological properties of South Australian Coal-Water Slurries, Coal Preparation," vol. 18(3), pp.185-199, 1997.

[12] C. Hicyılmaz, S. Ozun,and N.E. Altun, "Rheological Properties of Asphaltite-Water Slurries", Energy and Fuels, vol. 20, pp.2037-2045, 2006.

[13] Z. Wang, R. Zhang, Z. Jiang. and S. Jiang, " 1993. Preparation of Coal-Water Fuels from Coal Preparation Plant Fines", in J.J. Davis (ed.), Proceedings of the Sixth Australian Coal Preparation Conference, pp. 418-427, 1993

[14] J.G. Speight, Handbook of Coal Analysis. New Jersey, USA: John Wiley & Sons, Inc., 2005.

[15] Y. Shin and Y. Shen, "Preparation of coal slurry with organic solvents", Chemosphere, vol. 68, pp. 389-393, 2007.

[16] G. A. Nunez, M. I. Briceno, D.J. Daniel and T. Asa, "Colloidal coal in water suspensions", Energy Environ. Sci., vol. 3, pp.629-640, 2010.

[17] P. Li, D. Yang, H. Lou and X. Qiu, "Study on the stability of coal water slurry using dispersion-stability analyzer", J Fuel Chem Technol, vol. 36(5), pp. 524−529, 2008.

[18] R. Awang and C.Y. May, "Charcoal-Oil Mixtures as an Alternative Fuel: A Preliminary Study", American Journal of Applied Sciences, vol. 6 (3), pp. 393-395, 2009.

[19] R. Yavuz and S. Kucukbayrak, "Effect of particle size distribution on rheology of lignite-water slurry", Energy Sources, vol. 20(9), pp.787-794, 1998.

Correlation among vitrinite reflectance Ro%, pyrolysis parameters, and atomic H/C ratio: Implications for evaluating petroleum potential of coal and carbonaceous materials

Hsien-Tsung Lee[*], Li-Chung Sun

Department of Electrical and Information Technology, Nan Kai University of Technology, Nan Tou County, Taiwan

Email address:

t114@nkut.edu.tw (Hsien-Tsung Lee), sunlc@nkut.edu.tw (Li-Chung Sun)

Abstract: In this study, 26 samples from northwest Taiwan, 12 from Mainland China, 13 from Australia and 39 from literature were jointly examined to explore relationships among pyrolysis parameters, Vitrinite Reflectance Ro%, and Atomic H/C ratio. Samples of mixed high and low maturity coal were combined in proportions determined by the total quantity in the furnace prior to the Rock-Eval analysis and used to explore the correlation between the pyrolysis parameter, Tmax, and the vitrinite reflectance. These average values were then plotted against the corresponding Tmax results. The experimental results revealed that:(1) For low maturity coal samples that were mixed with different proportions of high maturity coal samples, the Tmax values fell within a range of low maturities. Alternativly, for samples containing the reworked sedimentary materials in the rock formation, the Tmax values were similar to the maturity of young material. (2) For sampling or Rock-Eval analysis of the high maturity materials, contamination with low maturity material should be avoided, even in very small amounts. (3) Afterproportional mixing, there was no evidence of a general linear relationship between the average of vitrinite reflectance, Ro%, and the corresponding Tmax value recorded. The atomic H/C ratio, as well as the BI, HI, QI, S1, and S2, generally decreases while the maturity (Tmax ($^\circ$C); vitrinite reflectance Ro%) increases. The atomic H/C ratio decreases slightly from 1.1 to 0.7 while maturity increased from Ro 0.55% to 0.85%. Samples with atomic H/C ratio within this range show significant change in certain other geochemical parameters (eg. BI, HI, QI, PI, S1, S2, S1+S2, Tmax). Organic matter in the samples studied is of type II/III kerogen based on the relationship between HI and Tmax. The hydrocarbon potential per unit organic carbon (S1+S2/TOC) of the organic matter in this study to be approximately 100~380, similar to the potential of humic coal used in general gas and oil production. This shows that organic matter in an oil window of Ro%=0.55 and atomic H/C=1.1 have reached a certain maturity and hydrocarbon potential. Overall, when the atomic H/C ratio increases, the BI, HI, QI, S1, and S2 also show an increasing trend; therefore, these parameters and atomic H/C ratio show a certain correlation.

Keywords: Atomic H/C Ratio, Vitrinite Reflectance Ro%, Kerogen, Rock-Eval Pyrolysis Tmax, Maceral Analysis, Reworked

1. Introduction

Despite the variety of forms in which organic matter evolves, the most important key in hydrocarbon production is maturity. The process of involves physical and chemical factors such as temperature, pressure, and time, but among them temperature plays the leading role. Bostick (1971, 1974) pointed out that organic matter in sedimentary strata can be used as the thermal index for contact or burial metamorphism, where the organic matter are heated (burial or tectonic movement) decomposed and then recombined. This process is irreversible and not affected by retrograde metamorphism; it can therefore record the most intensive thermal event in burial history (Tsai, 1988; Lee et al., 2010). Under the heating process, organic matter undergo two opposite reactions at the same time, i.e. condensation of molecular structure and partial bond breakage and degradation. Through time, the side chains and miscellaneous atomic groups in the kerogens break off, while the aromaticity increases due to condensation of aromatic entities and formation of polycyclic aromatic hydrocarbons and the orientation increases, leading to optical changes. In addition, the atomic H/C ratio and the maturity of organic matter changes, thus affecting hydrocarbon potential. In the evaluation of oil and gas potential, organic maturity is one of the most important criteria, and the assessment of organic maturity requires various techniques including fluorescence spectroscopy, HI, Tmax, Ro% and atomic H/C ratio. High atomic H/C ratio is observed since kerogens contain silicate minerals that cannot be removed in the kerogen separation procedure (Baskin, 1997). With regard to this problem, we referred to Dahl *et al.* (2004) in which TOC is graphed against S2 to calculate the Y-intercept and thus find the point where non-activated carbon content produces no pyrolyzed hydrocarbons; the H% (Y-axis) and C% (X-axis) obtained from the elemental analysis are graphed to find the Y-intercept (representing the hydrogen content when the organic carbon content is zero) to correct the H% value.

Sun et al (2001) used the Gini coefficient to calculate the distribution of the irvitrinite reflectance (Ro%) measurements. This allowed them to obtain quantified values of frequency distribution in order to illustrate the average vitrinite reflectance value of the measured samples and provide quantitative judgments on the deviation of the entire data set. Although this method transforms the vitrinite reflectance measurement into a frequency distribution histogram via numerical conversion, thus highlighting the gap between the average states, a number of problems may still occur. For instance, the method can cause overly broad vitrinite reflectance frequency distribution due to problems that may be caused by the individual who performs the measurement, or by poor functioning of the instrument itself. Additionally, the composition of the sample itself may cause problems, particularly if the sample has a range of different maturities, or if the sample itself is mixed with material of different maturities, such as reworked sedimentary material. The vitrinite reflectance measurement was also used by Kuo (1997) to explore oil and gas potential in western Taiwan. This study suggested a vitrinite reflectance Ro% frequency distribution of many samples that is larger at the two end points can indicate that the research sample is reworked sedimentary material, or that the sample has reworking environment significance. Additional problems with vitrinite reflectance (Ro%) measurements may also been countered when examining reworked sedimentary material using organic maturity indicators. In this situation it is easy to produce a bandwidth phenomenon. These situations raise the questions of; would using the good correlation between vitrinite reflectance and pyrolysis parameter Tmax (Tissot and Welte, 1984; Espitalié et al,1985) be able to deal with the bandwidth phenomenon? And what would the pyrolysis parameter Tmax be in the case of reworked sedimentary material samples?

A vitrinite reflectance (Ro %) value within some effective range is often regarded as a good index to evaluate organic maturity (Tissot, 1984; Hunt, 1996). Similarly, the pyrolysis parameter, Tmax, is another commonly used index for evaluation (Tissot,1984; Wu Liyan, 1986). Technically, both indices obtain the value of the sample maturity differently. Vitrinite reflectance is obtained by using physical optics measurements; it is mainly restricted to the identification of the material morphology by the observer. Although using the vitrinite reflectance frequency histogram can make up for some of the restrictions of the technique; when the vitrinite reflectance frequency distribution is overly broad, this problem persists and may incorrectly indicate that the sample itself is mixed with different maturity material. For instance, a past study used vitrinite reflectance values to explore the oil and gas potential of Western Taiwan. The study suggested that samples could be identified as reworked sedimentary material, or material with reworking environmental significance, if the frequency distribution of the vitrinite reflectance Ro% values was large at the two end points (Kuo Chenglung, 1997). On the other hand, the Rock-eval pyrolysis parameter, Tmax, is obtained using chemical heating analysis. It is an evaluation method developed by the French Institute of Petroleum (IFP), (Wu Liyan 1986). The biggest advantage of this technique is that it provides rapid data analysis; however the method can sometimes be restricted by the sensitivity of the analysis instrument.

Reworked sedimentary material refers to substances that are formed from sediment, fossil, rock fragments or other substances (Bates and Jackson, 1987). It is formed when materials of old strata deposit on younger strata. Vitrinite reflectance Ro% values tend to increase with increasing strata depth (Stach et al., 1982), thus making it a good index for the assessment of organic maturity (Tissot, 1984). Generally, for reworked sedimentary material, it is believed that when the Ro% value enters the oil window, which ranges from 0.65 to 1.35 (Hunt, 1996), the derived average of vitrinite reflectance Ro% does not accurately reflect the true state of organic maturity. Alternatively, the pyrolysis parameter, Tmax, is the recorded temperature of the S2peak value during the pyrolysis of hydrocarbons in the Rock-Eval instrument occurs (Kuo Chenglung, 1997; Walples, 1985). The hydrocarbon pyrolysis (S2) of reworked sedimentary material should reflect feature with at least two peaks, at this point, how will the pyrolysis parameters Tmax evolve variations? Past studies have shown that the vitrinite reflectance Ro% and the pyrolysis parameter, Tmax, of type-III kerogen have a very strong linear correlation (Teichmüller and Durand, 1983; Lang Dongsheng, 1999; Hou Dujie and Zhang Linye, 2003); however, the nature of this correlation is unknown for reworked sedimentary materials. This makes it difficult to conduct Rock-Eval analysis on affirmed reworked sedimentary material. This study aimed to clarify this relationship by adopting a simulation experiment. We used coal samples(which are enriched in organic matter) as a base sample, proportionally mixing coal samples with larger high-value and low-value differences in Ro%. We then subjected the mixed sample to Rock-Eval pyrolysis analysis in order to compare the differences with the original samples.

Obtaining geochemical data by heating a small amount of rock powder through pyrolysis process is an important geochemical tool in oil exploration. Nevertheless, the rock itself contains inorganic mineral matrices and the oil pollutants generated in the sampling process will reduce the hydrogen index (HI) value (Udo et al., 1986; Dahl et al., 2004) and affect the assessment for kerogen type and quality (Akinlua et al., 2005; Lee, 2010; Lee, 2011). As a result, when assessing the hydrocarbon potential of kerogens, one should not rely solely on the Rock-Eval pyrolysis parameters from pyrolysis analysis but also graph the atomic H/C ratio, TOC, and Ro% against one another to confirm and double check the results for accurate assessment. In terms of oil and gas potential, vitrinites are gas-prone, exinites are oil-prone, while inertinites have no oil and gas potential (Tissot&Welte, 1984). Stach et al. (1982) believed that vitrinite reflectance increases with the degree of heating because molecules move closer to each other when heated causing a higher reflectance. Rock-Eval pyrolysis is a fast, inexpensive analytical technique that provides useful data but requires only a small number of samples. This advantage allows the technology to be widely used in the analysis of organic matter samples. The pyrolysis results of Li et al. (2003) from the Dongying Depression in Bohai Bay Basin showed that with increased depth, Tmax and PI also exhibited an increase, while TOC and HI showed a decrease. Here, we focused mainly on lithofacies analyses (maceral analysis and measurement of vitrinite reflectance) and geochemical analyses (elemental analysis and Rock-Eval pyrolysis) for organic matter to understand their thermal maturity and assess their hydrocarbon potential. Subsequently, through exploring the relations between the relevant parameters (S1+S2, TOC, HI, atomic H/C ratio, Ro%, Tmax), we tried to shed light on their connections to hydrocarbon potential.

2. Samples and Methods

The coal samples for investigation included 26 from northwestern Taiwan, i.e. coal and carbonaceous shale samples from Nanchuang Formation, Shiti Formation, and Mushan Formation. The coal samples were mainly from beds formed through three sedimentary cycles in the Miocene, while the sedimentary environment for intercalated coal seams is of the regression type (Mao et al., 1988). On the other hand, 13 rock samples were collected from Australia, 12 from Mainland China, including samples from the Paleozoic, Mesozoic and Tertiary Cenozoic eras (Ge et al., 1993), and some from the Uinta and Permian Basins of the United States. According to the field study of Taylor &Ritts (2004) at the Green River Formation of Uinta Basin, Utah, the Green River Formation covers Flagstaff, comprising a continuous lake sediment. Lastly, by taking 39 sample data from literature, we studied how the variations of atomic H/C ratio of kerogens and pyrolysis parameters can be used to elucidate the impact of changes in atomic H/C ratio on the hydrocarbon potential of organic matter.

The material study starts from the collection of coal samples with maturity close to the early oil window, various

maceral groups were then prepared by using density centrifuge-separation. Coal briquettes were prepared according to the standard production procedure set forth in ASTM (1975). The coal samples were first crushed and filtered through a No.20 sieve, then mixed with glue and placed in a round steel mold to dry and harden. The hardened and dried briquettes were then polished (Ting, 1978; Sun, 2000). In accordance with the work of Su (2001), Sun (2000), and Dormans et al. (1957); zinc chloride was used as a density liquid for centrifuge-separations. Density of less than 1.25 denotes a concentrate of exinite, and density of 1.25–1.35 denotes a concentrate of vitrinite .Maceral analysis and vitrinite reflectance measurements were performed on polished pellets using a Leitz MPV Compact Microscope (light source 12 V, 100 W; wavelength 546 nm; refractive index of soak oil, Ne = 1.5180). Optical microscope was used to identify three main maceral groups (exinite, vitrinite, andinertinite), as well as inorganic minerals (pyrite and clay minerals) through point counting. According to ASTM standard (1980) and Bustin (1991), each sample was made into two pellets, and each pellet requires 200 point counts; a mean value should also be calculated. In addition, 100 random points (Rstd%) were measured in each pellet, and the mean value of the 100 points was calculated. Therefore, two mean values of the same sample were averaged to obtain the vitrinite reflectance (Ro%) value.

Geochemical analysis using elemental analysis (Elemental analyzer of Germany Heraeus Vario EL model, Accuracy ± 0.1%, Precision ±0.2%) to determine C, H, N, and S contents. The mineral of carbonate and silica-gel can be identified microscopically and removed by additional hydrochloric and hydrofluoric acid treatments. Samples were first transformed into the irgaseous forms (N_2, H_2O, CO_2, and SO_2), and then the segases (H_2O, CO_2, and SO_2) were adsorbed/desorbed through individual adsorption and desorption tubes. Finally, the weight percentage of C, H, N, and S wasobtained through a detection system. Rock-Eval pyrolysisis a standard pyrolysis method for assessing source rocks (Espitalie et al. 1977). In this study, Rock-Eval II was used for the pyrolysis analysis, and FID (flame ionization detector, Calibration sample IFP 55000, S_2 = 8.62 mgHC/g rock, S_3 = 1.00 mg CO_2/g rock, TOC = 2.86%,Tmax = 419°C) was used in the process to detect volatilized hydrocarbons (S_1) and organic matters from pyrolyzed hydrocarbons (S_2), as shown in Table 1.Organic matter have the same maturity but different hydrocarbon potentials, the maceral compositions were compared with the results of their pyrolysis and elemental analysis, so as to study the mechanism of hydrocarbon generation.

The content percentage of the three macerals (exinite, vitrinite and inertinite) from the maceral analysis was then used to estimate the atomic H/C ratios of the collected samples (Jones and Edison, 1978). The hydrogen and carbon content obtained from the elemental analysis could also be used to calculate the atomic H/C ratios. Furthermore, the hydrocarbon potential (S1+S2) and TOC values from the pyrolysis analysis could be used to calculate the hydrocarbon potential per unit organic carbon (S1+S2/TOC). By graphing H% (Y-axis) against C% (X-axis) from the elemental analysis, the obtained Y-intercept (representing the hydrogen content when the organic carbon content is zero) could be used to correct the H% value, as shown in Table 2. Next, by using the atomic H/C ratio, the pyrolysis parameters HI, S1, S2, S1+S2,S1/(S1+S2), S1/TOC, (S1+S2)/TOC, and Tmax were graphed against TOC And Ro% to explore the relations between atomic H/C ratios and the various pyrolysis parameters.

The mixed samples were prepared according to the weight ratio using coal sample swith high and low maturities from Mainland China (Jiao-Zuo Coal: Ro% of 3.34 ~3.77;Fu-Xin Coal: Ro% of 0.58 ~ 0.65). These samples were then subjected to Rock-Eval pyrolysis analysis. The analysis of the samples was performed by the Exploration & Development Research Institute of the CPC Corporation, Taiwan. 10mg of each mixed sample was inserted into the furnace. The mixed proportions of the samples were based on the sample number and proportion, as shown in Table 3.

This study used simulation experimentation. We used coal samples, which are organic matter enriched material, as the basis of the sample. The two types of coal samples used as simulation materials were Jiao-Zuo Coal (Ro% average value = 3.43)and Fu-Xin coal (Ro% average value = 0.61), both sourced from Mainland China. In accordance with the mixing proportion, the mixed samples were subjected to Rock-Eval pyrolysis analysis and compared with the original samples. At the same time, the vitrinite reflectance data points of these two coals with the same mixing proportion and random sampling were averaged (50 points each) after blending; for each mixed sample, average readings were taken as representative after 1,000 times measurements and the results were plotted against the corresponding Tmax results.The conductance of the Rock-Eval analysis and vitrinite reflectance measurements of the samples were entrusted to the Research Institute of Petroleum Exploration and Development (RIPED), CNPC. The sample numbers are as follows: 1. (H = Jiao-Zuo Coal original sample); 2. (H95: L5 mixed samples); 3. (H90: L10 mixed samples); 4.(H80: L20 mixed samples); 5 (H70: L30 mixed samples); 6. (L = Fu-Xin Coal original sample).

Table 1. The parameters when atomic H/C ratio=0.7~1.1[1]

Parameter[5]	All[2]	Taiwan[3]	Mainland China[3]	Australia[3]	Others[4]
Ro%	0.72~0.26	0.95~0.35	0.8~0.5	1.4(1 sample)	0.72~0.4
HI (mg/g)	25~445	75~340	52~165	50(1 sample)	25~445
S1+S2 (mg/g)	20~235	45~235	60~145	-----	20~220
BI (mg/g)	0.01~14	2~5	1~3	-----	0.01~14
QI (mg/g)	35~415	70~375	70~170	80(1 sample)	35~415
PI (mg/g)	0.005~0.072	0.03~0.06	0.01~0.035	0.36(1 sample)	0.005~0.072
T$_{max}$(°C)	395~510	414~480	418~435	418(1 sample)	395~510
TOC (%)	0.26~80	45~80	62~73	0.36(1 sample)	0.26~76
S2 (mg/g)	0.12~242	50~230	20~150	0.12(1 sample)	0.15~220
S1 (mg/g)	0.01~15	2~4	1~2.4	0.01(1 sample)	0.03~15

Note 1.Considering all samples and graphing H/C ratio against each parameter, the result shows that each parameter exhibits significant variations when H/C=0.7~1.1.

2."All" includes all samples from Taiwan, Mainland China, Australia and other reference sample data.

3."Taiwan" denotes the sample data from Taiwan; "Mainland China" denotes that from Mainland China; "Australia" denotes that from Australia.

4."Other" refers to the use of sample data from other references (Wang, 1998; Xiao et al., 1996; Xiao, 1997).

5. HI: S2 / TOC; QI: (S1+S2) / TOC; BI: S1 / TOC; PI: S1 / S1+S2

Table 2. The Ro, H/C ratio, HI, QI, TOC,V, E, and (S1+S2) values of samples analyzed in this study

Sample No.	S1+S2 (mg/g)	TOC (wt%)	QI (mg/g)	HI (mg/g)	V (%)	E (%)	H/C(1)	H/C(2)	Tmax (°C)	Ro (%)
A0	1.2	0.7	159.2	118.1	1.0	0.0	0.25	0.18	437	1.63
A1	0.3	0.4	67.6	40.2	1.0	0.0	0.26	0.25	407	1.61
A2	0.3	0.4	70.2	45.4	0.4	0.2	0.70	0.63	416	1.40
A3	3.2	1.7	184.1	168.3	2.4	0.2	2.54	2.15	434	1.14
A4	5.1	4.2	377.1	285.2	20.8	0.4	1.85	1.63	426	1.23
A5	4.1	1.9	213.3	199.1	1.8	1.8	2.11	1.82	433	1.02
A6	3.2	1.7	182.5	167.2	1.6	0.6	1.52	1.46	434	0.77
A7	3.6	3.8	282.2	220.5	18.8	0.4	1.48	1.34	426	0.86
A8	2.8	18.3	16.3	14.4	61.4	0.6	1.94	1.67	433	1.08
A9	1.2	2.0	59.6	50.6	4.6	1.0	1.75	1.72	432	0.84
01	112.6	75.2	149.8	148.1	60.6	5.0	0.78	0.73	434	0.74
02	84.2	62.6	134.5	133.4	95.0	1.0	1.04	0.83	431	0.53
03	140.3	67.7	207.2	205.3	68.8	2.4	0.89	0.62	429	0.58
14	81.1	44.3	183.2	177.1	53.2	1.2	1.18	0.47	430	0.39
15	37.7	18.6	202.9	196.4	30.2	1.4	1.36	0.28	425	0.52
16	68.3	43.2	158.0	155.2	66.2	0.6	1.08	0.57	427	0.52
17	1.6	2.2	73.7	70.5	3.8	0.0	1.86	0.03	433	0.36
18	177.6	79.2	224.2	221.2	93.2	1.2	0.90	0.81	437	0.74

Sample No.	S1+S2 (mg/g)	TOC (wt%)	QI (mg/g)	HI (mg/g)	V (%)	E (%)	H/C(1)	H/C(2)	Tmax (°C)	Ro (%)
19	214.9	79.3	271.1	286.1	84.0	11.4	0.90	0.86	435	0.68
21	152.6	79.8	191.2	189.1	93.2	3.0	0.80	0.84	437	0.75
22	104.6	64.1	163.1	161.0	81.6	3.6	0.85	0.76	441	0.73
23	32.6	58.1	56.2	54.1	95.4	0.4	1.07	0.82	377	0.23
24	227.3	71.2	319.4	314.7	33.2	1.2	1.10	0.31	431	0.36
25	8.4	1.5	545.7	512.0	11.6	0.2	0.16	0.11	440	0.31
26	155.0	28.6	541.0	511.5	12.8	0.0	1.36	0.12	428	0.28
27	164.1	35.5	462.3	437.5	0.6	16.8	1.44	0.22	430	0.21
28	152.7	29.0	526.4	498.5	4.0	10.2	1.43	0.17	428	0.18
29	70.2	66.6	105.3	104.2	87.0	4.6	0.84	0.8	429	0.54
30	124.3	71.9	172.8	171.2	88.2	9.4	0.96	0.88	429	0.59
31	155.5	45.9	338.7	334.1	80.0	10.8	1.18	0.86	450	0.58
32	354.3	77.6	456.3	454.0	69.4	24.8	1.27	0.93	453	0.39
33	159.2	44.0	361.4	354.4	6.6	1.8	1.38	0.06	444	0.45
34	114.6	72.8	157.3	155.4	82.6	15.4	0.89	0.91	429	0.68
35	93.5	74.4	125.7	124.0	71.4	4.8	0.74	0.78	432	0.59
36	17.5	73.4	23.8	23.0	78.0	2.6	0.70	0.74	445	1.24
37	54.8	71.1	77.0	74.2	61.2	6.8	0.77	0.77	432	0.72
38	17.9	7.3	245.6	239.0	1.4	0.8	6.68	0.02	426	0.17
39	0.4	82.2	0.5	0.1	–	–	–	–	488	4.60
40	71.2	68.9	103.4	101.7	88.0	2.8	0.83	0.81	430	0.54
41	49.6	55.9	88.8	85.5	80.0	5.6	0.77	0.75	477	1.57
42	45.3	50.4	90.0	87.1	73.0	3.4	0.68	0.67	479	1.56

Note 1. HI: S2 / TOC; QI: (S1+S2) / TOC; V: Vitrinite; E: Exinite

2. H/C(1): H/C atomic ratio (Elemental analysis); H/C(2): H/C atomic ratio (Maceral analysis)

Table 3. Preparation of Sample Proportions

Sample No.	1	2	3	4	5	6	7
H* (g)	0.010	0.002	0.004	0.005	0.006	0.008	
L* (g)		0.008	0.006	0.005	0.004	0.002	0.010
Total (g)	0.010	0.010	0.010	0.010	0.010	0.010	0.010

*H : Jiao-Zuo Coal   ※L: Fu-Xin Coal

Table 4. Rock-Eval Pyrolysis results of simulated samples

Sample No.	Tmax (°C)	S_1 (mgHC/g-coal)	S_2 (mgHC/g-coal)	$\dfrac{S_1}{S_2}$	$\dfrac{S_1}{S_1+S_2}$
1*	487	0.60	0.20	3.0	0.75
2	427	0.87	75.63	0.012	0.011
3	428	0.78	68.72	0.011	0.011
4	430	0.78	46.07	0.017	0.017
5	426	0.81	63.00	0.013	0.013
6	431	0.62	28.03	0.022	0.022
7※	428	1.00	75.00	0.013	0.013

*: No.1 as Jiao-Zuo Coal original sample
※: No.7 as Fu-Xin Coal original sample

3. Results and Discussions

3.1. Atomic H/C Ratios Vs. Various Parameters

When Ro%=0.55~0.85 (H/C=1.1~0.7), atomic H/C ratios exhibit a less noticeable decline. As the atomic H/C ratio increases, (S1+S2)/TOC value also increases. By graphing atomic H/C ratio against the following parameters HI, S1, S2, S1+S2, S1/(S1+S2), S1/TOC,(S1+S2)/TOC, Tmax and TOC, we observed that when atomic H/C ratio lies in the range of0.7~1.1, each parameter value displays significant changes (Table 1). According to the van Krevelen diagram for organic matter classification (Stach *et al.*, 1982), it is also observed that for different macerals, the H/C ratio of exinite is the highest, followed by vitrinite, then inertinite. The van Krevelen diagram can also be presented as Tmax versus HI graphs (Hunt,1996). The samples in this study are mainly type II and type III kerogens (Fig. 1). This is inline with the kerogen classification put forward by Magoon and Dow (1994), where Type II kerogens have H/C=1.2~1.5 and Type III kerogens have H/C less than 1.0 in. This also corresponds to the results of significant changes in the parameters when the atomic H/C ratiois in the range of 0.7~1.1.

By using Rock-Eval pyrolysis, (S1+S2) was graphed against TOC (Fig. 2). As shown in Fig. 2, the (S1+S2) and TOC values in the samples from Australia are lower, distributed around the proximity of the origin; otherwise the greater part of the samples lie in zone C. In assessing the hydrocarbon potential of source rocks according to hydrocarbon generation potential per unit organic carbon (S1+S2/TOC) (Wang, 1998), in zone A. (S1+S2/TOC)>600, denoting typical high potential oil shale; zone B (S1+S2/TOC) of 380~600 indicates excellent humic coal as source rocks; and zone C (S1+S2/TOC) of 100~380 indicates high-quality humic coals for general oil and gas production. Our results showed that samples in this study have the same hydrocarbon potential as high-quality humic coals used in general oil and gas production. As presented in Fig. 2, the hydrocarbon potential of zone A> B> C, while the more to the upper right-hand corner in each zone means higher potential. Ro% value is about 0.3~0.9, and Tmax ranges from 410oC~465oC, depicting the range from immature stage to mature stage. The higher the vitrinite reflectance (Ro%), the lower the atomic H/C ratio, and when Ro%=0.55~0.85, the decline is less noticeable (Fig. 3). For this type of kerogen (H/C=0.7~1.1), source rocks with maturity within this range have better hydrocarbon potential. By graphing HI against H/C (Fig. 4), a decrease of HI is observed with lowering H/C values. When H/C=0.7~1.1, HI exhibits a stable increasing trend, indicating the kerogenhas reached a certain maturity and has hydrocarbon potential. In Fig. 5 and 6, obvious correlation is detected between atomic H/C ratio and S1+S2 and (S1+S2)/TOC. When H/C=0.7~1.1, kerogens must

possess a certain hydrocarbon generating potential and hydrocarbon generating potential per unit organic carbon to have oil production possibility, and in this range, Tmax exhibits a greater range of change (Fig. 7), indicating mature hydrocarbon potential. From Fig, 8 and 9, we know that when H/C=0.7~1.1, TOC and S2values must reach a certain value for the source rocks to have oil production possibility.

For all the samples in this study, an increase in maturity leads to a decrease in HI and pyrolyzed hydrocarbon S2 value (Fig. 10 and 11), corresponding to the aforementioned situation when H/C=0.7~1.1, the kerogen in this range has lowered hydrocarbon potential when Ro% is greater than 0.8. When reflectance (Ro%) is below 0.55, HI shows significant irregular changes (Ro%=0.55~0.35→HI=80~520mg/g); when atomic H/C value is above 1.1, HI value has also show significant irregular changes (atomic H/C=1.1~1.4 →HI=160~520mg/g). From this result, we can conclude that organic matter in the oil window of Ro% =0.55% and atomic H/C=1.1 possess a certain maturity and hydrocarbon potential. In addition, if H% (Y-axis) is graphed against TOC% (X-axis) (Fig. 12), the obtained Y intercept (representing the hydrogen content when organic carbon content is zero) can be used to correct the H% value. By doing this, a better atomic H/C ratio is generated to enhance the accuracy of the assessed hydrocarbon potential.

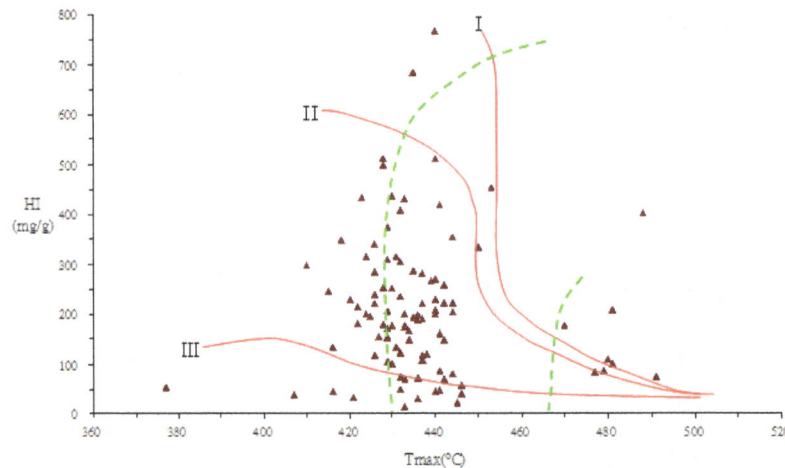

Fig. 1. *van Krevelen organic matter classification diagram.*

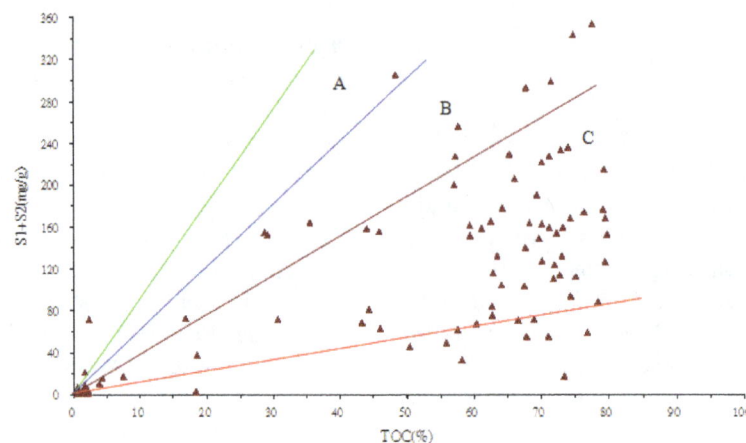

Fig. 2. *organic type and hydrocarbon generating potential zone diagram.*

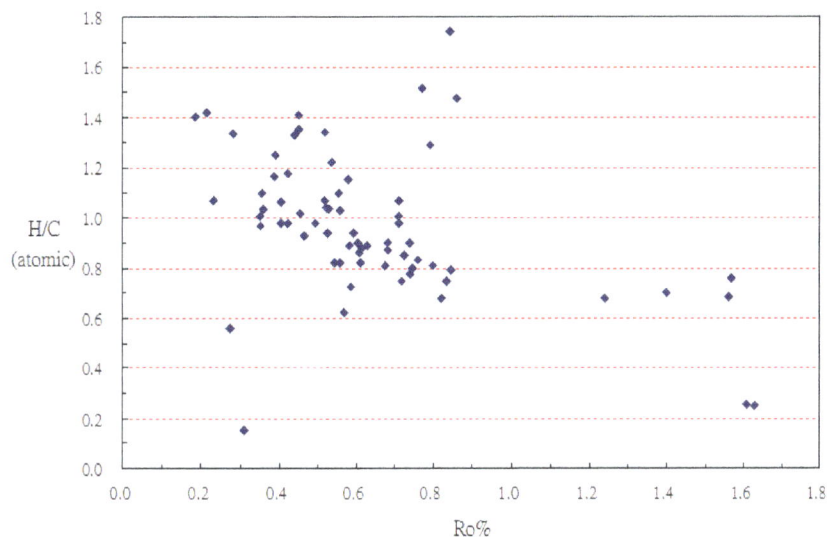

Fig. 3. *hydrocarbon atomic ratio (H/C) vs. vitrinite reflectance (%Ro).*

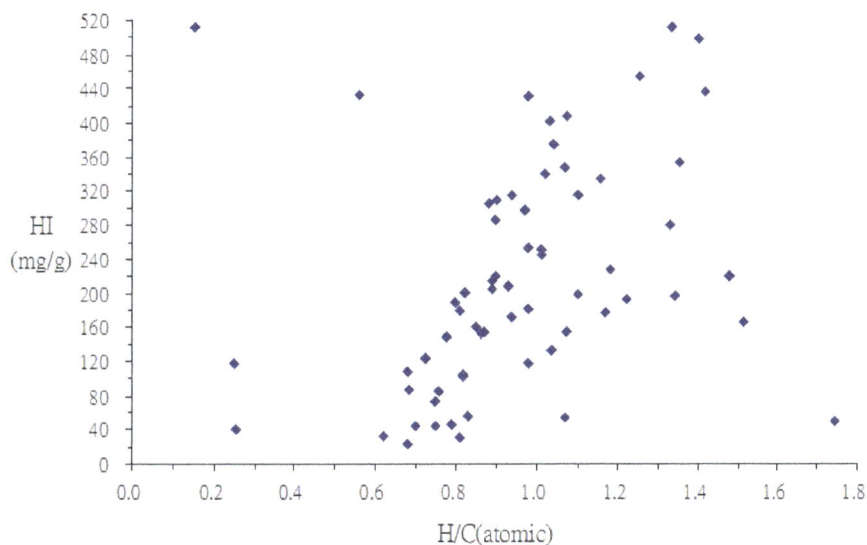

Fig. 4. *hydrogen index (HI) vs. hydrocarbon atomic ratio (H/C).*

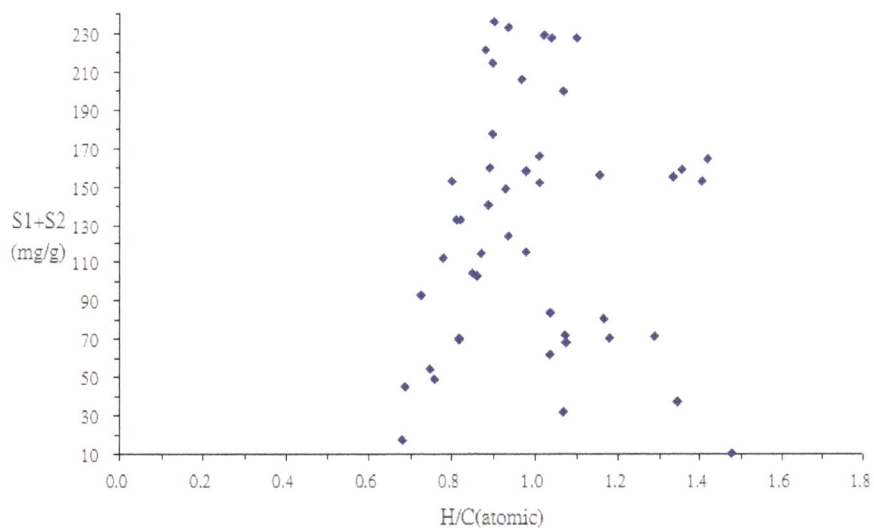

Fig. 5. *hydrocarbon generating potential (S1+S3)vs. hydrocarbon atomic ratio (H/C).*

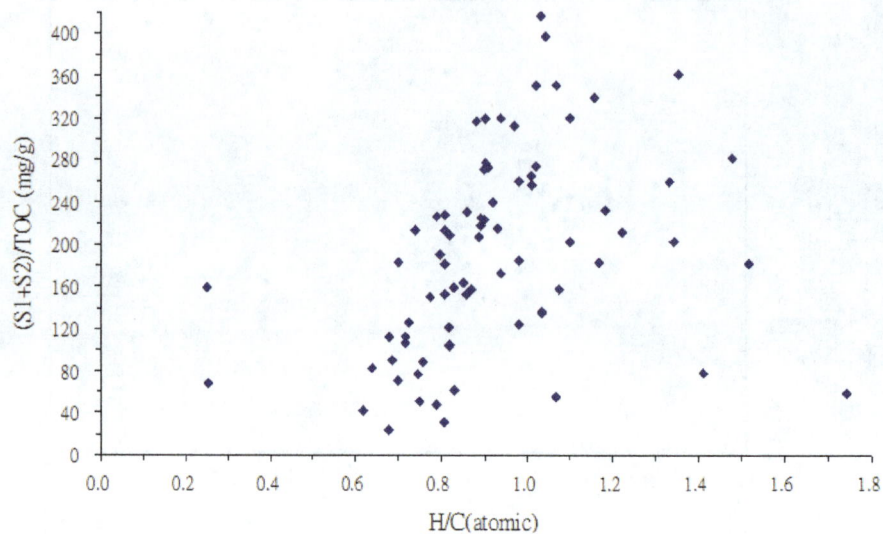

Fig. 6. *hydrocarbon generating potential per unit organic carbon vs. .hydrocarbon atomic ratio (H/C).*

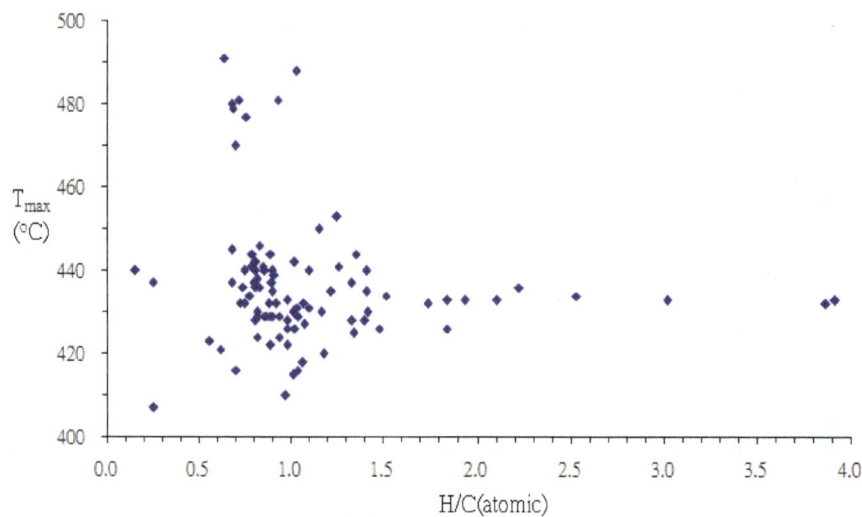

Fig. 7. *Tmax (°C) vs. hydrocarbon atomic ratio (H/C).*

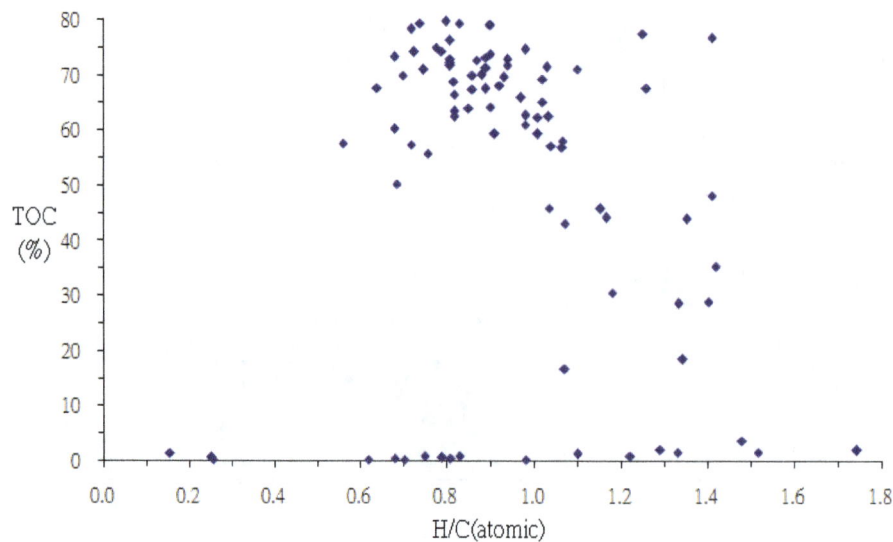

Fig. 8. *total organic carbon (TOC%) vs. hydrocarbon atomic ratio (H/C).*

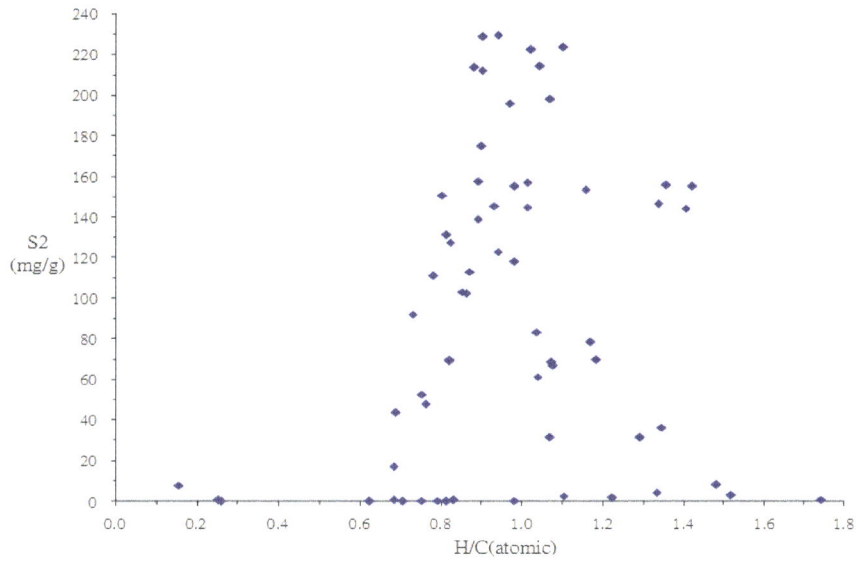

Fig. 9. *pyrolyzed hydrocarbons (S2) vs. hydrocarbon atomic ratio (H/C).*

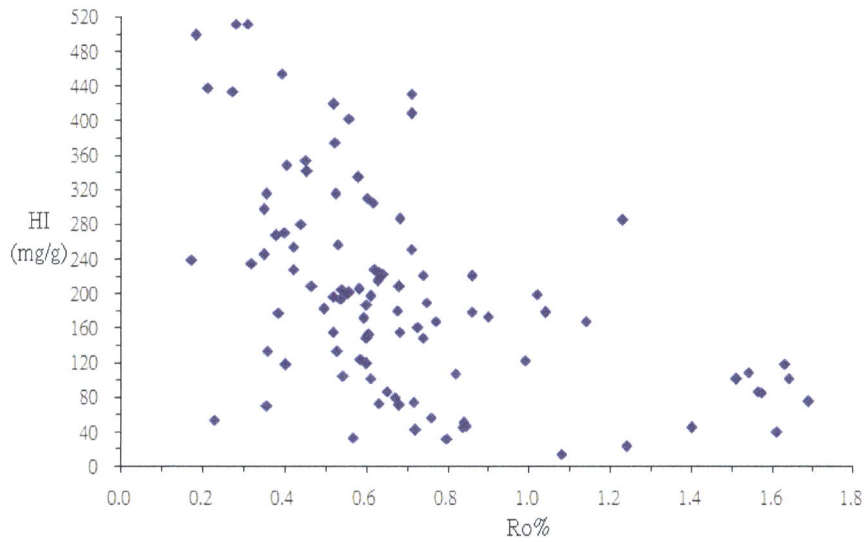

Fig. 10. *hydrogen index (HI) vs. vitrinite relflectance (%Ro).*

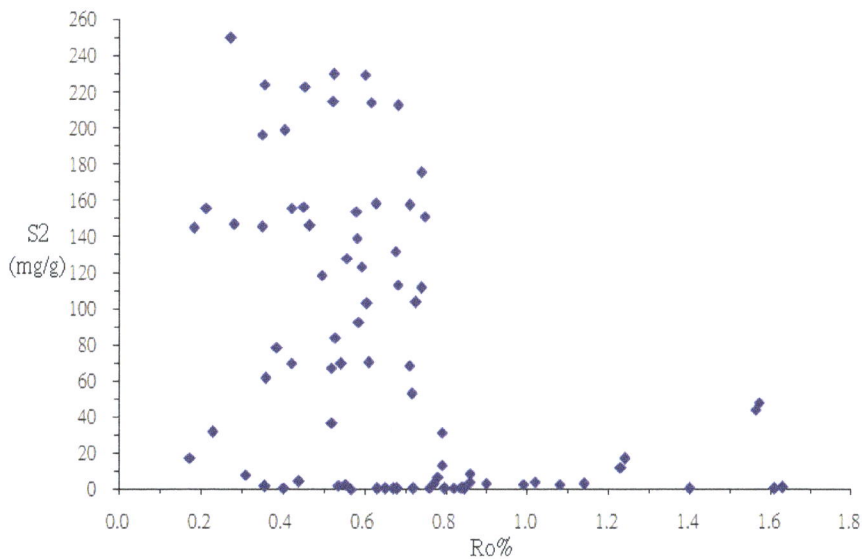

Fig. 11. *pyrolyzed hydrocarbons (S2) vs. vitrinite relflectance (%Ro).*

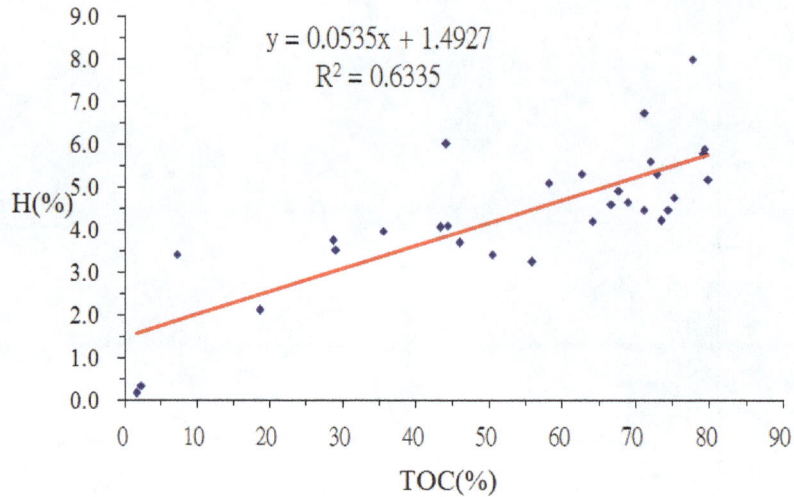

$$y = 0.0535x + 1.4927$$
$$R^2 = 0.6335$$

Fig. 12. hydrogen content (H%(wt%)) against total organic carbon (TOC%).

3.2. Hydrocarbon Potential Evaluation of Organic Materials

By graphing hydrogen index (HI) against atomic H/C ratio, we showed that when the atomic H/C ratio is greater than 0.4, the HI and the atomic H/C ratio exhibit a certain correlation (Fig. 13 and Fig. 14). However, when atomic H/C ratio less than 0.4 (sample No.17, 38, 33, 25, 26, 27, 28, 24 and 15), the H/C ratio versus HI graphs mapped from the results of maceral analysis and elemental analysis display significant differences. This can be attributed to the high mineral content and low maceral content in the samples (Fig. 15). When the sample has high mineral content, hydrous silicate would release hydrogen in the combustion process, resulting in abnormally high atomic H/C ratio (Table 2) in the elemental analysis (Fig. 16). In contrast, low maceral content gives rise to a lower atomic H/C ratio in the maceral analysis. In addition, all of the samples can only produce a certain amount of S2,therefore, when the total organic carbon (TOC) in the sample is low, it may produce a high HI value. This condition of low atomic H/C ratio and high HI is depicted in Fig. 13. For this reason, we should be more cautious when using HI and atomic H/C ratio, the so-called"hydrocarbon generating potential", to assess hydrocarbon potential of organic matter.

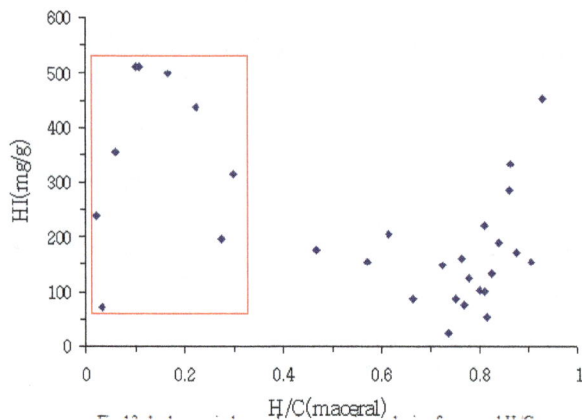

Fig. 14. hydrogen index and elemental analysis of H/C atomic ratio.

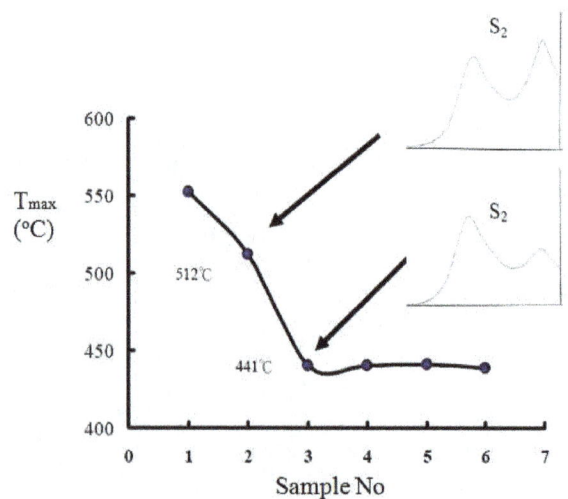

Fig.15. Rock-Eval Pyrolysis Parameter Tmax distribution of coal mixed samples.

Fig. 2 shows the hydrocarbon potential of zone A>B>C where the upperright-handcorner means higher potential. From our results, sample No. 32 Shuicheng2 in zone B has thehighest hydrocarbon potential, with low mineral content and high organic matter content(exinite and vitrinite), particularly in exinite (or H) content; therefore, besides an

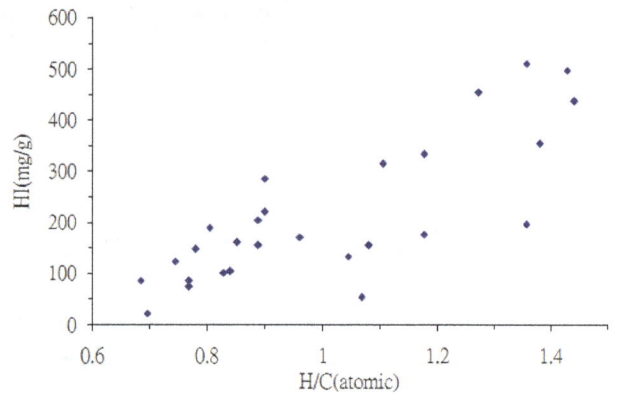

Fig. 13. hydrogen index and microscopic analysis of maceral H/C atomic ratio.

increase in TOC value, (S1+S2) also exhibits a significantincrease, resulting in an increase in thehydrocarbon potential of per unit organiccarbon (S1+S2/TOC). On the other hand, samplesNo. 26, 27 and 28 of the same zonehave high mineral content (above 80%) as their (S1+S2) and TOC values arerelatively lower; their (S1+S2/TOC) value still lie in zone B. With regardto samples below the zone C line (sample No. 42, 41, 23, 29, 40, 37, 36, 39 and 17), apartfrom the high mineral content (96.2%) in sample No. 17, making it close to the origin, the restof the samples have moderate mineral content, but their vitrinite content in the total organicmatter is relatively high (meaning a lower exinite content), resulting in low (S1+S2) valuesand high TOC. For this reason, their (S1+S2/TOC) values are below the zone C line.

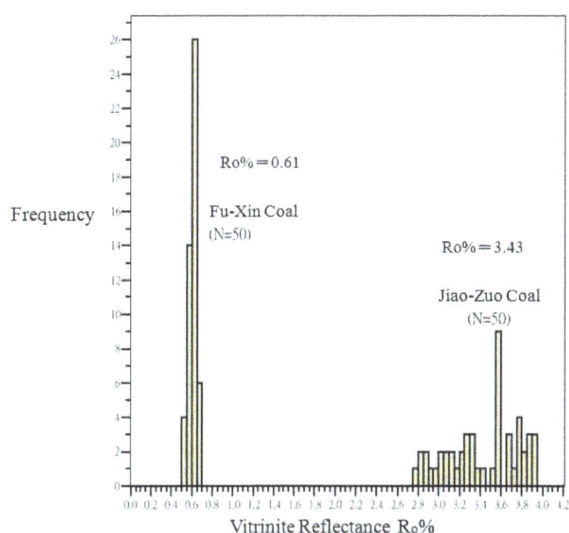

Fig. 16. Vitrinite Reflectance histogram of Jiao-Zuo and Fu-Xin coal samples.

3.3. Elemental Analysis and Maceral Analysis

Overall, the majority of samples contain vitrinites. In the samples from Taiwan, the maceral content of the ones collected from Shiti Formation is higher than the ones from Nanchuang Formation. The samples from Taiwan have less exinite and inertinite content, in which only sample No. 1 Lifung 1 had 29.6% maceral content which is higher than the rest. As for the samples from Mainland China, most samples contain vitrinites as well with contentranging from 1.8 to 24.8%, and the inertinite content could range from as little as 0% to32.0%. In terms of the overall trend, the vitrinite and inertinite contents in samples from Mainland China are both higher than the ones from Taiwan. Shale samples purchased from the United States contain mostly minerals, followed by vitrinites. With respect to heavy liquid separation, the results for samples from Yufeng are: when liquid density was less than 1.25,the   concentration of exinite is 11.4%, and when density liquid was greater than 1.35 the concentration of exinite is 4.8%. Results for Green River shales are: when density liquid wasless than 1.25, the concentration of exinite is 16.8%. After heavy liquid separation at a density of 1.25,

both samples exhibited a higher concentration of exinite. On the other hand, from themaceral analysis on the content percentage of the three major macerals (exinite, vitrinite andinertinite), we estimated the atomic H/C ratios of each sample as listed in Table 2. For aportion of the samples, the atomic H/C ratio obtained here differs significantly from there sults from the elemental analysis.

The main purpose of measuring the vitrinite reflectance is to use it as a maturity indicator. We classified the samples into various ranges of Ro% according to Tissot and Welte (1984). The measurement results are summarized as follows. The mean vitrinite reflectance of samples from Taiwan is between 0.3~0.7%, in which samples from Mushan Formation are in the mature stage, possessing crude oil and wet gas potential; samples from Shiti Formation are in the oil window stage; while most of the samples from Nanchuang Formation are in the immature stage. Most samples from Mainland China also have a vitrinite reflectance between 0.3~0.7%, but they exhibit a wide spectrum of reflectance. For example, sample No. 36 Shanxi has a wide reflectance from 0.5 to 2.0%. In contrast, samples from Taiwan have a narrower spectrum. Using heavy liquid separation to separate the coal samples from Yufeng, the vitrinite reflectance distribution displays a sideway extension to both sides; besides the lower reflectance in exinite concentrated samples, the rest show nosubstantial variations.

In the elemental analysis, most samples have carbon content (C%) ranging50~80%. In terms of hydrogen content (H%), sample No. 32 Shuicheng 2 from Mainland China has the highest (8.02 %), while Woodford Shale has the lowest (0.18%); the remaining samples range between 2~7%. Heavy liquid separation of concentrated macerals can be divided into twoparts: (1) for the separated samples from Yufeng, density less than 1.25 means higherhydrogen content (H%), with a maximum of 5.91%, and a high H/C ratio of 0.9, possibly dueto the concentration of exinites, (2) a similar trend is observed for the samples of Green River Shale where hydrogen content (H%=3.98) and atomic H/C ratio (1.44) are the highest for thosewith density less than 1.25. Rock-Eval pyrolysis results gave us data on Tmax, S1, S2and TOC for further calculation of HI and Pg (S1+S2, hydrocarbon potential). In terms of Tmax, all samples range between 377~488℃, with sample No. 39 Jiaozuo being the highestand sample No. 6 Mingde and the samples collected from Fuji, Hengshan being the lowest. S1 yields range from 0.07 to 8.81 mg HC/g rock, with the lowest being sample No. 17 fromMingde, Dongxia that contains mud. After heavy liquid separation, samples from Green Riverwith density less than 1.25 have the highest S1 yields. S2yields range from 0.08 (sample No.39 Jiaozuo) to 352.5 (sample No. 32 Shuicheng 2) mg HC/g rock. HI ranges between 0.1 to512 mgHC/gTOC, while the majority of samples lie within 200~300 mgHC/gTOC.

3.4. Tmax (°C) and Vitrinite Reflectance (Ro%) Results of Coal Samples Mixed

By using two coal samples with larger differences in

maturity, mixed samples wereprepared in calculated proportions and then subjected to Rock-Eval pyrolysis analysisin order to explore the change and impact on Tmax. The experimental results (Table 4): The Tmax values of the original samples with high and low maturities were490 °C and 428 °C, respectively; the average Tmax value of the other five mixedsamples was 428.4 °C. This demonstrated that all of the mixed samplespresented Tmax values similar to the original sample with low maturity. The resultsalso indicated that regardless of the mixing proportion of low maturity sample to highmaturity sample, all of the mixed samples presented Tmax characteristics of lowmaturity samples.

The pyrolysis parameter, Tmax, is determined mainly by the S2 peak position(hydrocarbons derived from kerogen pyrolysis) (Espitalié et al., 1977; 1985). In the experimental results, when the high maturity original sample was mixed proportionally with low maturity sample, the S2 values of the former sample did not significantly affect the S2 value of the original low maturity sample (Table 4). Thisnon-significant change of the peak value temperature was due to the low S2 of thehigh maturity sample. After mixing, despite the increase in the proportion of high maturity sample(Table 3), the overall S2 value exhibited a downward trend (Table 4). However, the temperature of S2 peak was still determined by the low maturity sample, such that, when the Rock-Eval pyrolysis instrument was producing curves it took the S2 of low maturity material as the main contribution and merged with it the S2 of high mature material (low volume), giving smooth results.

This study also used the proportional mixing of two different kinds of maturity coal samples to simulate the changes in the Tmax and vitrinite reflectance of reworked sedimentary material. After mixing and being subjected to Rock-Eval pyrolysis analysis, the vitrinite reflectance measurement was performed on the original coal samples. The experimental results are outlined in Table 5 and discussed below:

(1) Rock-Eval pyrolysis analysis: In the results of this pyrolysis analysis, Jiao-ZuoCoal is the main body of the mixing proportion of the two coals as Jiao-Zuo coal is a high maturity coal sample. The amount of pyrolyzed hydrocarbon S2 waslower, whereas the Tmax value should be higher. However, the Tmax values are determined by the peak of S2(Bates, 1987; Teichmüller and Durand, 1983).Therefore, by using the low-maturity Fu-Xin coal sample, wich has a higheramount of pyrolysis hydrocarbon S2, as the main component of the mixed sample the Tmax value of high maturity Jiao-Zuo coal cannot be easily observed.

TheRock-Eval pyrolysis results obtained after proportional mixing of the two coal samples showed (Fig. 15): 1. The high maturity Jiao-Zuo Coal (H) had a Tmax value of 552 °C and the low maturity Fu-Xin coal sample (L) had a Tmax value of439 °C. When the mixing proportion was H: L = 95:05 (Tmax = 512 °C), abimodal phenomenon appeared and the result was closer to the Tmax of Jiao-Zuo coal (high maturity). 2. When the mixing proportion was H: L = 90:10 (Tmax =440 ° C) bimodal phenomenon also appeared, however is the result was closer to the Tmax of Fu-Xin coal (low maturity). When the mixing proportion of H was less than 90%, all of the Tmax values recorded were similar to the low maturity Fu-Xin coal.

(2) Vitrinite reflectance: 1. 50 points of measurement were assessed to obtain the vitrinite reflectance measurement of the Jiao-Zuo coal sample and the Fu-Xin coal sample and the average Ro% values were respectively 3.43 and 0.61 (Fig. 16). 2. According to the original mixing proportion, as determined by the pyrolysis analysis and using random sampling, 50 measured points respectively that complied with the proportion were taken and then averaged. Each sample with mixing proportion was taken in this manner for 1000 times and then the average was calculated as the vitrinite reflectance results after proportional mixing. The results for mixed samples are as follows (Fig. 17): Sample no. 2 (mixed sample)had an Ro% of 3.32; sample no. 3 (mixed sample) had an Ro% of 3.15; sample no.4 (mixed sample) had an Ro% of 2.87; sample no. 5 (mixed sample) had an Ro% of 2.59. Besides, each Ro% value of randomly mixed proportion was stable. As the vitrinite reflectance measured point of low maturity increased, the vitrinite reflectance results of mixed samples showed a linear decline. Not quite unexpectedly one obtains a approximate "mixing line" for the Vitrinite Reflectance values: Ro% (xH)= xH*3.5 + (1-xH)*0.6in the figure 17 (xH: the mixing proportion of high maturity Jiao-Zuo original coal sample). Nevertheless, several mixed samples continued to show high maturity characteristics. 3. The vitrinite reflectance results of the original samples and mixed samples, as well as the corresponding Tmax values, were placed on a plot of good correlation (Lang Dongsheng et al, 1999). The results display that in more extreme cases (mixed samples comprising of substrates with highly different maturities), the relationship of mixed samples deviates from the general linear trajectory (Fig. 18). This suggests that when extracting strata samples that contain reworked sedimentary materials, the Tmax measurement of the sample is likely to reflect the maturity of the younger materials present.

Table 5. Rock-Eval Pyrolysis and Vitrinite Reflectance results obtained from mixed coal samples

Sample No.	Tmax (°C)	S_1(mg/g)	S_2(mg/g)	Ro%
No.1 (Jiao-Zuo Original Coal Sample, H)	552	0.41	2.40	3.43
No.2 H95:L5 Mixed Sample	512	0.34	6.19	3.32*
No.3 H90:L10 Mixed Sample	440	0.32	9.45	3.15
No.4 H80:L20 Mixed Sample	440	0.50	17.99	2.87
No.5 H70:L30 Mixed Sample	441	0.70	26.53	2.59
No.6 (Fu-Xin Original Coal Sample, L)	439	1.84	86.12	0.61

*:Ro% of Sample No.2 is derived from the results of H96:L4.

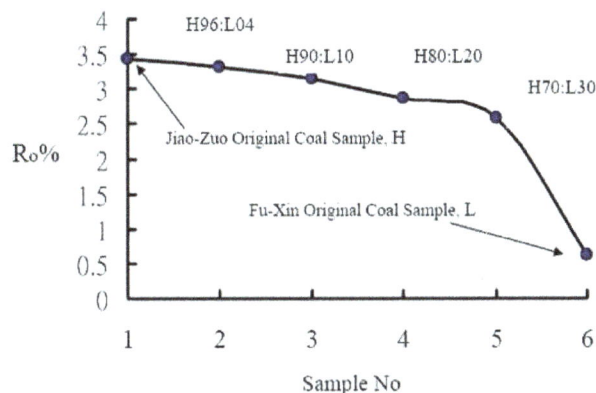

Fig. 17. *Vitrinite reflectance Ro% of simulation different proportion of coal mixed samples. Show the approximate "mixing line" for the Vitrinite Reflectance values: Ro%(xH)= xH*3.5 + (1-xH)*0.6(xH: the mixing proportion of high maturity Jiao-Zuo original coal sample).*

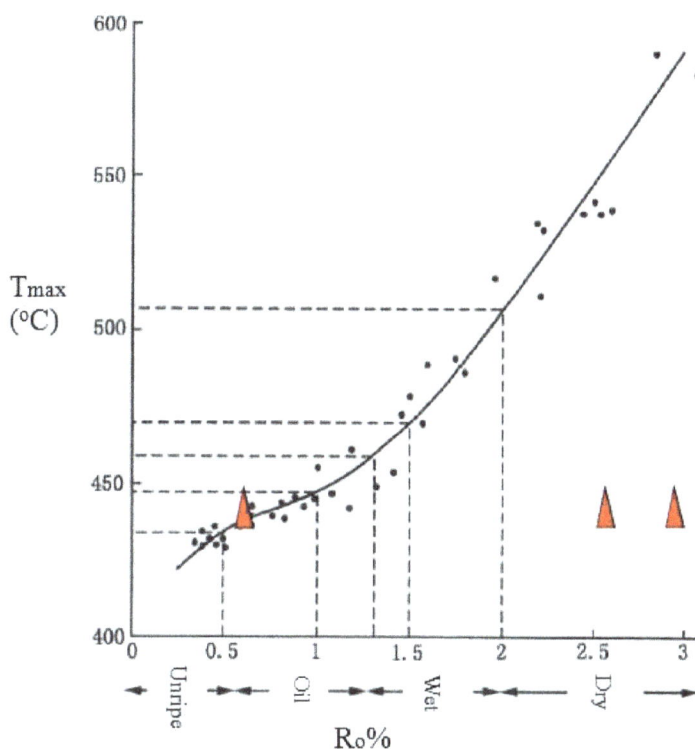

Fig.18.Correlation between Tmax temperature and vitrinite reflectance for type-III kerogen (Adapted from TeichmÜller and Durand, 1983).

3.5. Vitrinite Reflectance (Ro%) Versus Tmax (°C)

For samples separated with heavy liquid, those with density less than1.25 have more exinite than the original sample, thus giving them lower Tmax and Ro% values; those with density greater than 1.35 have more inertinite than the original, and since inertinites are macerals that are heated and oxidized before burial, they result in higher Tmax and Ro%. Generally speaking, the washing process leads to the loss of free hydrocarbon, thus a decreasein S1. The value of S2 shows the pyrolyzed hydrocarbon content, and since hydrogen index(HI) is the ratio of S2 and TOC, the value of S2 also serves as reference hydrogen content. In view of the above, macerals with density less than 1.25 contain more exinite (i.e. higher hydrogen content) and will produce a higher volume of hydrocarbons; therefore, the higher the S2

value, the higher the peak value, meaning a better maceral composition for higher hydrocarbon potential. Macerals with density is greater than 1.35 indicates less hydrogen content. The van Krevelen diagram for organic matter classification (Stach *et al.*, 1982) can also be used to represent the different macerals, where exinite has the highest H/C ratio, followed by vitrinite, then inertinite. The van Krevelen diagram can also be presented as Tmax versus HI graph (Hunt, 1996). The samples in this study are mostly type II and type III kerogens. From immature to mature stage, Ro% value ranges from 0.3 to 0.7, while Tmax ranges from 420oC~465oC. The atomic H/C ratio decreases with the increase of vitrinite reflectance (Ro%), but when Ro%=0.55~0.85, the decline is less obvious, as shown in Fig. 3.

Thermal maturation is the most important process in hydrocarbon production. In the evaluation of oil and gas potential, organic maturity is one of the most important parameters to be considered. This is the reason why "hydrocarbon potential" and "maturity" are the two major topics in oil and gas potential assessment studies. According to the evaluation parameters for source rocks set by Peters and Cassa (1994); when Wt.% TOC values is greater than 2.0 and Rock-Eval S2 value greaterthan10.0, the source rocks are rich in organic matter; when the atomic H/C ratio is greater than 1.0 and hydrogen index (HI, mgHC/gTOC) value is more than 200, the organic matter has oil-producing potential; and when Ro% is greater than 0.6 and Tmax (°C) value is more than 430, the organic matter are at the beginning of the mature stage for hydrocarbon production.

In this study, most samples are abundant in organic matter content, while the majority has gas producing potential ranging from immature stage to mature stage when hydrocarbon production begins to happen. Drawing from the obtained parameters of both lithofacies and geochemical analyses, the organic matter from Lifung have gas producing potential (less oil)and have reached the maturity when hydrocarbons begin to form; while the organic matter from Shuichengbian have oil-producing potential and have reached the maturity when hydrocarbons are about to form.

In the assessment of thermal maturity, vitrinite reflectance (Ro%) and Tmaxare the most frequently used indicators. Both of them increase as the maturity increases. As shown in Fig.19 where Tmax is graphed against the average vitrinite reflectance (Ro%), we found these two factors display a positive correlation. The sample with the highest exinite content, No. 32Shuicheng 2, has a S2 value as high as 352.5 mg HC/g rock. From the results of the analysis, it is also known that with the increase of vitrinite reflectance (Ro%), S2 generally shows a declining trend, which is similar to the findings of Rimmer et al. (1993). After Rock-Eval pyrolysis, the samples in this study display little correlation between the S2 value and Ro% and Tmax, thus we were unable to discuss the reflectance suppression, maximum pyrolysis temperature suppression, and reworked sediments. However, if Ro% is graphed against Tmax,the more similar the sedimentary environment and sedimentary time, the better the correlation between Tmax and Ro% (all samples, r^2=0.51; samples from Taiwan,r^2=0.78; samples from Mainland China, r^2=0.73; all meet the significant level α=0.01) (Table 6); it may be possible to use these results to examine suppression and reworked sediments.

Table 6. *The correlation coefficient (r) of Tmax versus Ro%*

Number of Samples	r-value	r^2-value
90(All Samples)	0.810	0.656
26(Samples from Taiwan)	0.885	0.782
12(Samples from Mainland China)	0.852	0.726

(All meet the significant level á=0.01)

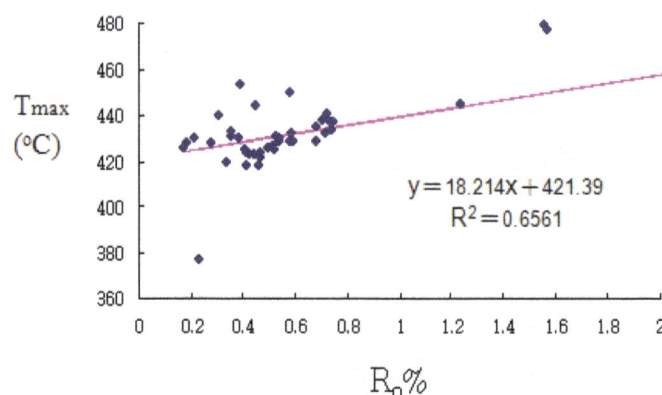

$$y = 18.214X + 421.39$$
$$R^2 = 0.6561$$

Fig. 19. *linear regression of T_{max} (°C) vs. vitrinite reflectance (Ro%).*

4. Conclusions

Hydrogen index (HI) and atomic H/C ratio generally show a certain correlation. When themineral content of the sample is high and total organic carbon (TOC)content is low, ahigh HI and low atomic H/C ratio may result. Therefore, when using HI and atomic H/C ratio to assess hydrocarbon potential of organic matter, a morecautious approach should be adopted. In addition, when H% is graphed against C%, using the obtained Y-intercept to correct the H% will give

better results for the samples from the same region or in the same sample group. Vitrinite reflectance (Ro%) increases as the organic material becomes more mature; at the same time, the atomic H/C ratio declines, but the decline becomes less noticeable when Ro%=0.5~0.7. When organic matter have atomic H/C ratio>1.0 and HI (mgHC/gTOC)>200, they possess oil-producing potential, whereas Tmax (°C)>430 indicates that the organic matter has reached the mature stage that hydrocarbon production is about to take place.

The results of proportionally mixing two coal samples with different maturities, and then subjecting them to Rock-Eval pyrolysis analysis, vitrinite reflectance measurements and mixing treatment showed that in low maturity materials (i.e.younger rock formations), regardless of the proportion of mixed high maturity material (i.e. reworked sediments of old rock formations), the Tmax value remained similar to that of low maturity samples (i.e. younger strata maturity). This implies that when a strata sample contains reworked sedimentary material, the Tmax may reflect the maturity of young material. Therefore, during the sampling or pyrolysis analysis on relatively high maturity material, low-maturity material contamination should be avoided whenever possible, even in very small amounts; otherwise the Tmax value is likely to be biased towards that of the low maturity contaminant. Our results also indicate that after mixing, there is a distinctively non-linear relationship between the average value of vitrinite reflectance Ro% and the corresponding Tmax values measured. When Tmax is graphed against Ro%, the more similar the sedimentary environment and sedimentary time, the better the correlation between Tmax and Ro% (all samples,r^2=0.51;samples from Taiwan, r^2=0.78; samples from Mainland China, r^2=0.73).

As thermal maturity indicator (vitrinite reflectance, Ro%) increases, HI, S_1, S_2, and H/C values decrease; when the atomic H/C ratio has a Ro%=0.55~0.85 (H/C=1.1~0.7), the decline is less noticeable. When the atomic H/C ratio increases, (S_1+S_2)/TOC value also increases. It was observed that when atomic H/C ratio is in the range of 0.7~1.1, each parameter value displays large-scale changes. When the vitrinite reflectance (Ro%) is below 0.55, the value of the corresponding HI showsirre gular substantial changes (Ro% =0.55~0.35→HI=80~520mg/g). However, the decline is less noticeable in the range of Ro%=0.55~0.85. Whereas atomic H/C ratio lies in the range of 0.7~1.1 and HI (mgHC/gTOC)>200, organic matter are expected to possess oil-producing potential. On the other hand, when the atomic H/C value is in the range of 0.7~1.1, the value of the corresponding HI shows a stable increasing trend (atomic H/C=0.7~1.1→HI=120~520mg /g). This shows that organic matter inan oil window of Ro%=0.55 and atomic H/C=1.1 have reached a certain maturity and hydrocarbon potential. Tmax (°C)>430 also indicates that the organic matter has reached the stage of hydrocarbon generation starting to take place. When the atomic H/C ratio=0.7~1.1, HI, S_1, S_2, S_1+S_2, S_1/(S_1+S_2), S_1/TOC, (S_1+S_2)/TOC, Tmax and TOC exhibit significant large-scale variations, demonstrating the existence of correlations between them.

Acknowledgements

We sincerely thank Prof. Jin KuiLi from the Graduate School of China University ofMining and Technology (CUMT) (Beijing) for providing the coal samples used in thisexperiment. At the same time, we would also like to thank the Research Institute ofPetroleum Exploration and Development, CNPC, for providing adequate assistancewith the Rock-Eval pyrolysis analysis and vitrinite reflectance measurements. Notably,we are grateful toMr. Hsieh MF from the Exploration &Development Research Institute of CPC Corporation, Taiwan (Maio-Li) for theassistance on pyrolysis experiments. Also appreciate the financial support from theNational Science Council for the research projects of NSC-95-2116-M-252-001, NSC-99-2116-M-252-001 and NSC-101-2116-M-252-001.

References

[1] Akinlua A, Ajay TR, Jarvie DM, Adeleke BB (2005) A re-appraisal of the application of Rock-Eval pyrolysis to source rock studies in the Niger Delta. Journal of Petroleum Geology 28(1):39-48

[2] ASTM (1975) Standard D-2797, ASTM Standard manual Part, vol 26, ASTM. Philadelphia,pp350-354

[3] ASTM (1980) Standard D-2797, Microscopical determination of volume percent of physical components in a polished specimen of coal. ASTM,Philadelphia

[4] Baskin DK (1997) Atomic H/C ratio of kerogen as an estimate of thermal maturity and organic matter conversion, AAPG Bulletin 81(9):1437-1450

[5] BatesRL,JacksonJA (1987) Glossary of Geology [M], American Geological Institute, Alexandria, Virginia, 565~565

[6] BostickNH (1971) Thermal alteration of clastic organic particles as a indicator of contact and burical metamorphism in sedimentary rocks. Geoscience and Man3:83-93

[7] BostickNH (1974)Phytoclasts as Indicator of thermal metamorphism, Franciscan Assemblage and Great Vally Sequence (upper Mosozoic),California. Spec. Pap. Geol. Soc. Amer153:1-17

[8] Dahl B,Bojesen-Koefoed J,Holm A,Justwan H,Rasmussen E, Thomsen E (2004) A new approach to interpreting Rock-Eval S2 and TOC data for kerogen quality assessment. Organic Geochemistry 35:1461-1477

[9] DormansHNM, Huntjens FJ, van Krevelen DW (1957) Chemical structure and properties of coal XX-composition of the individual macerals (vitrinites, fusinite, micrinites and exinites). Fuel36:321-339

[10] Espitalie J, La Porte JL, MadecM, Marquis F, Le Plat P,Paulet J, Boutefeu A (1977)Methoderapide de caracterisation des roches meres de leur potential petrolier et de leurdegred'evolution. Revue l'Inst Francais du Petrole32(1):23–42

[11] Espitalie'J, Deroo G, Marquis F (1985) La pyrolyse Rock-Eval et ses applications. Revue Institut Franc- ais du Pe' trole Part I40:563–578, Part II40:755–784

[12] Ge T-S, Chen Y-X, Yu T-X, Wang G-S, Niu Z-R, Shen S-W, Li Y-L, Song S-Y, Wu T-S, Zhao Q-B, Xiong C-Y (1993)Liaohe Oil Field:Petroleum geology of China3: 4 74-494

[13] Hou D-J, ZhangL-Y (2003) Practical geochemistry illustrated handbook, [M]. Petroleum Industry Press,pp63

[14] HuntMJ(1996) Petroleum Geochemistry and Geology(2rd ed.), W. H. Freeman and Company, New York, pp1~743

[15] Hutton AC, Cook AC (1980) Inffluence of alginite on the reflectance of vitrinite from Joadja, NSW, and some other coals and oil shales containing alginiate. *Fule* 59, p711-714.

[16] Jones RW, Edison TA (1978) Microscopic observations of kerogen related to geochemical parameters with emphasis on thermal maturation, in D.F. Oltz, ed., Low temperature metamorphism of kerogen and clay minerals: SEPM Pacific Section, Los Angeles, October, p.1-12.

[17] Kuo C-L (1997) Application of vitrinite reflectance on oil and gas exploration in Western Taiwan [D]: [dissertation]. Taipei: Department of Geosciences, National TaiwanUniversity, Doctoral Dissertation PP 1-302

[18] LangD-S, Jin C-Z, Guo G-Y (1999) The evaluation technique of pyrolysis andgas chromatography of reservoir fluids, [M] Petroleum Industry Press, Pg117.Cited from: HouDujie, Zhang Linye, Practical geochemistry illustrated handbook [M].Petroleum Industry Press, 2003, pp63

[19] Lee, H-T (2011) Analysis and characterization of samples from sedimentary strata with correlations to indicate the potential for hydrocarbons, Environmental Earth Sciences (formerly Environmental Geology)64(7):1713-1728

[20] Lee, H-T, TsaiLY, Sun L-C (2010) Relationships among Geochemical Indices of Coal and Carbonaceous Materials and Implication for Hydrocarbon Potential Evaluation. Environmental Earth Sciences60(3):559-572

[21] Lee, H-T (2010) Statistical analysis of relations between petrographic and geochemical indices of oil generation potential in organic matter, Carbonates and Evaporites 25(4):303-323

[22] LiS,Pang X, Li M, Jin Z (2003) Geochemistry of petroleum systems in the Niuzhuang South Slope of Bohai Bay Basin—part 1: source rock characterization. Organic Geochemistry34:389–412

[23] Mao EW, Zhou J-N, Wang J-G, Zhang S-J (1988) Depositional environment of the Nanchuang Formation in the Miaoliand Hsinchu area, Taiwan. Petroleum Geology of Taiwan24:37-50

[24] Pang X, Li M, Li S, Jin Z (2003) Geochemistry of petroleum systems in the Niuzhuang South Slope of Bohai Bay Basin,Part 2: evidence for significant contribution of mature source rocks to "immature oils" in the Bamianhe field. Organic Geochemistry34:931-950

[25] Peters KE, Cassa MR (1994) Applied source-rock geochemistry, In: Magoon, L.B., Dow, W.G. (Eds.), The Petroleum System—From Source to Trap. American Association of Petroleum Geologists Memoir 60:93–120

[26] Rimmer SM, Cantrell DJ, Gooding PJ (1993) Rock-Eval pyrolysis and vitrinite reflectance trends in the Cleveland Shale member of the Ohio Shale, eastern Kentucky. Organic Geochemistry20(6):735-745

[27] Stach E, Mackowsky M-Th, Teichmüller M, Taylor GH, Chandra D, Teichmüller R (1982)Stach's Textbbok of Coal Petrology(3rd ed.)[M].Berlin, Gebruder Borntraeger, pp 1~535

[28] Su PJ(2001) The influence of maceral composition on Rock-Eval Pyrolysis, Masters Dissertation, Graduate Institute of Applied Geology at National Central University, p 73

[29] SunL Z (2000) Measurement of vitrinite reflectance suppression- coal sample separation from Yufeng, Taiwan as an example, Ph.D. Dissertation of Institute of Geophysics at National Central University, p 81

[30] Taylor AW, Ritts BD (2004)Mesoscale heterogeneity of fluval-lacustrine reservoir analogues: examples from the Eocene Green River and Colton Formations, Uinta Basin, Utah, U.S.A. Journal of Petroleum Geology27:3-26

[31] TeichmÜller M, Durand B (1983) Fluorescence microscopical rank studies on liptinites and vitrinite in peat and coals, and comparison with the results of the Rock-Eval pyrolysis[J]. Int. J. Coal Geol 2:197~230

[32] Teichmüller M (1974) Generation of petroleum-like substance in coal seams as seen under the microscope. In: Tissot B, Bieener F (eds) (1973) Advances in organic geochemistry. Rueil-Malmaison, France, pp 379- 407

[33] Ting FTC (1978) Petrographic techniques in coal analysis, In: Karr, C., Jr. (ed.)：Analytical Methods for Coal and Coal Products. Academic Press, Inc., New York, pp3-26

[34] Tissot BP, Welte DH (1984) Petroleum formation and occurrence；a new approach to oil gas exploration, Springer-Verlag, Berlin, Heidelberg, New York, p 699

[35] Tsai LY(1988)Paragenetic implications of vitrinite in Chilung-Taipei coal fields,Ti-Chih8(1-2):63-70

[36] Udo OT, Ekweozor CM, Okogun JI (1986) Organic petrographic and programmed pyrolysis studies of sediments from Northwestern Niger delta, Nigeria, Journal Min. Geol 24:85-96

[37] Walples DW (1985) Geochemistry in petroleum exploration [M]. D. Reidel Publishing Co. Dordrecht BostonL ancasterpp 1~232

[38] Wang CJ(1998) "Folded-fan" evaluation of coal's hydrocarbon potential,Geochemistry27(5):483-492

[39] Wu L-Y, Gu X-Z, Sheng Z-W, Fan C-L, Tong Z-Y, Cheng K-M (1986) Quick evaluation quantitatively on the pyrogenation of source rocks [M]Beijing: Science Press,pp1 ~ 198

Comparative Analysis of the Fuel Properties of Ethylester Biodiesels from *Cyperus esculentus, Sesamum indicum* and *Colocynthus vulgaris* Seed Oils

Anekwe Ozioma Juliana, Ajiwe Vincent Ishmael Egbulefu

Department of Pure and Industrial Chemistry, Nnamdi Azikiwe University, Awka, Anambra State, Nigeria

Email address:

oziane2002@yahoo.com (A. O. Juliana), Vaj_04@yahoo.com (A. V. I. Egbulefu)

Abstract: The relative abundance of some Nigerian seed oils coupled with the little knowledge of their biodiesel usage prompted the need for this research. Biodiesel production is a very modern and technological area for researchers due to the relevance that it is winning everyday because of the increase in the petroleum price and the environmental advantages. In this work, studies were carried out to investigate the fuel properties of *Cyperus esculentus, Colocynthus vulgaris, Sesamum indicum* ethylesters and their corresponding ethyl ester blends. Ethylesters of these oils were prepared by H_2SO_4 catalysed transesterification reaction between the oils and ethanol. The fuel properties such as kinematic viscosity, flash points, pour point and water crackle were determined. GC-MS was used to determine the fatty acid profile of the transesterified oils. The fuel properties correlated very well with the fatty acid compositions. The results showed that the oils were rich in saturated fatty acids (66.67%), appreciable value of monounsaturated fatty acids (33.33%) and little percent (33.33%) polyunsaturated fatty acids and absence of polyunsaturated fatty acids in *Cyperus esculentus* and *Sesamum indicum* esters. The fatty acid chain lengths that were predominant in *Cyperus esculentus, Colocynthus vulgaris, Sesamum indicum* were C16, C18 and C20. The fuel properties of *Colocynthus vulgaris, Cyperus esculentus* and *Sesamum indicum* biodiesel blends were better when compared to the oil and petrodiesel in terms of flash points, viscosity and pour points, hence the optimum engine performance of both oil and petrodiesel could be improved by use of these biodiesel feedstock.

Keywords: Ethylbiodiesels, *Cyperus esculentus, Colocynthus vulgaris, Sesamum indicum*, Comparative Analysis

1. Introduction

Escalating crude oil prices and environmental awareness have increased interest in the use of renewable fuel sources. One area of attention is the upgrading of vegetable oils for use as a fuel or fuel additive. Besides being a renewable source, the use of vegetable oils has benefits economically and environmentally. Such oils are CO_2 neutral and contain little, if any sulfur, nitrogen and metals, which are major pollutants in current fuel emissions (Katikaneni *et al*; 1998). The possibility also exists for the reuse of current vegetable oil wastes such as wastes from fast food restaurants. As such, oils come from plants that can be easily grown; its production can be localized and adjusted according to demand. The conversion of the oil to fuel can therefore bring benefits to the community economically as well as making them no longer reliant on outside sources.

Over the years, vegetable oils have been substituted for diesel for use in engines but this has led to problems such as carbon deposits, oil ring sticking and gelling of the lubricating oil (Ma and Hanna, 1999). Because of such problems, research in this area has been centered on the conversion of these oils to a form that is similar to current fuels. One such fuel, which is currently gaining much attention, is biodiesel. This is a variety of ester-based oxygenated fuels made from vegetable oils or animal fats.

There are several methods for the conversion of vegetable oils to biodiesel of which the most common is the transesterification process, in which an alcohol is reacted with the oil to form esters and glycerol (Ma and Hanna, 1999). The esters are separated and commonly used as a mixture with petroleum diesel (20:80) to minimize engine modification requirements. (Altin *et al*; 2001), showed that vegetable oil methyl esters gave performance and emission characteristics close to petroleum diesel. The main problems associated with the increased use of this fuel are the costs of the oil and its processing. Also, the marketing of this product is limited to diesel engine applications.

Vegetable oils are predominantly made up of triacylglycerols with a small amount of minor compounds (2-5%) (Cert *et al*; 2000).Triacylglycerols are made up of one glycerol molecule joined to 3 fatty acids by an ester link.

2. Materials and Method

The seeds of *Cyperus esculentus* L., *Colocynthus vulgaris* and *Sesamum indicum* L. seeds were collected, winnowed to remove sand and other particles, sun dried, ground using a milling machine and stored separately in dry air tight containers prior to use. Oil from the ground sample was thoroughly extracted with petroleum ether using a soxhlet extractor and separated using rotary evaporator apparatus. The extracted oil samples were transesterified into ethylesters by a method described in Ajiwe *et al*; 2006. Different dilutions of the ethylesters were made with petroleum diesel in the ratio of 0:100, 10:90, 20:80, 30:70, 40:60, 50:50, 60:40, 70:30, 80:20, 90:10 and 100:0. The fatty acid profiles of the ethylesters were investigated with GC-MS auto sampler and auto analyzer model fused with HP-Innowax fused with capillary column (60x60mmi.d) and helium as the carrier gas (flow rate =1.61ml/min.). The ethylesters and their diesel blends were analyzed for parameters such as flash point, pour point, kinematic viscosity, using standard American Society for Testing and Materials (ASTM) procedures previously adopted (Ajiwe *et al.*, 2006). The chemicals used in the experiment include, sulfuric acid (98 %wt), sodium hydroxide pellet of 98.2% purity and ethanol (analytical grade reagents).

3. Results and Discussion

Table 1. *Fatty acid profile of Cyperus esculentus ethyl ester seed oil.*

s/no	Common name	Systematic name	Shorthand name	R.A (%)
1	Myristic acid	Tetradecanoic acid	14:0	0.75
2	-	9-hexadecenoic acid		1.12
3	Palmitic acid	Hexadecanoic acid	16:0	24.15
4	Linoleic acid	9,12-octadecadienoic acid	18:2	18.04
5	Oleic acid	9-octadecenoic acid	18:1	36.31
6	Stearic acid	Octadecanoic acid	18:0	17.02
7	-	Heptadecanoic acid	18:1	2.61

R.A = relative abundance

Table 2. *Fatty acid profile of Colocynthus vulgaris ethyl ester seed oil.*

s/no	Common name	Systematic name	Shorthand name	R.A (%)
1	-	Pentadecanoic acid		22.42
2	Linoleic acid	9,12-octadecadienoic acid	16:0	54.86
3	Stearic acid	Octadecanoic acid	18:0	22.73

R.A = relative abundance

Table 3. *Fatty acid profile of Sesamum indicum ethylester seed oil.*

s/no	Common name	Systematic name	Shorthand name	R.A (%)
1	Palmitic acid	Hexadecanoic acid	16:0	2.76
2		Hexadecanoic acid		27.76
3	Cis vaccenic acid	11-octadecadienoic acid	18:1	9.26
4	Oleic acid	9-octadecenoic acid	18:1	41.09
5		pentadecenoic acid		16.10
6		heptadecanoic acid		3.03

R.A = relative abundance

Table 4. *Saturation levels of fatty acids components of the ethylesters.*

Sample	%SFA	%MUSFA	%PUSFA
Sesamum indicum	66.67	33.33	-
Cyperus esculentus	57.14	28.57	-
Colocynthus vulgaris	66.67	-	33.33

Table 5. *Average chain length of the fatty acids components of the ethylesters.*

Chain length	*Cyperus esculentus*	*Colocynthus vulgaris*	*Sesamum indicum*
%C16	100	-	-
%C17	-	33.33	66.67
%C18	66.67	-	33.33
%C19	33.33	-	66.67

Table 1 showed that the major fatty acids present in *Cyperus esculentus* ethylesters were myristic acid (0.75%), palmitic acid (24.15%), stearic acid (17.02%) and heptadecanoic acid (2.61) as the saturated fatty acids while 9-hexadecenoic acid (1.12%) and oleic acid (36.31%) as the monounsaturated fatty acids, linoleic acid (18.04%) as the polyunsaturated fatty acid.

The presence of four saturated fatty acid , 2 monounsaturated fatty acids and one polyunsaturated indicated that *Cyperus esculentus* ethylester can have good oxidation stability without the aid of additives to improve oxidation stability since the palmitic acid present in it is a natural anti oxidant.

Table 2 showed the major fatty acids present in *Colocynthus vulgaris* ethyl esters as pentadecanoic (22.42%) and stearic acid (22.73%) as the only saturated fatty acids while linoleic acid (54.86%) as the only polyunsaturated fatty acid and the absence of monounsaturated fatty acids. The presence of polyunsaturated fatty acids indicated that that it has a poor oxidative stability and the possibility of polymerising at high temperature. It would therefore require an anti oxidant to prevent lipid peroxidation.

Table 3 showed the major fatty acids present in *Sesamum indicum* ethyl esters as palmitic acid (22.76%), pentadecenoic acid (16.1%) and heptadecanoic acid (3.03%)as saturated fatty acids, while cis vaccenic acid (9.26%) and oleic acid (41.09%) as the monounsaturated fatty acid, with the absence of polyunsaturated fatty acid. The absence of polyunsaturated fatty acid and the presence of palmitic acid (a natural antioxidant) showed that it had a good oxidation stability and low possibility of polymerizing at high temperature.

Table 4 showed that *Sesamum indicum* and *Colocynthus vulgaris* ethylesters had greater percentage of saturated fatty acids than *Cyperus esculentus* ethylesters. *Sesamum indicum* and *Cyperus esculentus* ethylester contained no polyunsaturated fatty acid while *Colocynthus vulgaris* ethylesters contained no monounsaturated fatty acid.

Table 5 showed chain length by composition of *Sesamum indicum* ethyl esters as C_{17} (2), C_{18} (1), C_{19} (2), C_{20} (1), *Cyperus esculentus* ethylester as C_{16} (1), C_{18} (2), C_{19} 1(), C_{20} (3) while *Colocynthus vulgaris* ethyl esters had chain length of C_{17} (1) and C_{20} (2).

3.1. Variation of Flash Point of the Ethylesters and Their Diesel Blends

Fig 1 showed the variation of flash points of *Colocynthus vulgaris*, *Cyperus esculentus* and *Sesamum indicum* ethylesters and their corresponding diesel blends. From the plot it was observed that *Colocynthus vulgaris* ethylesters and its diesel blends had the highest value of flash points, followed by *Sesamum indicum* then *Cyperus esculentus*. The high flash point value of *Colocynthus vulgaris* ethylesters indicated that the samples contained relatively low volatile components when compared to *Cyperus esculentus* and *Sesamum indicum*. From the plot, it was also observed that the flash point tends to increase with increase in concentration of the esters in the biodiesel blends.

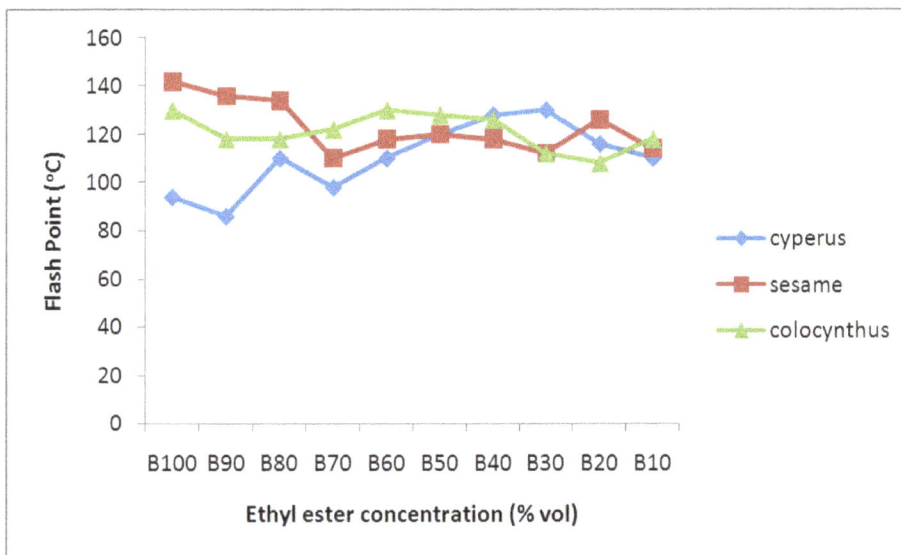

Fig. 1. *plot of flash point verus Ethylester concentrations in ethylesters/diesel blends.*

3.2. Variation of Pour Point of the Ethylesters and Their Diesel Blends

From the plot in fig 2, *Cyperus esculentus* ethyl ester had the poorest pour point; this result could be explained further from the result obtained in Table 1 which showed that

Cyperus esculentus ethylester contained high amount of saturated fatty acids (57.14%). Cold flow properties could worsen with increase in degree of saturation of fatty acids (Knothe, 2007; Keith 2001, Shrestha *et al*; 2008).

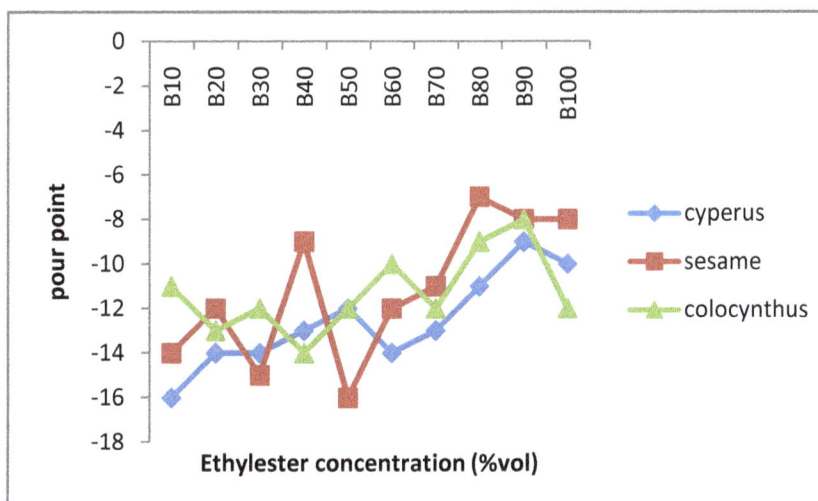

Fig. 2. *Plot of pour point versus Ethylester concentrations in ethylesters/diesel blends.*

Dilution of the ethylesters decreased the pour point of the biodiesels. The ethylesters and their diesel blends were liquid at room temperature (27°C) and will have no cold flow problems in countries where room temperature is always above 10°C.

Colocyntyhus vulgaris ethyl ester showed better cold flow properties than others.

3.3. Variation of Viscosity of the Ethylesters and Their Diesel Blends

From the plot in Fig 3, *Colocynthus vulgaris* ethylester had the highest viscosity on dilution with diesel when compared to the other ethylesters at40°C, followed by *Sesamum indicum* methylesters then *Cyperus esculentus* ethylesters. This showed that the interaction between fossil diesel and these esters were the same and that the diesel had no or little interaction with the esters.

There was a progressive decrease in viscosity of the ethylesters blends (B10-B100), this showed that blending petrodiesel with biodiesel improves the viscosity of the blend thereby ensuring their compatibility with modern diesel engines which have fuel injection systems and are sensitive to viscosity change (Jaichander *et al*; 2011).

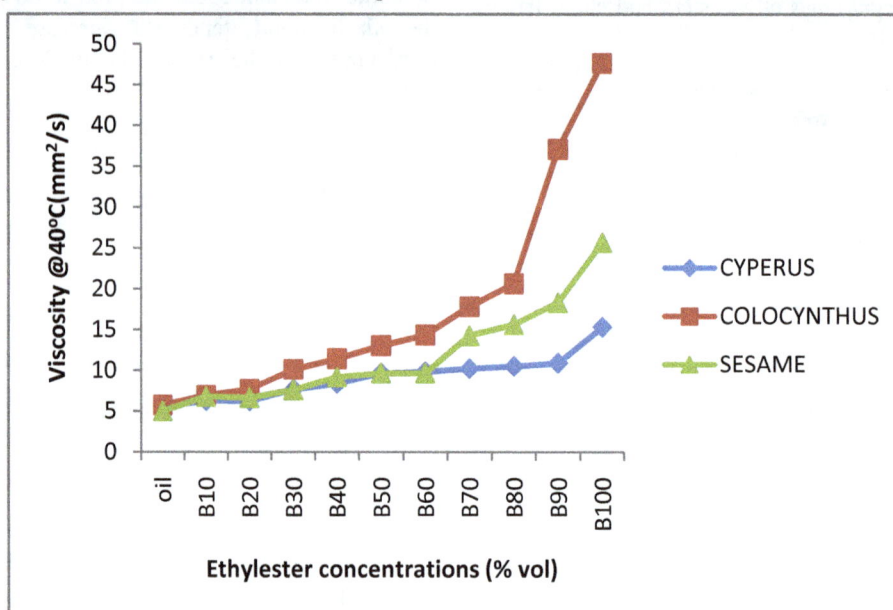

Fig. 3. *plot of viscosity @ 40°C versus ethylester concentration in ethylester/diesel blends.*

Table 6. *Grading of the oils, ethylesters and its diesel blends based on their viscosities according to ASTM standards.*

Ethylesters	1D	2D	4D
@ 40°C			
Colocynthus	-	-	B10-B80
Sesamum	-	-	B10-B90
Cyperus	-	-	B10-B100

From Table 6 above, it was observed that *Colocynthus vulgaris* ethylesters B90-B100, *Sesamum indicum* B100 at 40°C did not fall into any grade while others fell into grade 4D diesel class (heavy diesel grades) showing that they more likely to be used as diesel fuels after heating and cooling.

4. Conclusion

Colocynthus vulgaris, Cyperus esculentus and *Sesamum indicum* ethylester were rich in saturated fatty acid than in monounsaturated fatty acids and poor in polyunsaturated indicated that these biodiesel can have good oxidation stability. The fuel properties of *Colocynthus vulgaris Cyperus esculentus* and *Sesamum indicum* biodiesel blends were better when compared to the oil and petrodiesel in terms of flash points, viscosity and pour points, hence the optimum engine performance of both oil and petrodiesel could be improved by use of these biodiesel feedstock.

Acknowledgements

We wish to express our profound gratitude to Department of Pure and Industrial Chemistry, Nnamdi Azikiwe University Awka and NARICT Zaria for their research laboratory facilities.

References

[1] Ajiwe, V.I.E., Mbonu, S.O., and Enukorah, E.A.O., (2006): Biofuel from *Irvinga gabonesis* seed oil and its methylesters, proceedings of the NASEF'06 International Conference, Awka, Solar Energy society of Nigeria, pp. 247-252.

[2] Ajiwe, V.I.E., Okeke, C.A., Agbo, H.U, (1995): Extraction and utilisation of *Afzelia africana* seed oil; Bioresources technology, Elsevier Science Limited, Britain, 53, 89-90.

[3] Altin, R., Çetinkaya, S, Yücesu,H.S., (2001) The potential of using vegetable oil fuels as fuel for diesel engines. Energy Conversion and Management, 42: 529-538.

[4] Cert, A., Moreda, W., Pérez-Camino, M.C., (2000) Chromatographic analysis of minor constituents in vegetable oils. Journal of Chromatography, 881: 131-148.

[5] Jaichandar, S., and Annamaiia, K. (2011) The Status of Biodiesel as an Alternative Fuel for Diesel Engine- An Overview, Journal of Sustainable Energy & Environment, 2, 71-75.

[6] Katikaneni, S. P. R., Adjaye, J.D., Idem, R.O., Bakhshi, N. N., (1998) Performance studies of various cracking catalysts in the conversion of canola oil to fuels and chemicals in a fluidized-bed reactor. JAOCS, 75: 381-391.

[7] Kaufmann, H. P., Funke, S., (1938) The field of fats. LIX. The viscometry of fats (ZurViskometrie de fette(Studien auf demfettgebiet, 59. Mitteilung.)).Fette u seifen, 45(5):255-62.

[8] Keith, A. (2001) Handmade Projects (http://joumeytoforever. org/keith_cv.htm).

[9] Knothe, G., Steidley, K. R., (2005) Kinematic viscosity of biodiesel fuel components and related compounds, Influence of compound structure and comparison to petrodiesel fuel components. Elsevier journal fuel, 84:1059-1065.

[10] Ma, F., Hanna, M.A., (1999) Biodiesel production: a review. Bioresource Technology 70: 1-15.

[11] Shrestha, D.S., Van, G., and Thompson, J., (2008) Effectiveness of Cold Flow Additives on Various Biodiesels, Diesel and Their Blends; Transactions of the ASASE, 51 (4):1365-1370.

4

Saccharification of *Ulva lactuca* via *Pseudoalteromonas piscicida* for biofuel production

El-Naggar M. M.[1], Abdul-Raouf U. M.[2], Ibrahim H. A. H.[1], El-Sayed W. M. M.[3, *]

[1]Microbiology Lab., Environ. Div., National Institute of Oceanography and Fisheries (NIOF), Alexandria, Egypt

[2]Botany and Microbiology Department, Faculty of Science, Al-Azhar University-Assiut Branch. Egypt

[3]Microbiology Lab., Environ. Div., National Institute of Oceanography and Fisheries (NIOF), Hurghada, Egypt

Email address:

mmelnaggar@yahoo.com (M. M. El-Naggar), oabdulraouf@yahoo.com (U. M. Abdul-Raouf),
drhassan1973@yahoo.com (H. A. H. Ibrahim), walled_mohamed78@yahoo.com (W. M. M. El-Sayed)

Abstract: *Pseudoalteromonas piscicida* WM21 was isolated from seawater at Hurghada, Red Sea, Egypt. It was promising to hydrolyze the polysaccharides of *Ulva lactuca*. *Ulva lactuca* contained 44% carbohydrates, 5% lipids, 16% proteins, 12% Fibers and 23% ash. Optimization of reducing sugars production by *P. piscicida* WM21 was investigated using Plackett-Burmman design. The main effect data as well as the *t*-test results suggested that the beef extract and inoculum size are the most effective variables that controlled the reducing sugar produced by *P. piscicida*. Considerable positive effects of the high levels of substrate concentration and low levels of incubation period were also suggested. On the other hand, variations within the examined levels of pH levels, NaCl and peptone recorded slight effects. While the main effect data as well as the *t*-test results suggested that the substrate concentration and incubation period were the most effective variables that controlled amylase activity produced by *P. piscicida*. To evaluate the accuracy of the applied Plackett-Burman statistical design, a verification experiment was carried out. The predicted near optimum and far from optimum levels of the independent variables were examined and compared to the basal condition settings. The applied near optimum condition, resulted in approximately 56 mg/g increase in reducing sugar with 6 mm amylase activity by *P. piscicida* when compared to the basal medium formulation, while the conditions predicted to be far from optimal recorded approximately 45 mg/g decreases in reducing sugar with 3 mm amylase activity. These results supported the predictions of the applied Plackett-Burman experiment for enhancement of reducing sugar production by marine microorganisms.

Keywords: Reducing sugar, *Ulva lactuca*, *Pseudoalteromonas piscicida*, Saccharification process, Biofuel

1. Introduction

Conversion of biomass from marine algae into ethanol could be economically feasible since some algae hydrolysate can contain more total carbohydrate and hexose sugars than some terrestrial, lignocellulosic biomass feedstock (Philippidis *et al.*, 1993; Chynoweth, 2002; Sluiter, 2006; John *et al.*, 2011). The green macroalgae (chlorophyceae), like *Ulva lactuca* has been considered as a potential aquatic energy crop as early as in the Aquatic Species Program in the USA back in 1978–1996, due to its high potential growth rates and high content of carbohydrates (Ryther *et al.*, 1984). The process of breaking a complex carbohydrate into its monosaccharide components is called saccharification

process. Marine bacteria act a vital role in production of industrial enzymes. Marine bacterial enzymes have several advantages for industrial utilization (Ventosa and Nieto, 1995; Hong *et al.*, 2013; Koppram *et al.*, 2013). Statistical experimental designs are powerful tools for searching the key factors rapidly from a multivariable system and minimizing the error in determining the effect of parameters and the results are achieved in an economical manner (El-Helow and El-Ahawany, 1999; Xiong *et al.*, 2008). The application of statistically based experimental designs is a more efficient approach to deal with a large number of variables (Ooijkaas *et al.*, 1998). Using the Plackett-Burman experimental designs has resulted in increased optimization of fermentation titers, and an ability to predict the presence of mixtures and to select substitutes for complex medium ingredients (Monaghan and Koupal, 1989). The Plackett-Burman experimental design, a fractional factorial design (Plackett and Burman; 1946, Yu et al., 1997) was applied in this research to reflect the relative importance of various environmental factors on the saccharification process in liquid cultures. The current study was suggested for enhancing the ability of isolated marine bacteria to hydrolyze the polysaccharides of *Ulva lactuca* by Plackett-Burman experimental designs.

2. Material and Methods

2.1. Sample Collection

Sea water and sediment samples were collected from marine environment at Hurghada coastline, Red Sea, Egypt, summer 2011. While *Ulva lacttuca* were harvested from Red Sea, Egypt coastline, to be used as substrates of reducing sugars production.

2.2. Screening of Hydrolysis Enzymes Producing Marine Bacteria

Isolation of marine bacteria from the seawater and sediment samples was performed by serial dilution and spread plate method. A volume of 1 ml of each dilution was transferred aseptically to starch nutrient agar plates containing: starch; 10 g/L and nutrient agar; 23 g/L. The plates were incubated at 30°C ± 2°C for 24 h. Tooth picking technique was used to test the ability of isolated bacteria to produce amylase (Margesin et al., 2003). The plates were stabbed consequently using a sterile clean tooth pick in each time with a single colony of each of the tested bacteria. After 3 days the stabbed plates were flooded with iodine solution for detecting the amylase enzyme (Horikoshi, 1999).

2.3. Phenotypic Characterization of the Promising Marine Bacterial Isolate

Phenotypic characteristics such as Gram's staining, motility, cultural characteristics, catalase, oxidase and IMViC test of the marine bacterial isolate was studied by adopting standard procedures. Effect of sodium chloride, pH level, and temperature on growth was tested.

2.4. Electron Microscope Investigation

For scanning electron microscopy (SEM), bacterial cells, grown in S.N.B medium, were harvested by mild centrifugation, washed with phosphate buffer pH 8 and fixed with 2% glutaraldehyde followed by 1% osmium tetroxide treatment. After completion of fixation, sample was washed in buffer solution, and the washed cells were dehydrated in ascending order of ethanol concentrations. The sample was dried completely in a critical point dryer, and finally coated with gold in JEOL-JFG1100 E ion-sputter-coater. The

specimen was viewed in JEOL-JSM 5300 scanning electron microscope operated at 20 kV with a beam specimen angle of 45°.

2.5. Genotypic Characterization the Promising Marine Bacterial Isolate

The promising bacterial isolate was cultured in SN liquid medium for 2 days and genomic DNAs were extracted with genomic DNA extraction protocol of GeneJet genomic DNA purification Kit (Fermentas). PCR using Maxima Hot Start PCR Master Mix (Fermentas) and PCR clean up to the PCR product was performed using GeneJET™ PCR Purification Kit (fermentas). The sequencing to the PCR product on GATC Company was made by using ABI 3730xl DNA sequencer by using universal primaries (16S 27F and 16S 1492R) Table 1.

Table 1. *The primers used in PCR amplification and sequencing.*

Primers	Sequence (5` to 3`)
16S 27F	AGAGTTTGATCCTGGCTCAG
16S 1492R	GGTTACCTTGTTACGACTT

2.6. Chemical Composition of Ulva Lactuca

Humidity, organic matter and ash were determined by standard methods (Pádua et al., 2004). Protein content was measured with Kjeldahl method using a factor of N = 6.25 (AOAC, 2000; method 976.05). Lipid content was determined with the Soxhlet method (AOAC, 2000; method 920.39). Subtraction of the sum of humidity, protein, lipids, fibers and ash values from 100 was the carbohydrate contents in percentage (Pádua et al., 2004).

2.7. Pretreatment of Ulva Lactuca

The algal sample was dried at 70°C over night. All dried algae were milled using a laboratory hammer mill in order to obtain a chip size less than 0.2 mm. the dilute acid pretreatment of biomass was optimized at 121°C in an autoclave for with sulfuric acid concentrations 1N, w/v for one hour (Taherzadeh and Karimi, 2008).

2.8. The Saccharification Medium

The saccharification process was carried out using medium containing pretreated substrate. A volume of 100 ml of sterile starch nutrient broth medium was added to 3 g from pretreated substrate as sole carbon source and inoculated with 3 ml freshly prepared inoculum of bacteria. The flasks were loaded on a rotary shaker incubator at a speed of 120 ± 2 rpm at 30 ± 2°C for 24 h. After incubation, the production broths were centrifuged at 5000 rpm for 15 min. The supernatants were collected for determining the reducing sugars concentration and enzymatic activity.

2.9. Estimation of Reducing Sugars

The amount of reducing sugars was estimated by dinitrosalicylic acid (DNSA) method (Miller, 1959).

2.10. Optimization of Saccharification Process by Experimental Design

Seven independent variables were screened in nine combinations organized according to the Plackett-Burman design matrix. For each variable, a high (+) and low (-) level was tested. The factors tested were given in Table 2. All trials were performed in duplicates and the averages of observation results were treated as the responses. The main effect of each variable was determined using the following equation:

$$E_{xj}= (M_{i+} -M_{i-})/N$$

Where E_{xj} is the variable main effect, M_{i+} and M_{j-} are enzyme activity & reducing sugar (glucose conc.) in trials where the independent variable (xi) was present in high and low levels, respectively, and N is the number of trials divided by 2. A main effect figure with a positive sign indicates that the high level of this variable is nearer to optimum. Using Microsoft Excel, statistical *t*-values for equal unpaired samples were calculated for determination of variable significance (Plackett and Burman; 1946, Yu *et al.*, 1997).

Table 2. Factors examined as independent variables affecting the production of enzymes production by P. piscicida and its levels in the Plackett-Burman experiment.

Factor	Symbol	Level		
		-1	0	+1
Substrate conc. (g/L)	SC	10	30	50
Incubation period (h)	IP	24	72	120
NaCl (g/L)	NC	20	30	40
Peptone (g/L)	Pep	3	5	7
Beef extract (g/L)	BE	1	3	5
pH	pH	7	8	9
Inoculum size (ml)	IS	1	3	5

2.11. Verification of Plackett-Burman Experiment

In order to validate the obtained results and to evaluate the accuracy of the applied Plackett-Burman statistical design, a verification experiment was carried out in triplicates. According to the main effect results, the predicted near optimum and far from optimum levels of the independent variables were examined and compared to the basal condition settings. The enzymes activity and reducing sugar concentration were then estimated as described before (Plackett and Burman; 1946, Yu *et al.*, 1997).

2.12. Reducing Sugar Estimation by High Performance Liquid Chromatography (HPLC)

The concentration of monosaccharides obtained from predicted near optimum conditions were measured by HPLC. The Shimpack SPR-Ca column (Shimadzu, Japan) at 80°C with IR-detector was used. The mobile phase was distilled water at a flow rate of 0.5 ml/min. The samples were filtered with 0.45 μm of cellulose acetate filter and 10 μl of injection volume was added.

3. Results

3.1. Characterization of the Promising Marine Bacterial Isolate

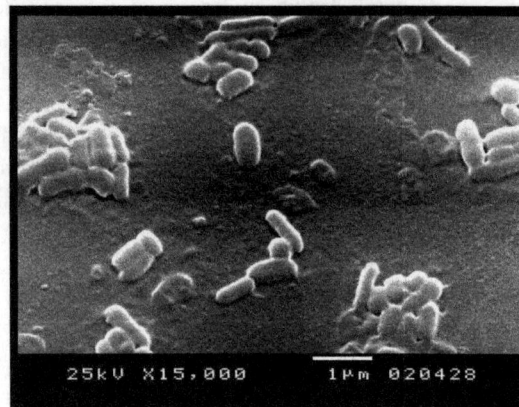

Figure 1. Scanning electron microscopic feature of bacterial isolate WM21 shown as rods.

The most promising marine bacterial isolate (WM21) was submitted to the phenotypic characterization through morphological, physiological and biochemical tests and submitted to genotypic characterization through 16S rDNA technique.

It grew at 20-45°C, pH range (6-10) and at all examined sodium chloride tested concentrations (0-10%) Also catalase and oxidase are positive. Moreover, produce gelatinase, amylase, agarase, lipase (Tween 80), and lysine decarboxylase. In addition to, it was utilizing of glucose.

It has been found that the bacterial isolate WM21 had 96% identical counterpart with respect to 16S rRNA sequence. Sequence of the isolate WM21 was affiliated according to their 16S rDNA to the genus Pseudoalteromonas. However, the isolate WM21 showed 96% sequence homology to *Pseudoalteromonas piscicida*.

Table 3. Accession number of the experimental 16S rDNA sequence and similarity percentage to the closest known species.

Accession no.	Most related Species	Similarity (%)
359805839	*Pseudoalteromonas piscicida*	96

3.2. Chemical Cmposition of Algal Substrate

The chemical analysis for the algal species; *U. lactuca* was carried out and the content of organic matter included the total carbohydrates, lipids, proteins and ash percentages were detected (Table 4).

Table 4. The chemical composition of different algal substrates.

Composition (%)	U. lactuca
Moisture content	10
Organic matter	77
Carbohydrate	44
Lipid	5
Protein	16
Fibers	12
Ash	23

3.3. Optimization of Saccharification Process by Plackett-Burman Design

Seven factors were examined as independent variables affecting the production of reducing sugars by *P. piscicida*. The symbols and level of these factors were presented in Table 2, while their distribution in nine major trails according to fitted design and their responses were shown in Table 5. Data conducted that the highest concentration of reducing sugar recorded 144 mg/g by trial No.6 with maximization of 41 mg/g comparing to the basal condition 103 mg/g. The main effect data presented in Table 6 as well as the t-test results in suggested that the beef extract is the most effective variable that controlled the reducing sugar value by *P. piscicida*. The main effect of each variable, based on the reducing sugar concentration, was estimated as the difference between both averages of measurements made at the high level (+) and the low level (-) and represented graphically in Figure 2. The main effect data presented in Table 6 as well as the *t*-test results suggested that the beef extract and inoculum size are the most effective variables that controlled the reducing sugar produced by *P. piscicida*. According to these results, the lower inoculum size (1ml) and the higher beef extract (5g/L) are nearer to optimum than their opposite

levels. Considerable positive effects of the high levels of substrate concentration and low levels of incubation period are also suggested by Figure 2. On the other hand, variations within the examined levels of pH levels, NaCl and peptone recorded slight effects. While the main effect data presented in Table 6 as well as the *t*-test results suggested that the substrate concentration and incubation period are the most effective variables that controlled amylase activity produced by *P. piscicida*. According to these results, the lower incubation period (24 h) and the higher substrate concentration (5%) are nearer to optimum than their opposite levels. Considerable of the low levels of pH and the high levels of peptone are positive effect also suggested by Figure 3. On the other hand, variations within the examined levels of beef extract, NaCl, and inoculum size recorded slight effects. The interacting effects of beef extract and substrate concentration as described in three-dimensional graph (Figure 4; A), suggest that, within the examined ranges the higher beef extract accompanied by the higher substrate concentration would markedly increase the reducing sugar expressed by the experimental bacterium; *P. piscicida*. However, the interaction beef extract and inoculum size (Figure 4; B) with respect to production of the reducing sugar appeared to be high.

Table 5. Applied Plackett-Burman design for seven cultural variables and the experimental results of reducing sugar produced by P. piscicida.

Trial No.	Independent variables							Reducing sugar conc. (mg/g)	Amylase (mm)
	SC	IP	NC	Pep	BE	pH	IS		
1	-	-	-	+	+	+	-	110	21
2	+	-	-	-	-	+	+	92	28
3	-	+	-	-	+	-	+	90	18
4	+	+	-	+	-	-	-	106	29
5	-	-	+	+	-	-	+	80	23
6	+	-	+	-	+	-	-	**144**	**32**
7	-	+	+	-	-	+	-	70	15
8	+	+	+	+	+	+	+	116	27
9	0	0	0	0	0	0	0	103	29

Table 6. Statistical analyses of the Plackett-Burman experimental results.

Variable	Reducing sugar conc. (mg/g)		Amylase (mm)	
	Main effect	*t*-value[1]	Main effect	*t*-value[1]
Substrate conc. (%)	27	1.943	9.75	2.015
Incubation period (h)	-11	2.015	-3.75	2.015
NaCl (g/L)	3	2.013	0.25	1.943
Peptone (g/L)	4	2.013	1.75	2.131
Beef extract (g/L)	28	2.015	0.75	1.943
pH level	-8	1.943	-2.75	1.943
Inoculum size (ml)	-13	2.013	-0.25	2.015

[1]*t*-value significant at the 1% level = 3.70
t-value significant at the 5% level = 2.446
t-value significant at the 10% level = 1.94
t-value significant at the 20% level =1.372
Standard *t*-values are obtained from Statistical Methods (Snedecor and Cochran 1989).

The interacting effects of substrate concentration and incubation period as described in three-dimensional graph (Figure 5; A) suggest that, within the examined ranges, the

shorter incubation period accompanied by the higher substrate concentration would markedly increase the amylase activity expressed by the experimental bacterium; *P. piscicida* to activity of the amylase appeared to be very slight. However, the interaction of substrate concentration and pH level (Figure 5; B) with respect to activity of the amylase appeared to be very slight.

3.4. Verification of Plackett-Burman Experiment

In order to validate the obtained results and to evaluate the accuracy of the applied Plackett-Burman statistical design, a verification experiment was carried out in triplicates. The predicted near optimum and far from optimum levels of the independent variables were examined and compared to the basal condition settings. The Reducing sugar concentration and amylase activity of out using *P. piscicida* observations are shown in Table 7. The applied near optimum condition, resulted in approximately 56 mg/g increase in reducing sugar and 6 mm amylase activity by *P. piscicida* when compared to the basal medium formulation. On the other hand, the condition predicted to be far from optimal recorded approximately 45 mg/g decreases in reducing sugar and 3 mm amylase activity. These results support the predictions predicted from the applied Plackett-Burman experiment.

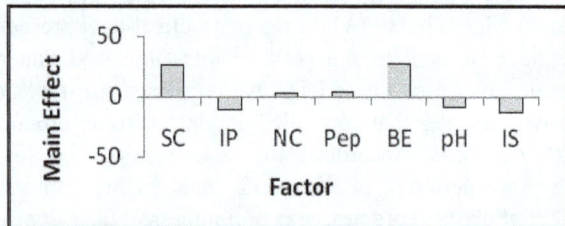

Figure 2. Elucidation of the factors affecting the production of reducing sugars by P. piscicida.

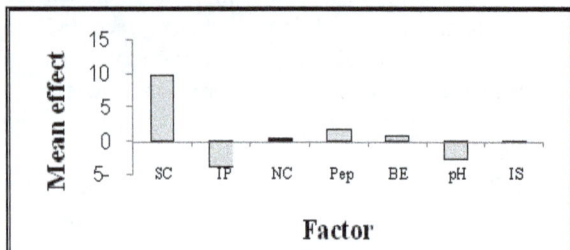

Figure 3. Elucidation of the factors affecting the amylase activity by P. piscicida.

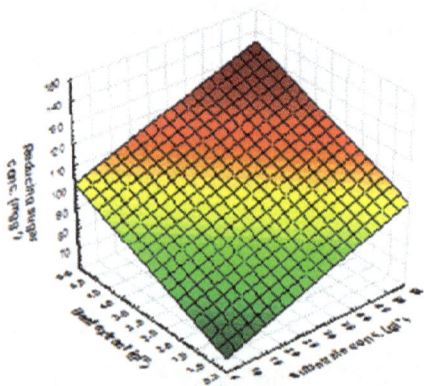

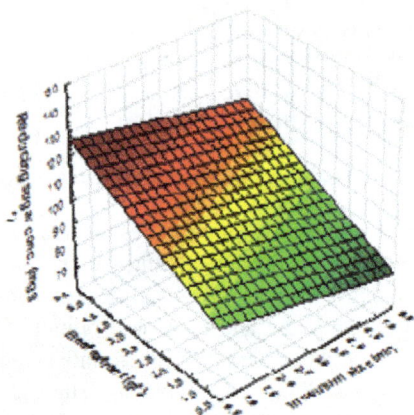

Figure 4. The response 3D Surface Polt analysis show the interaction of the different beef extract with different substrate concentration (A) and different inoculum size (B) in relation to reducing sugar mg/g by P. piscicida.

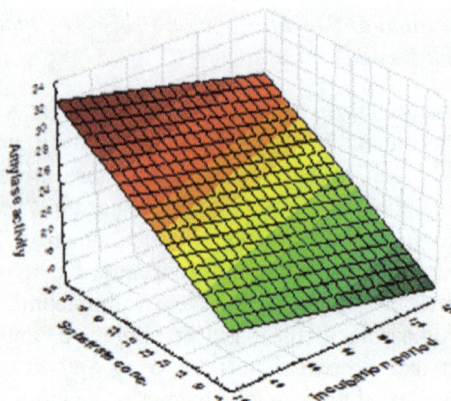

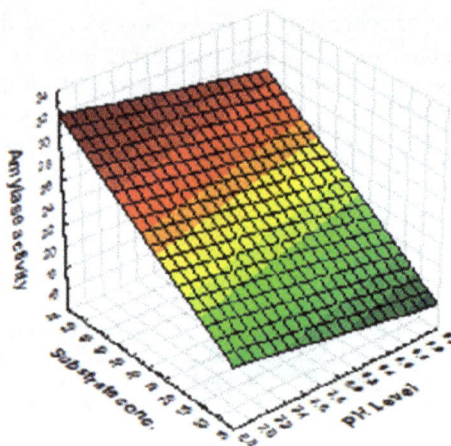

Figure 5. The response 3D Surface Polt analysis show the interaction of the different substrate concentration with different incubation period (upper) and different pH level (lower) in relation to amylase activity by P. piscicida.

Table 7. Verification of the Plackett-Burman experimental results that carried out using P. piscicida.

Response (Average)	Basal medium	Near optimum medium	Far from optimum medium
Reducing sugar conc. (mg/g)	102.3±0.007	158±0.005	57±0.005
Amylase (mm)	29.3±0.003	35±0.005	26.3±0.005

3.5. The Reducing Sugar Analysis by HPLC

The monosaccharides and their concentration were estimated in near optimum medium using HPLC Glucose and galactose were detected at 144 mg/g and 24 mg/g respectively (Figure 6).

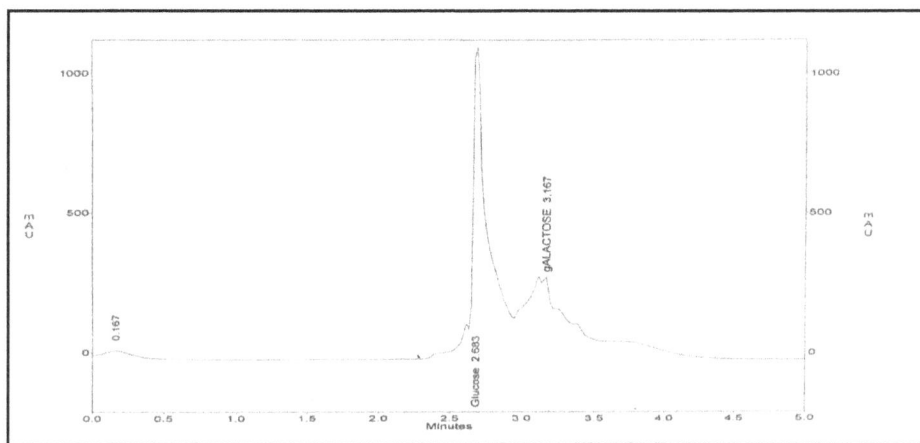

Figure 6. HPLC chromatogram showing glucose and galactose concentrations produced from U. lactuca hydrolysis by P. piscicida at retention time 2.68 and 3.17 min, respectively.

4. Discussion

Macroalgae are gaining some attention as an alternative renewable source of biomass for the production of bioethanol (Borowitzka, 1992). Some algae were represented to have high contents of carbohydrate that can be used as substrate for bioethanol production (Singh and Olsen, 2011). They also contain a low concentration of lignin (Wi et al., 2009) or no lignin at all (Ge *et al.*, 2011). Therefore, the biomass of marine algae can be converted economically into simple sugars and then to bioethanol (Chynoweth, 2002; Sluiter, 2006; John *et al.*, 2011).

The green macroalga; *Ulva lactuca* is appropriate substrate that contain 44% carbohydrates.

The marine bacterium WM21 exhibited promising enzymatic hydrolysis of polysaccharides. So, it was selected to be identified using phenotypic and genotypic analyses, respectively. It was gram-negative rod, grew at 20-45°C with pH range (6-10) and sodium chloride concentrations (2-10%). Also, catalase and oxidase were positive. Moreover, it produced gelatinase, amylase, agarase, lipase (Tween 80), and lysine decarboxylase. In addition to, it was utilizing of glucose. This isolate genotypically was *Pseudoalteromonas piscicida* with similarity 96%.

Actually, some workers isolated *Pseudoalteromonas sp.* and *P. piscicida* from different marine resources. For examples, Nelson and Ghiorse (1999) isolated *P. piscicida* from the diseased eggs of two damselfish (Pomacentridae) species; *Amphiprion clarkii* and *Amblyglyphidodon curacao*. As well as, Lau *et al.* (2005) isolated a Gram-negative, non-spore-forming, short rod-shaped bacterium (UST010723-006T) from the surface of the sponge Mycale adherents in Hong Kong waters. Cells of UST010723-006 did not have flagella and were non-motile. Colonies were pale orange in color, 2–4 mm in diameter, convex with a smooth surface and an entire translucent margin. Gas bubbles were observed in the colonies and also in the agar matrix underneath and adjacent to the colonies. UST010723-006 was heterotrophic, strictly aerobic and required NaCl for growth (2–6%). It grew at pH 5–10 and

between 12 and 44°C. Phylogenetic analysis of the 16S rRNA gene sequence placed UST010723-006T within the genus *Pseudoalteromonas* of the c-subclass of the Proteobacteria. These data supported the affiliation of UST010723-006T to the genus *Pseudoalteromonas*. The closest relatives were *Pseudoalteromonas luteviolacea*, *P. phenolica*, *P. rubra* and *P. ruthenica* with similarity values ranging from 95.4 to 96.8%. Molecular evidence, together with phenotypic characteristics, suggests that UST010723-006T constitutes a novel species within the genus Pseudoalteromonas. The name *Pseudoalteromonas spongiae* sp. is proposed for this bacterium.

Tao *et al.* (2008) examined the ability of *P. piscicida* and/or closely related species for hydrolytic enzymes secretion especially; amylase and agarase. They isolated marine bacteria producing extracellular α-amylase from seawater and identified as member of *Pseudoalterimonas* species. Matsumoto *et al.* (2003) inoculated *Pseudoalterimonas undina* NKMB 0074 into suspensions containing the green microalgae NKG 120701 cells and increasingly reduced suspended sugars with incubation time. Terrestrial amylase and glucoamylase were inactive in saline suspension. They concluded that the marine amylase is necessary in saline conditions for successful saccharification of marine microalgae.

In a complementary step of optimization saccharification, the concentrations of medium components were simultaneously investigated using Plackett-Burman experiment design. Occasionally, seven factors were examined as independent variables affecting the production of reducing sugars by *P. piscicida*. Data conducted that the highest concentration of reducing sugar recorded 144 mg/g by trial No.6 with maximization of 41 mg/g comparing to the basal condition (103 mg/g) in trail No. 9. And when applied the near optimum media we reached to reducing sugar production about 158 mg/g.

By comparing these data (158 mg/g equals 7.9 mg/ml) with the reducing sugar ratio in other investigations, it was observed that our yield is was much more than that achieved by Malek *et al.* (1988) which was (5.8 mg/ml) obtained with a

Cytophaga sp., using sugar cane bagasse as growth and hydrolysis substrates. While, Sunarti *et al.* (2010) obtained lower hydrolytic activity produced from cellulose fraction by isolate C4-4, which liberated (3.5 mg/ml) of total sugar. Yanagisawaa *et al.* (2011) used successive saccharification with an enzyme that was effectively used to obtain high concentrations of glucose from these seaweeds. They hypothesized that agar weed contained both galactan and glucan. For this reason, it was possible to obtain a high concentration of ethanol from agar weed using combined saccharification, which is the acid hydrolysis of galactan to produce galactose followed by the enzymatic hydrolysis of glucan to obtain glucose. Begum and Alimon (2011) pretreated three lignocellulosic substrates viz. sugarcane bagasse, sawdust and water hyacinth with alkali and enzyme and studied their effect on bioconversion agricultural and industrial wastes to chemical feedstock. They found that the maximum degree of conversion of substrate by *Aspergillus oryzae* ITCC (0.415%) and improved specific substrate consumption (0.99 g substrate/g dry biomass) was exhibited in sugarcane bagasse after alkali treatment at 96 h. they observed that alkali-treatment and enzyme-treatment, water hyacinth was the best for cellulase induction and showed maximum endoglucanase activity of 11.42 U/ml. Reducing sugar yield ranged from 1.12 mg/ml for enzyme treated sawdust at 48 h to 7.53 mg/ml for alkali treated sugarcane bagasse at 96 h. Alkali-treated sugarcane bagasse gave the highest saccharification rate of 9.03% after 96 h. The most resistant substrate was sawdust which produced 5.92% saccharification by alkaline treatment.

However, the current study supports the following points:

1. Marine macroalgae (*U. lactuca*) can be used as promising substrates for the bioethanol production.
2. *Pseudoalteromonas piscicida* have ability to hydrolysis of *U. lactuca* substrate for production of reducing sugars.
3. Optimization of saccharification process enhanced the reducing sugars production.

References

[1] AOAC (Association of Analytical Chemists), (2000). Official Methods of Analysis of Association of Analytical Chemist. Horwitz, W., Gaithersburg, Maryland, USA.

[2] Austin, B. (1988). The marine environment. In: Marine Microbiology, (pp.1-11). Cambridge: Cambridge University Press.

[3] Begum, M.F. and Alimon, A.R. (2011). Assessment of some wild *Aspergillus* species for cellulase production and characterization. African Journal of Microbiology Research, 5(27), 4739-4747.

[4] Borowitzka, M. (1992). Algal biotechnology products and processes – matching science and economics. J. Appl. Phycol; 4, 267–79.

[5] Box, G.E.P., and Draper, N. R. (1987). Empirical Model Building and Response Surfaces, John Wiley & Sons, New York, NY.

[6] Carolissen-Mackacy, V., Arendse, G. and Hastings, J.W. (1997). Purification of bacteriocins of lactic acid bacteria: problems and proteins. Food Microbiol; 34, 1-16.

[7] Chynoweth, D.P. (2002). Review of Biomethane from Marine Biomass. Department of Agricultural and Biological Engineering, University of Florida, Gainesville, Florida, USA.

[8] El-Helow, E.R. and El-Ahawany, A.M. (1999). Lichenase production by catabolite repression resistant *Bacillus subtilis* mutants: Optimization and formation of an agro-industrial by product medium. Enz. Microbiol. Technol; 24, 325-331.

[9] Fritze, D., Flossdorf J. and Claus, D. (1990). Taxonomy of alkaliphilic Bacillus strains. International Journal of Systematic Bacteriology, 40, 92-97.

[10] Ganzon-Fortes, E.T. (1991). Characteristics and economic importance of seaweeds. In: Proceedings of the seaweed research training and workshop for project leaders. Philippine Council for Aquatic and Marine Research and Development.

[11] Ge, L., Wang, P. and Mou, H. (2011). Study on saccharification techniques of seaweed wastes for the transformation of ethanol. Renewable Energy, 36, 84-89.

[12] Hong, L.S., Ibrahim, D. and Omar I.C. (2013). Effect of physical parameters on second generation bio-ethanol production from oil palm frond by *Saccharomyces cerevisiae*. BioResource, 8(1), 969-980.

[13] Horikoshi, K. (1999). Introduction: Definition of Alkaliphilic Organisms. In Alkaliphiles. (Kodansha Ltd., Tokyo, 1999b), p.1.

[14] Hu, Z., Lin, B.K., Xu, Y., Zhong, M.Q. and Liu, G.M. (2009) Production and purification of agarase from a marine agarolytic bacterium *Agarivorans* sp. HZ105. Journal of Applied Microbiology, 181–190.

[15] John, R.P., Anisha, G.S., Nampoothiri, K.M. and Pandey, A., (2011). Micro and macroalgal biomass: a renewable source for bioethanol. Bioresour Technol; 102, 186–93.

[16] Koppram, R., Nielsen, F., Albers, E., Lambert, A., Wännström, S., Welin, L., Zacchi, G. and Olsson, L. (2013). Simultaneous saccharification and co-fermentation for bioethanol production using corncobs at lab, PDU and demo scales. Biotechnology for Biofuels, 6, 2-10.

[17] Lau, S.C.K., Tsoi, M.M.Y., Li, X., Dobretsov, S., Plakhotnikova, Y. Wong, P-K. and Qian, P-Y. (2005). *Pseudoalteromonas spongiae* sp. nov., a novel member of the c-Proteobacteria isolated from the sponge Mycale adhaerens in Hong Kong waters. International Journal of Systematic and Evolutionary Microbiology, 55, 1593–1596.

[18] Malek, , N.A., , Q.M. and N. (1988). Bacterial cellulases and saccharification of lignocellulosic materials. Enzyme and Microbial Technology, 2, 750–753.

[19] Margesin, R., Gander, S., Zacke, G., Gounot, A.M. and Schinner, F. (2003). Hydrocarbon degradation and enzyme activities of cold-adapted bacteria and yeasts. Extremophiles, 7, 451–458.

[20] Matsumoto, M., Yokouchi, H., Suzuki, N., Ohata, H. and Matsunaga, T. (2003). Saccharification of Marine Microalgae Using Marine Bacteria for Ethanol Production. Applied Biochemistry and Biotechnology, 4(6): 105–108.

[21] Miller, G.L., (1959). Use of dinitrosalicylic acid for determination of reducing sugar. Analytical Chemistry, 31, 426–428.

[22] Monaghan, R.L. and Koupal, L.R. (1989). Use of the Placket and Burman technique in a discovery program for new natural products. Novel Microbes: Products for Medicine and Agriculture, Chapter 2, 25-32.

[23] Myers, R.H. and Montgomery, D.C. (1995). Response surface methodology: process and product optimization using designed experiments. John Wiley and Sons Inc., New York, N.Y.

[24] Nelson, E.J. and Ghiorse, W.C. (1999). Isolation and identification of *Pseudoalteromonas piscicida* strain Cura-d associated with diseased damselfish (Pomacentridae) eggs. Journal of Fish Diseases, 22, 253-260.

[25] Ooijkaas, L.P., Wilkinson, E.C., Tramper, J. and Buitelaar, R.M., (1998). Medium optimization for spore production of Conithyrium minitans using statistically-Based experimental designs. Biotechlnol. Bioeng. 64 (1), 92-100.

[26] Pádua, M., Fontoura, P.S.G. and Mathias, A.L. (2004). Chemical Composition of *Ulvaria oxysperma* (Kützing) Bliding, *Ulva lactuca* (Linnaeus) and *Ulva fascita* (Delile). 47(1), 49-55.

[27] Park, J.I., Woo, H.C. and Lee, J.H. (2008). Production of bio-energy from marine algae: status and perspectives. Korean Chem. Eng. Res; 46 (5), 833–844.

[28] Philippidis, G..P, Smith, T.K., and Wyman, C.E. (1993). "Study of the Enzymatic Hydrolysis of Cellulose for Production of Fuel Ethanol by the Simultaneous Saccharification and Fermentation Process." Biotechnology and Bioengineering, 41, 846-853.

[29] Plackett, R.L. and Burman, J.P. (1946). The design of optimum multifactorial experiments. Biomrtrika, 33, 305-325.

[30] Ryther, J.H., Debusk, T.A. and Blakeslee, M. (1984). Cultivation and conversion of marine macroalgae (*Gracilaria* and *Ulva*). In: SERI/STR-231-2360, pp. 1–88.

[31] Singh, A. and Olsen, S.I. (2011). A critical review of biochemical conversion, sustainability and life cycle assessment of algal biofuels. Applied Energy, 88, 3548–3555.

[32] Sluiter, A. (2006). Determination of structural carbohydrates and lignin in biomass. Vol. Version 2006. National Renewable Energy Laboratory, USA. http://www.nrel.gov/biomass/analytical_procedures.html.

[33] Sun, Y. and Cheng, J. (2002). Hydrolysis of lignocellulosic materials for ethanol production: A review. Bioresour. Technol. 83(1): 1-11.

[34] Sunarti, T.C., Meryandini, A., Sofiyanto, M.E. and Richana, N. (2010). Saccharification of corncob using cellulolytic bacteria for bioethanol production. BIOTROPIA, 17(2).

[35] Taherzadeh, M.J. and Karimi, K. (2008). Pretreatment of Lignocellulosic Wastes to Improve Ethanol and Biogas Production: A Review. International Journal of Molecular.

[36] Tao, X., Jang, M.S., Kim, K.S., Yu, Z. and Lee, Y.C. (2008). Molecular cloning, expression and characterization of alpha-amylase gene from a marine bacterium *Pseudoalteromonas* sp. MY-1. Indian J Biochem Biophys.;45(5), 305-309.

[37] Todorov, S., Giotcheva, B., Douset, X., Onno, B. and Ivanov, K. (2000). Influence of growth medium on bacteriocin production in Lactobacillus ptantarum ST31. Biotechnol. Equip; 14, 50-55.

[38] Todorov, S.D. and Dicks, L.M.T. (2004). Effect of medium components on bacteriocin production by *Lactobacillus pentosus* ST15/BR, a strain isolated from beer produced by the fermentation of maize, barley and soy flour. World J. Microbiol. Biotechnol; 20, 643-650.

[39] Trono, J.G.C. and Ganzon-Fortes, E. (1988). Philippine seaweeds. Philippines: National Book-store Inc.

[40] Ventosa, A. and Nieto, J.J. (1995). Biotechnological applications and potentialities of halophilic microorganisms. World J. Microbiol. Biotechnol; 11, 85–94.

[41] Wi, S.G., Kim, H.J., Mahadevan, S.A., Yang, D.J. and Bae, H.J., (2009). The potential value of the seaweed Ceylon moss (*Gelidium amansii*) as an alternative bioenergy resource. Bioresour Technol; 100, 6658–6660.

[42] Xiong, W., Li, X., Xiang, J. and Wu, Q. (2008). High-density fermentation of microalga *Chlorella protothecoides* in bioreactor for microbio-diesel production. Appl. Microbiol. Biotechnol; 78, 29–36.

[43] Yanagisawaa, M., Nakamuraa, K., Arigab, O. and Nakasakia, K. (2011). Production of high concentrations of bioethanol from seaweeds that contain easily hydrolyzable polysaccharides. Process Biochemistry, 46, 2111–2116.

[44] Yokoyama, M.Y. and Guimarães, O., (1975). Determinação dos teores de Na, K, O e proteínas em algas marinhas. Acta Biológica Paranaense, 4 (1/2), 19-24.

[45] Yu, X.S., Hallett, G., Sheppard, J. and Watson, A.K. (1997). Application of the Plackett-Burman experimental design to evaluate nutritional requirements for the production of *Colletotrichum coccodes* spores. Appl. Microbiol. Biotechnol; 47, 301-305.

5

A Variable Step Size MPPT Method for Stand-Alone PV Energy Systems

Tahar Tafticht, Yamina Azzoug

Electrical Engineering Department, College of Engineering, Majmaah University, Majmaah, KSA

Email address:

tahar.tafticht@uqtr.com (T. Tafticht)

Abstract: This paper presents an optimal algorithm control of photovoltaic generator system with battery energy storage. The system is the combination of photovoltaic (PV) array and battery storage via a common dc bus. The system components have substantially different voltage-current characteristics and they are integrated on the DC bus through power conditioning devices for optimal operation by using the developed Maximum Power Point Tracking (MPPT) control method. Using this method, it is possible to adapt the load to the PV modules and to follow the MPP howsoever the weather conditions may vary. This algorithm is based on perturbation and Observation (P&O) method with initial measurement of the tracking reference for estimating the step size to get the optimal operating point. The results show that the approach improves clearly the tracking efficiency of the maximum power available at the output of the PV modules and reduces the oscillations around the maximum power point.

Keywords: Photovoltaic, Battery Energy Storage, MPPT Method

1. Introduction

PV energy is a clean energy which has had a fast development for the last two decades. However, the intermittent character of the solar irradiation requires the use of some energy storage device so as to be able to continuously supply loads from the PV generators. PV energy seems to be an interesting alternative that makes possible to control the load demand energy and to produce not polluting fuel usable in transportation or the buildings, and thus to diversify the energy markets. There are some previous works on PV energy systems [1-11]. Several publications tackle the problem concerning the optimal design and the sizing of the PV systems [1-5]. The majority of these publications show that the PV power system therefore has higher availability to deliver continuous power and results in a better utilization of power conversion and control equipment than either of the individual sources. In addition, the cost of PV systems, based on long terms of data hourly, daily, and yearly was analyzed [6-11]. The price of PV energy remains high and the energetic efficiency of the installations is relatively low. In order to be useable in a broad reach of applications, and to satisfy the economic constraints, the conversion chain of this energy must be robust and reliable. It must also present a better efficiency and be realized at low cost. For that, it is necessary to extract the maximum power from renewable sources. The operating power of renewable sources depends on the sun intensity, the temperature and especially output voltage for the PV modules. If the power transfer between renewable sources and the load is not optimal, the total efficiency of PV energy system will be largely affected. The renewable sources can be operated at the maximum power operating point for various conditions by adjusting the PV modules output voltage optimally.

This paper presents a new MPPT method used to achieve the maximum power control of PV system. By acquiring the output power of PV system, the control searches the maximum power point and tracks this point in the event of change of the meteorological conditions, in order to reduce the error between the operating power and the maximum power. The application of the MPPT method to PV systems with battery energy storage shows that it is possible to increase significantly the efficiency of storage because of the improved control of the MPP during changing ambient weather conditions. Using this method, it is possible to adapt the load to the PV modules and to follow the MPP howsoever the weather conditions may vary. This paper presents an optimal algorithm control of photovoltaic generator system with battery energy storage. This algorithm is based on P&O method with initial measurement of the tracking reference for estimating the step size to get the optimal operating point. The new algorithm reduces the oscillations around the maximum power point, and increases the average efficiency of the MPPT obtained.

2. System Configuration

Figure 1 presents the diagram of the different components of the hybrid PV system. The system consists of a 1 kW (peak) solar PV array as primary energy sources. The excess energy with respect to load demand has been stored as batteries storage. The hybrid system components have substantially different voltage-current characteristics and are integrated through the developed power conditioning devices on a 48V DC bus, which allows power to be managed between input power, energy storage and load. The PV generator is controlled by a DC/DC buck converter, where the duty ratio is used as the control means to track the maximum power point. The power generated from the PV generator is supplied to the common DC bus. The different sensors are used to record real time voltages and currents of PV generator, common DC bus, load, and DC-DC converter duty ratio.

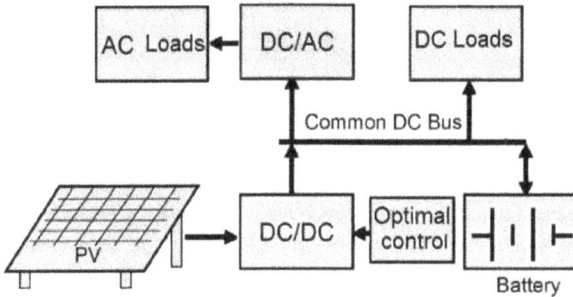

Fig. 1. Schematic diagram for the PV system.

3. PV Characteristics

The equivalent electric diagram of a PV array is shown in figure 2. The PV array's electric characteristic under solar radiation is given by the equation (1) in terms of output current (Ipv) and voltage (Vpv) [12]:

$$I_{pv} = I_L - I_O \left[\exp\left(\frac{V_{pv} + R_s.I_{pv}}{V_T} \right) - 1 \right] - \frac{V_{pv} + R_S.I_{pv}}{R_P} \quad (1)$$

Where $I_L = N_p.I_{ph}$ corresponds to the light-generated current of the solar array with Iph being the cell light-generated current and N_p represents the number of parallel modules.

$I_o = N_p.I_{os}$ corresponds to the reverse-saturation current of the solar array, with I_{os} the cell reverse-saturation current

$V_T = (N_s.n.K_B.T)/q$ is the thermal voltage, where N_s is the

number of cells connected in series, K_B is the Boltzmann's constant, n is the ideality factor, T is the cell temperature, and q is the electron charge;

R_s is the series parasitic resistance of a solar array, Rp its shunt parasitic resistance. The load resistance will be represented by R_{Load} .

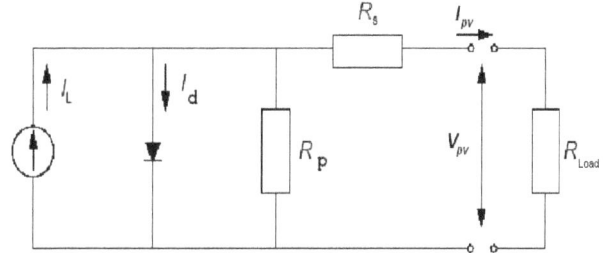

Fig. 2. Equivalent electric diagram of a PV array.

An ideal PV module is one for which R_s is zero and R_p is infinitely large. The output current and voltage are then:

$$I_{pv} = I_L - I_O \left[\exp\left(\frac{V_{pv}}{V_T} \right) - 1 \right] \quad (2)$$

$$V_{pv} = V_T \ln[(I_L - I_{pv})/I_0 + 1] \quad (3)$$

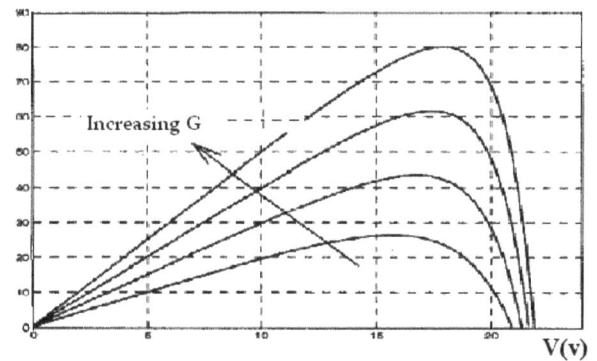

a) Solar radiation influence

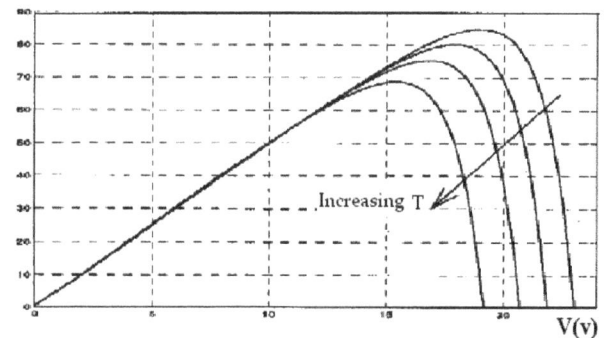

b) Temperature influence

Fig. 3. Solar radiation and temperature influences on the P–V characteristics.

The PV array characteristic presents three important points:

the short circuit current (I_{sc}), the open voltage (V_{oc}) and the optimum power (P_{op}) delivered by the PV array to an optimum load (R_{op}). In this case the PV modules operate at their MPP. Figures 3 a and b give the Power-Voltage (P-V) characteristics of a PV module for varying values of solar radiation and temperature. The short circuit current is clearly strongly proportional to the solar radiation: more radiation, more current, and also more maximum output power. On the other hand the temperature dependence is inverse: an increase in temperature causes a reduction of the open circuit voltage (when sufficiently high) and hence also of the maximum output power. Hence these opposite effects of the variations of solar radiation and temperature on the maximum output power make it important to track the maximum power point efficiently.

4. System Operation

Figure 4 shows the MPPT buck converter diagram for PV optimum power transfer. The switch S of the buck converter is a MOSFET transistor with a low internal resistance Ron. The MOSFET is controlled by a PWM signal generation circuit that uses a microcontroller. For searching the MPP and tracking this point in order to reduce the error between the operating power and the maximum power, in the event of change of the weather conditions, the control of the buck converter track periodically the operating point of the PV module. By acquiring the output voltage and current of PV modules, the control used this information to increase or decrease the duty cycle of the buck converter to change the operating point of the PV array. The operation of the buck converter is described in our previous paper [12] and Load voltage is given by:

$$\frac{V_{Load}}{V_{pv}} = \frac{I_{pv}}{I_{Load}} = \frac{t_{on}}{T} = d \qquad (4)$$

Where t_{on} is turn on period of S switch and T is it period.

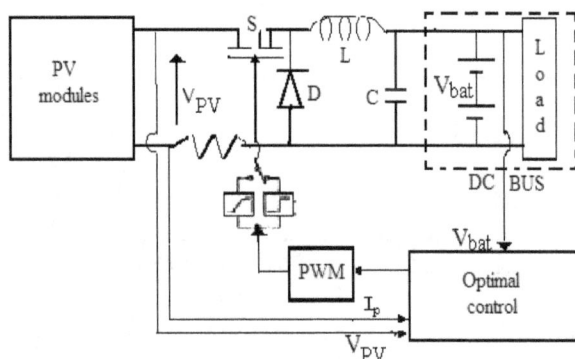

Fig. 4. MPPT Buck converter diagram.

5. PV Optimal Control

In order to extract the maximum power, an analysis was provided to understand the probable displacement of the operating point in the two operating zones of PV Generator.

Figure 5 represents the typical curve of power variation according to the operating voltage. This figure shows that there are two operating zones: the first is located on the right side of the MPP where dP/dv < 0 and the second on the left side of the MPP where dP/dv > 0.

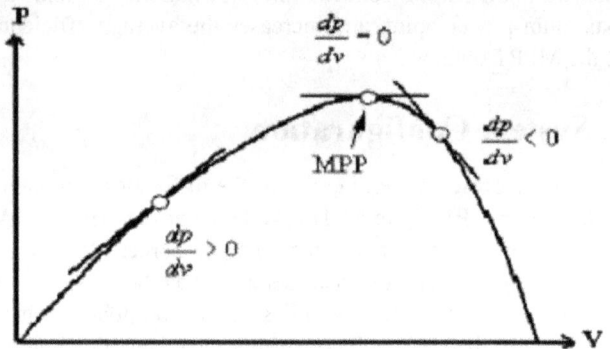

Fig. 5. Probable displacement of the operating point.

Figure 6 gives the algorithm of the proposed MPPT control method, where the information on solar radiation or temperature is not required. For searching the maximum power operating point and tracking this point in order to reduce the error between the operating power and the maximum power if the solar radiation or temperature changes, the control of the buck converter perturbs periodically the operating point of the PV Generator. By measuring the output voltage and current of PV Generator, the control uses this information to change the operating point of the PV Generator by increasing or decrease the duty cycle of the buck converter. After the perturbation, there is a displacement of the operating point from (k-1) to (k). Four cases of perturbation from operating point are distinguished:

1st case: If P(k)>P(k-1) and V(k)>V(k-1), the power increases after perturbation. This indicates that the MPP research is oriented to the good direction. So, the search of the MPP continues in the same direction and reaches the operating point (k+1) by increasing the duty cycle by Δd.

2nd case: If P(k)<P(k-1) and V(k)<V(k-1), the power decreases after perturbation. This indicates that the MPP search is oriented to the bad direction. The MPP search direction must be changed and the duty cycle is increased by two Δd to reach the operating point (k+1).

3rd case: If P(k)>P(k-1) and V(k)<V(k-1), the power increases after perturbation. This indicates that the MPP search is oriented to the good direction. Then, the MPP search direction must be maintained and the duty cycle is decreased by one Δd to reach the operating point (k+1).

4rd case: If P(k)<P(k-1) and V(k)>V(k-1), the power decreased. This indicates that the MPP search is oriented to bad direction. The MPP search direction must be changed and the duty cycle is increased by two Δd to reach the operating point (k+1).

To estimate the MPP reference value, the power wand the voltage of the PV modules is measured cyclically at the output solar panels during the opening of the switch of the buck converter. This voltage allows the MPP reference value to be calculated from equation 5, and consequently the value of the converter's duty cycle (step size) to be adjusted by equation 4. The proposed method scans sequentially the voltage range applicable to the load. At each iteration, the power consumed by the load is calculated. To find the maximum power, the derivative of output power (ΔP) is used. The optimal operating point corresponds to $\Delta P = 0$.

$$\Delta P = \frac{P_i - P_{i-1}}{V_i - V_{i-1}} \qquad (5)$$

6. Results and Discussion

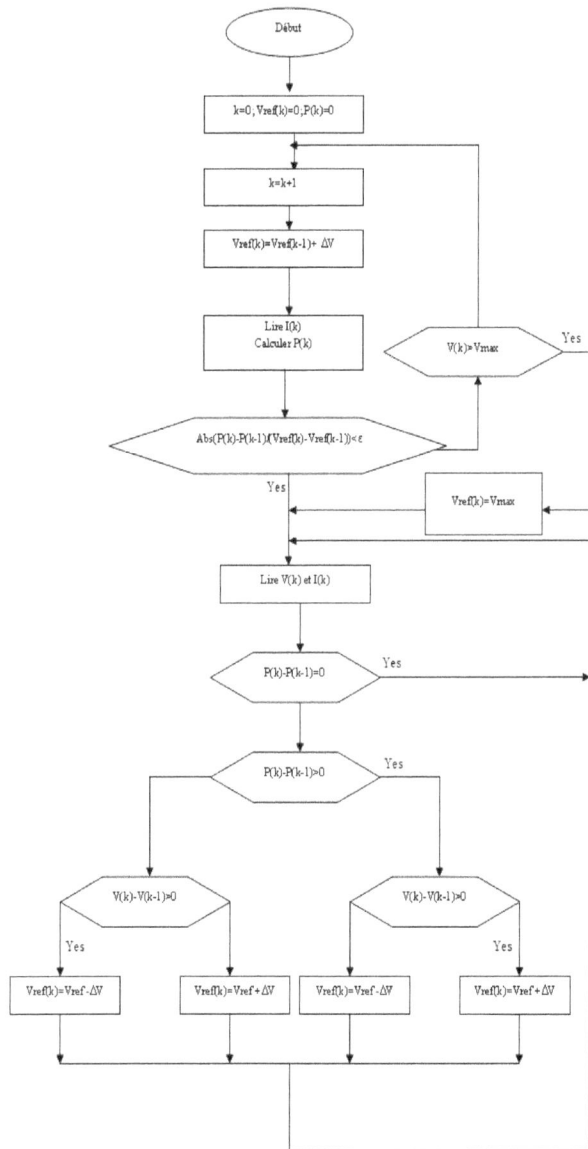

Fig. 6. *Algorithm of the proposed MPPT control method.*

The MPPT control of a PV generator system described above was implemented in a circuit based on a micro-controller PIC18F242 and tested. Figures 7 and 8 give the results obtained with the proposed approach. It is noticed that classic P&O method does not instantly locate the MPP and it

has also an oscillation around the MPP (Fig. 7). The new approach reduces the oscillations around the MPP (Fig. 8). This results in almost stable PV voltage and current.

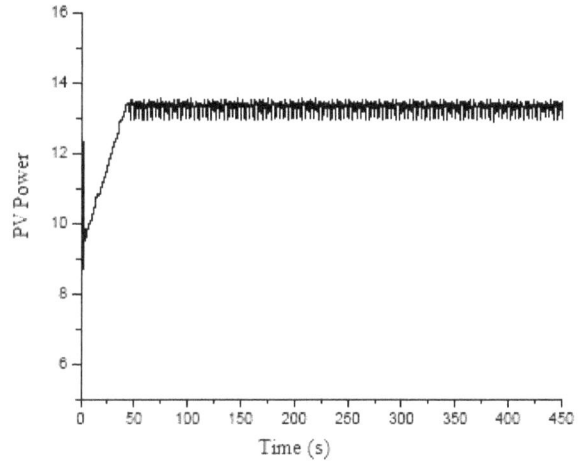

Fig. 7. *Experimental results of the classic P&O method.*

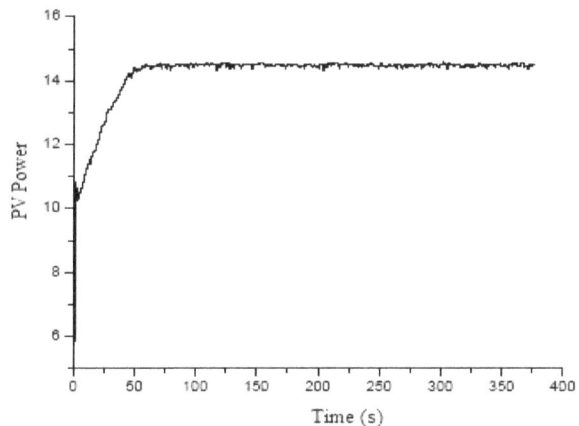

Fig. 8. *Experimental results of the proposed method.*

7. Conclusion

This paper has presented a MPPT Control for efficient operating of PV energy system taking into consideration the operational features of the system. An operational algorithm has been proposed for that purpose, where the information about the temperature and solar radiation is not required. The results show that the approach reduces clearly the oscillations around the maximum power point.

References

[1]   A. Ashraf, L. Ran and J. Bumby "Simulation and Control of a Hybrid PV-Wind System", Power Electronics, Machines and Drives, 2008. PEMD 2008, York, UK, 2-4 April 2008.

[2]   F. Valencaga, P. F. Puleston, and P. E. Battaiotto, "Power control of a solar/wind generation system without wind measurement: A passivity/ sliding mode approach," *IEEE Trans. Energy Convers.*, vol. 18, no. 4, pp. 501–507, Dec. 2003.

[3] B. S. Borowy and Z. M. Salameh, "Dynamic response to a stand-alone wind energy conversion system with battery energy storage to a wind gust," *IEEE Trans. Energy Convers.*, vol. 12, no. 1, pp. 73–78, Mar. 1997.

[4] R. Chedid and S. Rahman, "Unit sizing and control of hybrid wind-solar power systems," *IEEE Trans. Energy Convers.*, vol. 12, no. 1, pp. 79–85, Mar. 1997.

[5] W. D. Kellogg, M. H. Nehrir, G. Venkataramanan, and V. Greez, "Generation unit sizing and cost analysis for stand-alone wind, photovoltaic, and hybrid wind/PV systems," *IEEE Trans. Energy Convers.*, vol. 13, no. 1, pp. 70–75, Mar. 1998.

[6] R. Chedid and S. Rahman, "A decision support technique for the design of hybrid solar–wind power systems," *IEEE Trans. Energy Convers.*, vol. 13, no. 1, pp. 76–83, Mar. 1998.

[7] D. Das *and al.*, "An optimal design of a grid connected hybrid wind/photovoltaic/fuel cell system for distributed energy production," in *Proc. 32nd Conf. IECON*, Nov. 6–10, 2005, pp. 2499–2504.

[8] F. Giraud and Z. M. Salameh, "Steady-state performance of a grid connected rooftop hybrid wind–photovoltaic power system with battery storage," *IEEE Trans. Energy Convers.*, vol. 16, no. 1, pp. 1–7, Mar. 2001.

[9] J. T. Bialasiewicz, E. Muljadi, and R. G. Nix, "Simulation-based analysis of dynamics and control of autonomous wind–diesel hybrid power systems," *Int. J. Power Energy Syst.*, vol. 22, no. 1, pp. 24–33, 2002.

[10] E. Muljadi and J. T. Bialasiewicz, "Hybrid power system with a controlled energy storage," in *Proc. 29th Annu. Conf. IECON*, Nov. 2–6, 2003, vol. 2, pp. 1296–1301.

[11] F. Valenciaga and P. F. Puleston, "Supervisor control for a standalone hybrid generation system using wind and photovoltaic energy," *IEEE Trans. Energy Convers.*, vol. 20, no. 2, pp. 398–405, Jun. 2005.

[12] T. Tafticht, K. Agbossou, A. Chériti and M.L.Doumbia "An Improved Maximum Power Point Tracking Method for Photovoltaic Systems", International Journal of Renewable Energy, Vol.33, Issue 7, July 2008.

[13] T. Tafticht, K. Agbossou, A. Chériti and M. L. Doumbia.

[14] "Output Power Maximization of a Permanent Magnet Synchronous Generator Based Stand-alone Wind Turbine", IEEE International Symposium on Industrial Electronics, ISIE'06, ETS-Downtown Montréal (Québec), 9-13 July 2006.

[15] T. Noguchi, S. Togachi and R. Nakamoto "Short-current pulse-based maximum-power-point tracking method for multiple photovoltaic and converter module system", *IEEE Trans Ind Electron.*, vol. 49, no. 1, pp. 217–23, Feb.2002.

[16] M. A. S. Masoum *and al.*, "Theoretical and experimental analyses of photovoltaic systems with voltage and current-based maximum power-point tracking" *IEEE Trans Energy Convers.*, vol. 17, no. 4, pp. 514-522, Dec. 2002.

6

Dye-Sensitized Solar Cells Using Natural Dyes Extracted from Roselle (*Hibiscus Sabdariffa*) Flowers and Pawpaw (*Carica Papaya*) Leaves as Sensitizers

Danladi Eli[1, *], Muhammad Ahmad[2], Idodo Maxwell[1], Danladi Ezra[3], Aungwa Francis[1], Sunday Sarki[2]

[1]Department of Physics, Nigerian Defence Academy, Kaduna, Nigeria
[2]Department of Physics, Kaduna State University, Kaduna, Nigeria
[3]Department of Agricultural Science, Kaduna State University, Kaduna, Nigeria

Email address:

danladielibako@gmail.com (D. Eli)

Abstract: Dye-sensitized solar cells (DSSCs) were fabricated using natural dyes extracted from roselle flowers and carica papaya leaves extract as photosensitizers. The photovoltaic perfomance of the DSSCs were evaluated under 100 mAcm^{-2} light intensity. The roselle (*Hibiscus Sabdariffa*) extract sensitized solar cell gave a short circuit current density (*Jsc*) of 0.180 mAcm^{-2}, an open circuit voltage (*Voc*) of 0.470 V, a fill factor (*FF*) of 0.552, and an overall solar energy conversion efficiency (*η*) of 0.046%. Also, the pawpaw leaves extract sensitized cell gave a *Jsc* of 0.094 mAcm^{-2}, *Voc* of 0.433 V, *FF* of 0.544 yielding a conversion efficiency of 0.022%. The cell sensitized by the roselle extract shows better sensitization, which was in agreement with the broadest spectrum of the extract adsorbed on TiO$_2$ film. The sensitization performance related to interaction between the dye and TiO$_2$ surface is discussed.

Keywords: DSSCs, Natural Dyes, Dye Extracts, Sensitization

1. Introduction

Attempts to create photo electrochemical solar cells by mimicking nature's photosynthesis started in the 1970s with early efforts involving the covering of crystals of semiconductor titanium dioxide with a layer of chlorophyll [1].

Historically the dye-sensitization dates back to the 19th century, when photography was invented. The work of Vogel in Berlin after 1873 can be considered the first significant study of dye-sensitization of semiconductors, where silver halide emulsions were sensitized by dyes to produce black and white photographic films referred in [2].

However, owing to the electrons' reluctance to move through the layer of pigment, the efficiency of the first solar cells sensitized in this way was about 0.01%. In the late 1980s, Scientists discovered that nanotechnology could overcome the problem [1]. The use of dye-sensitization in photovoltaic remained however rather unsuccessful until a breakthrough at the early 1990's in the Laboratory of Photonics and Interfaces in the École Polytechnique Fédérale de Lausanne (EPFL) Switzerland. By the successful combination of nanostructured electrodes and efficient charge injection dyes, professor Grätzel and his co-workers developed a solar cell with energy conversion efficiency exceeding 7% in 1991 [3] and 10% in 1993 [4]. This solar cell is called the dye sensitized nanostructured solar cell or the Grätzel cell named after its inventor.

Dye-sensitized solar cells (DSSCs) are the third generation of photovoltaic devices which is used for the conversion of visible light into electric energy. These new type of solar cells are based on the photosensitization produced by the dyes on wide band-gap mesoporous metal oxide semiconductors; this sensitization is produced by the dye absorption of part of the visible light spectrum [5].

The sensitization approach enables the generation of electricity with irradiation of energy lower than energy of the bandgap of the semiconductor. The progress of such devices occurred with the development of nanostructured semiconductor films onto which light absorbing dye molecules are adsorbed. Many efforts have focused on sensitizer dye, since dye plays a key role in harvesting sunlight and transforming solar energy into electric energy. Several organic dyes and organic metal complexes have been employed to sensitize nanocrystalline TiO_2 semiconductors [6].

Naturally most flowers and leaves show various colours and contain several pigments which are easily extracted and then employed in DSSCs [7]. The leaves of most green plants dye has been investigated in many studies [8, 9, 10]. Anthocyanins are natural compounds that give colour to fruits and plants [11]. Roselle extract is rich in anthocyanins. It was reported that anthocyanin obtained from roselle are delphinidin and cyanidin complexes [12, 13]. The chemical structure of cyanidin and delphinidin in the *Hibiscus Sabdariffa* dye is shown in figure 4.

In this research work, anthocyanins extracts of roselle flowers, and chrorophyll extract from carica papaya leaves were the natural dyes used as sensitizers in the formed DSSCs. The performances of the formed DSSCs shows that, the roselle extract has higher photosensitized performance as compared to the pawpaw leaves extract.

The *Hibiscus sabdariffa* dye (commercially known as Roselle) belongs to the family *Malvacae* and is present in abundance through-out the world and has attained prominence as a jute substitute [14].

Pawpaw, belongs to the family of caricaceae. It is not a tree but a herbaceous succulent plants that posses' self-supporting stems of spongy and soft wood [15].

2. General Composition, Function and Parameters of the DSSCs

The cell is composed of four elements, namely, the transparent conducting and counter conducting electrodes, the nanostructured wide bandgap semiconducting layer, the dye molecules (sensitizer), and the electrolyte. The transparent conducting electrode and counter-electrode are coated with a thin conductive and transparent film (such as fluorine-doped tin dioxide (SnO_2)). The TiO_2 surface is stained with a dye. TiO_2 nanocrystals are used rather than a continuous layer to maximize surface area for light absorption. Between the electrodes is an electrolyte. Upon absorption of photons, dye molecules are excited as shown schematically in Figure 1. Once an electron is injected into the conduction band of the wide bandgap semiconductor nanostructured TiO_2 film, the dye molecule (photosensitizer) becomes oxidized. The injected electron is transported between the TiO_2 nanoparticles and then extracted to a load where the work done is delivered as an electrical energy. The electrons flow through the TiO_2 onto the electrode, through an electric circuit, and then to the counter electrode. The electrolyte carries electrons back to the dye from the counter electrode. Electrolytes containing redox ions is used as an electron mediator between the TiO_2 photoelectrode and the coated counter electrode. Therefore, the oxidized dye molecules (photosensitizer) are regenerated by receiving electrons from the ion redox mediator that get oxidized.

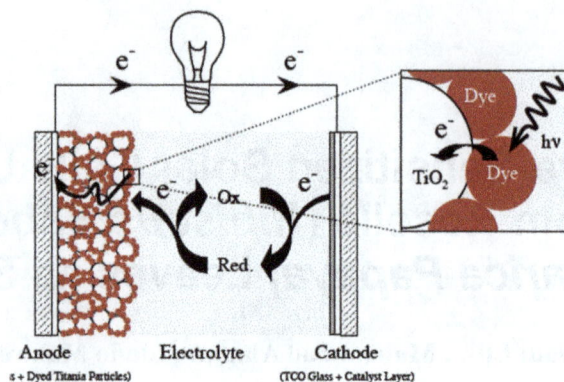

Fig. 1. *Schematic of the structure of the dye sensitized solar cell.*

3. Materials and Methods

3.1. Preparation of the Natural Dye

The leaves of *carica papaya* were grinded to small particles using blender with water as extracting solvent, and flowers of *Hibiscus Sabdariffa* were air dried till they became invariant in weight.

The dried flowers of *Hibiscus Sabdariffa* were left uncrushed because previous attempts proved failure to extract the dye from crushed samples due to jellification [16]. The method of heating in water (*Aq.*) was used to extract the dye. Distilled water was the solvent for aqueous extraction. 5 g of the sample (*Hibiscus Sabdariffa*) Flower was measured using analytical scale balance and dipped in 50 ml of the solvent heated to 100°C for 30 minutes after which solid residues were filtered out to obtain clear dye solutions.

3.2. DSSCs Assembling

The photoanode was prepared by first depositing a blocking layer on the fluorine doped tin oxide (FTO) glass (solaronix), followed by the nanocrystalline TiO_2 (solaronix). The blocking layer was deposited from a 2.5 wt% TiO_2 precursor and was applied to the FTO glass substrate by spin coating and subsequently sintered at 400°C for 30 minutes. The 9 μ m thick nanocrystalline TiO_2 layer was deposited by screen printing. It was then sintered in air for 30 minutes at 500°C. The counter electrode was prepared by screen printing a platinum catalyst gel coating onto the FTO glass. It was then dried at 100°C and fired at 400°C for 30 minutes.

The sintered photoanode was sensitized by immersion in the sensitizer solution at room temperature overnight. The cells were assembled by pressing the photoanode against the platinum-coated counter electrodes slightly offset to each other to enable electrical connection to the conductive side of the electrodes. Between the electrodes, a 60 μ m space was retained using two layers of a thermostat hot melt sealing foil. Sealing was done by keeping the structure in a hot-pressed at 100°C for 1 minute. the liquid electrolyte constituted by 50 mmols of tri-iodide/iodide in acetonitrile was introduced by injection into the cell gap through a channel previously fabricated at opposite sides of the hot melt adhesive, the channel was then sealed.

3.3. Characterization and Measurement

The current-voltage (J-V) data was obtained using a keithley 2400 source meter under AM1.5 (100 mw/cm^2) illumination from a Newport A solar simulator. Scanning electron micrographs of the nanocrystalline TiO$_2$ films are taken with Carl Zeis SEM. The absorption spectra of the dyes was recorded on Ava-spec-2048 spectrophotometer. The cell active area was 0.5 cm^2. Thickness measurement was obtained with a Dektac 150 surface profiler. X-ray microanalysis was carriedout with INCA EDX analyzer.

4. Results and Discussion

Fig. 2a. shows the SEM image of TiO$_2$ nanoparticles fabricated using screen printing method. The SEM micrograph shows that the TiO$_2$ nanoparticles produced have a mean particle size of about 20 nm. It also reveals that the

surface is porous and has agglomeration.

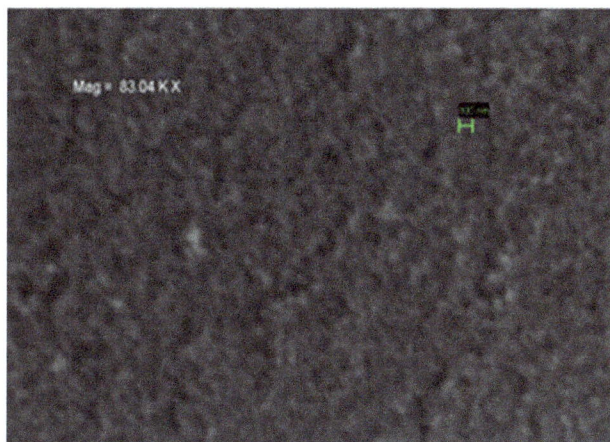

Fig. 2a. The Scanning electron microscope surface morphology of TiO$_2$ sample.

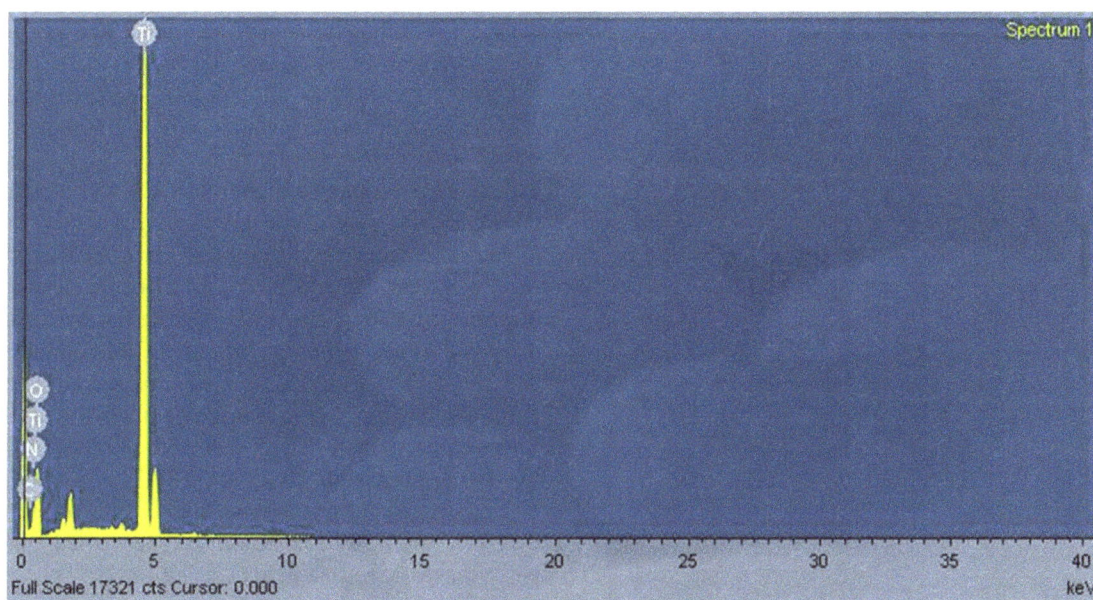

Fig. 2b. EDX image showing the elements present in the TiO$_2$ compound.

Fig. 2b presents the EDX Image of TiO$_2$. The elements present in the TiO$_2$ are Titania, Oxygen and Nitrogen. Nitrogen is present due to the blower that was used to dry the TiO$_2$ semiconductor.

Fig. 3a shows the UV-vis absorption spectra of roselle extract, and pawpaw leaves extract. It was found that the absorption peak of roselle extract is about 540 nm (Fig 3b) while it was deduced that the carica papaya dye absorbs photons best at a wave length peak of 370 nm (Fig 3a). The difference in the absorption characteristics is due to the different type of pigments (anthocyanin for roselle and chlorophyll for pawpaw) and colors of the extracts. After immersion of the TiO$_2$-coated electrode (photoanode) in the extracts, observable colors of TiO$_2$ films turned deep purple for the roselle extract but the film turned light green for the pawpaw extract. In the case of roselle extract, an absorption peak of the photoanode is broader than that of the dye

solution (Fig. 3(c)), with a shift to a higher wavelength (from 540 to 560 nm). The difference in the absorption peak is due to the binding of anthocyanin in the extract to the TiO$_2$ surface [17].

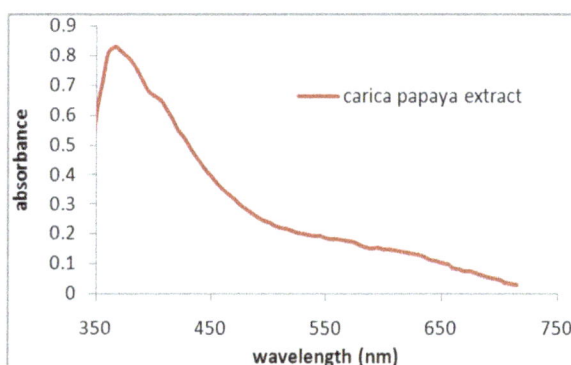

Fig. 3a. Absorption spectra of papaya extract.

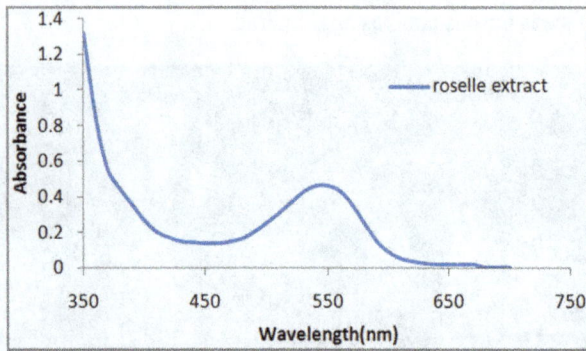

Fig. 3b. *Absortion spectra of roselle extract.*

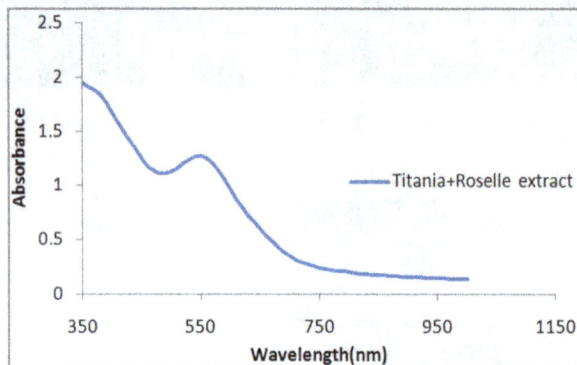

Fig. 3c. *Absortion spectra of titania immersed in roselle extract.*

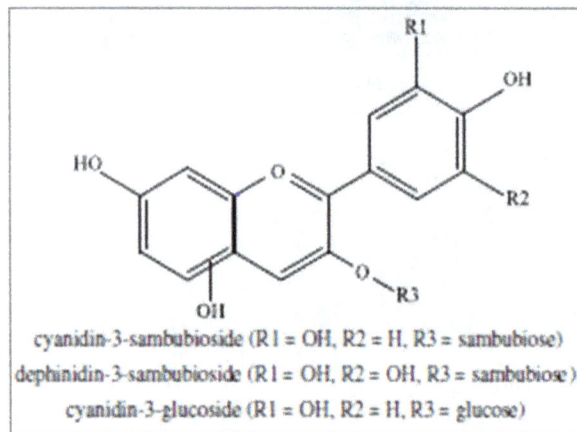

cyanidin-3-sambubioside (R1 = OH, R2 = H, R3 = sambubiose)
dephinidin-3-sambubioside (R1 = OH, R2 = OH, R3 = sambubiose)
cyanidin-3-glucoside (R1 = OH, R2 = H, R3 = glucose)

Fig. 4. *Chemical structures of: cyanidin and delphinidin in roselle dye.*

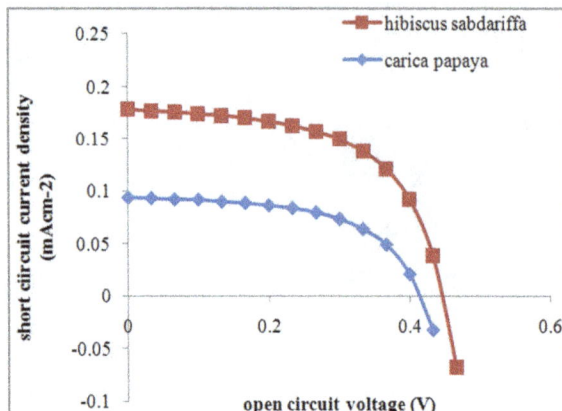

Fig. 5. *Photocurrent density-voltage (J-V) curve under 100mWcm^{-2} light intensity.*

Fig. 5 shows the *J–V* (current density–voltage) curves of the roselle and pawpaw leaf extract under illumination.

Based on the *J-V* curve, the *fill factor* (*FF*) and *solar cell efficiency* (η) were determined using equations (1) and (2) respectively.

$$FF = \frac{P_{max}}{P_{in}} = \frac{J_{max} \times V_{max}}{J_{SC} \times V_{OC}} \qquad (1)$$

$$\eta = \frac{FF \times J_{SC} \times V_{OC}}{P_{IRRADIANCE}}.100\% \qquad (2)$$

Where V_{max} = maximum voltage (V);
J_{max} = maximum current density (mA/cm^2);
J_{sc} = short circuit current density (mA/cm^2);
V_{oc} = open circuit voltage (V) and
$P_{IRRADIANCE}$ = light intensity (mW/cm^2)

From the effective absorption area of 0.5 cm^2 of the DSSCs, the averaged values of the light-to-current conversion efficiencies (η) of the DSSCs, shortcircuit photocurrent (J_{SC}), open-circuit voltage (V_{OC}), and fill factor (*FF*) were recorded as presented in *Table 1*. Obviously, the efficiency of the cell sensitized by the roselle extract was significantly higher than that sensitized by the pawpaw leaf extract. This is due to a higher intensity and broader range of the light absorption of the extract on TiO$_2$ (Fig. 3c), and the higher interaction between TiO$_2$ and anthocyanin in the roselle extract which leads to a better charge transfer. Moreover, anthocyanin in the roselle extract (cyanidin and delphinidin) has a shorter distance between the dye skeleton and the point connected to TiO$_2$ surface compared to the chrorophyll extract from *carica papaya* leaves that shows lower charge transfer effect due to the limited absorbance of the dye in the visible spectrum [18]. This could facilitate an electron transfer from anthocyanin in the roselle extract to the TiO$_2$ surface and could be accounted for a better performance of roselle extract sensitization [19].

Table 1. *Performance characteristics of DSSC under 100mWcm^{-2}.*

Sample	J_{SC}(mAcm^{-2})	V_{OC}(V)	FF	η (%)
Roselle	0.180	0.47	0.552	0.046
Pawpaw	0.094	0.43	0.544	0.022

When the results in this study is compared to a studiy performed under similar conditions by Ahmed *et., al* [20] who found *Jsc* of 0.17 mAcm^{-2}, *Voc* of 0.46 V, *FF* of 0.41 and η of 0.033% for *Hibiscus Sabdariffa* respectively, our result is alitle higher. Also when compared to Kimpa *et.al* [21], where the conversion efficiency of the DSSC prepared by pawpaw leaf extract was 0.20%, with V_{OC} of 0.50 V, J_{SC} of 0.649 mA/cm^2 and *FF* of 0.605, their result has better perfrmance. These differences may be attributed to differences in concentrations of phytoconstituents in different parts of the plants for the *Hibiscus Sabdariffa* dye [16], and due to the extracting solvent used for the case of pawpaw (distilled water for our studies and ethanol for Kimpa *et., al*).

5. Conclusions

Sensitization of a dye sensitized solar cell with extracts from roselle flowers was demonstrated. A similar cell sensitized with *carica papaya* leaves (containing a mixture of carothenoids and appreciable concentration of phenolic acid) extract shows a lower *Jsc, Voc, FF* and *η*. Using water as extracting solvent, the energy conversion efficiency (*η*) of the cells consisting of roselle extract and pawpaw leaves extract was 0.046%, and 0.022%, respectively. The roselle extract has higher photosensitized performance as compared to the pawpaw leaves extract. This is due to the better charge transfer between the roselle dye molecule and the TiO$_2$ surface which is related to a dye structure, [18].

Acknowledgement

The authors are grateful to the physics advanced laboratory, Sheda Science and Technology Complex (SHESTCO), Abuja, Nigeria for the use of their research facilities.

References

[1] Pagliaro, M., Palmmisano, G., and Ciriminna, R. (2000). Working Principles for Dye-Sensitized Solar Cells and Future Applications. *Photovoltaics International journal, third edition. third edition, 47-50.*

[2] Grätzel, M. and McEnvoy, A. J. (1994). Principles and Application of Dye Sensitized Nanocrystalline Solar Cells (DSC). Laboratory of Photonics and Interfaces in the École Polytechnique Fédérale de Lausanne (EPFL), Switzerland.

[3] O'Regan, B. and Grätzel, M. (1991). A low-cost, High-Efficiency Solar Cell Based on Dye-Sensitized Colloidal TiO$_2$ Film. *Nature 353, (6346): 737-740.*

[4] Nazerruddin, M. K; Kay, A.; Ridicio, I.; Humphry-Baker, R.; Mueller, E.; Liska, P.; Vlachopoulos, N. & Gratzel, M. (1993). *J.* Conversion of Light to Electricity on Nanocrystalline TiO$_2$ Electrodes. *Amer. Chem. Soc.* Vol. 115, pp. 6382-6390.

[5] Hernandez-Martinez, A. R., Estevez, M., Vargas, S., Quintanilla, F., and Rodriguez, R. (2011). New Dye-Sensitized Solar Cells Obtained from Extracted Bracts of Bougainvillea Glabra and Spectabilis Betalain Pigments by Different Purification Processes. *Int. J. Mol. Sci. 12, 5565-5576.*

[6] Lai, W. H., Sub, Y. H., Teoh, L. G., and Hona, M. H. (2007). Commercial and Natural Dyes as Photosensitizers for a Water-Based Dye-Sensitized Solar Cell Loaded with Gold Nanoparticles. *Journal of Materials Sciences and Applications, 195, 307–313.*

[7] G. Calogero and G. D. Marco, "Red Sicilian Orange and Purple Eggplant Fruits as Natural Sensitizers for Dye-Sensitized Solar Cells," *Solar Energy Material Solar Cell*, Vol. 92, No. 11, 2008, pp. 1341-1346.

[8] M. Gratzel, "Dye-Sensitized Solar Cell," *Journal of Photochemistry & Photobiology C: Photochemistry Reviews*, Vol. 4, No. 2, 2003, pp. 145-153.

[9] K. Tennakone, G. R. R. A. Kumara, A. R. Kumarasinghe, P. M. Sirimanne and K. G. U. Wijayantha, "Efficient Photosensitization of Nanocrystalline TiO$_2$ Films by Tannins and Related Phenolic Substances," *Journal of Photo- chemistry & Photobiology A: Chemistry*, Vol. 94, No. 2-3, 1996, pp. 217-220.

[10] D. Zhang, S. M. Lanier, J. A. Downing, J. L. Avent, J. Lumc and J. L. McHale, "Betalain Pigments for Dye- Sensitized Solar Cells," *Journal of Photochemistry Photo- biology A: Chemistry*, Vol. 195, No. 1, 2008, pp. 72-80.

[11] M. Rossetto, P. Vanzani, F. Mattivi, M. Lunelli, M. Scarpa and A. Rigo, "Synergistic Antioxidant Effect of Catechin and Malvidin 3-Glucoside on Free Radical-Ini- tiated Peroxidation of Linoleic Acid in Micelles," *Ar- chives of Biochemistry and Biophysics*, Vol. 408, No. 2, 2002, pp. 239-245.

[12] T. Frank, J. Clin. Pharmacol. 45 (2005) 203.

[13] N. Terahara, N. Saito, T. Honda, K. Tokis, Y. Osajima, Phytochemistry 29 (1990) 949.

[14] Oday A. Hammadi, Noor I. Naji, Effect of Acidic Environment on the Spectral Properties of Hibiscus sabdariffa Organic Dye used in Dye-Sensitized Solar Cells, *iraqi journal of applied physics*, Vol. 10, No. 2, 2014, pp 27-31.

[15] D. Gross, "Papaya: A Tantalising Taste of the Tropics. Maricopa County Master Gardener Volunteer information, University of Arizona CooperativeExtension,"2003. www.papayamaricopa-hort@ag.arizo.edu.

[16] Barness Chirazo Mphande, and Alexander Pogrebnoi, Outdoor Photoelectrochemical Characterization of Dyes from *Acalypha wilkesiana 'Haleakala'* and *Hibiscus sabdariffa* as Dye Solar Cells Sensitizers. *British Journal of Applied Science & Technology 7(2): 195-204, 2015.*

[17] K. Vinodgopal, X. Hua, R. L. Dalgren, A. G. Lappin, L. K. Patterson, P. V. Kamat, J. Phys. Chem. 99 (1995) 10883.

[18] K. Wongcharee, V. Meeyoo and S. Chavadej, "Dye-Sensitized Solar Cell Using Natural Dyes Extracted from Rosella and Blue Pea Flowers," *Solar Energy Material Solar Cells*, Vol. 91, No. 7, 2007, pp. 566-571.

[19] K. Sayama, S. Tsukagoshi, T. Mori, K. Hara, Y. Ohga, A. Shipou, Y. Abe, S. Suga, H. Arakawa, Sol. Energy Mater. Sol. Cells 80 (2003) 47.

[20] Ahmed TO, Akusu PO, Alu N, Abdullahi MB. Dye-sensitized solar cell (DSC) based on titania nanoparticles and hibiscus sabdariffa. British Journal of Applied Science and Technology. 2013; 3(4): 840-846.

[21] M. Isah Kimpa, M. Momoh, K. Uthman Isah, H. Nawawi Yahya and M. Muhammed Ndamitso, "Photoelectric Characterization of Dye Sensitized Solar Cells Using Natural Dye from Pawpaw Leaf and Flame Tree Flower as Sensitizers," *Materials Sciences and Applications*, Vol. 3 No. 5, 2012, pp. 281-286. doi: 10.4236/msa.2012.35041.

Jordanian Oil Shales: Variability, Processing Technologies, and Utilization Options

Hani Muhaisen Alnawafleh[1, 2, *]**, Feras Younis Fraige**[1, 3]**, Laila Abdullah Al-khatib**[1]**, Mohammad Khaleel Dweirj**[1]

[1]Faculty of Engineering, Al-Hussein Bin Talal University, Ma'an, Jordan
[2]Faculty of Engineering, Tafila Technical University, Tafila, Jordan
[3]Faculty of Engineering, King Saud University, Al-Muzahmiayah Branch, Riyadh, Kingdom of Saudi Arabia

Email address:
hanialnawafleh@ahu.edu.jo (H. M. Alnawafleh)

Abstract: Jordan has a huge Oil Shale (OS) reserves with about 50 billion tons are located in the central part. Oil Shale is considered the only potentially discovered fossil fuel in Jordan. Jordanian OS is characterized by its good quality, near surface deposits with low stripping ratio. The main deposits are located in central Jordan with good road network and effectively thin population. On general, the reported physical and chemical properties suggest vertical variability pattern within the same deposits and lateral variability between the deposits. Extraction behavior also confirms such variability. The dependency of Jordan on imported crude oil and gas put extra pressure on the Jordanian economy as a result of the heavy and yet growing energy bill. The Jordanian government shows interest in commercializing its potential OS reserves and signed many agreements and memo of understanding with many external companies' expert in OS processing, such as Shell international. In this work, the variability of Jordanian OS will be investigated with reference to the current status on their processing and utilization options.

Keywords: Oil Shale, Jordan, Variability, Utilization, Processing

1. Introduction

Jordan has a huge oil shale (OS) reserves spread over the country. More than 50 billion tons are located in the central part (Fig. 1), and with unlimited OS quantities are reported in the north of the country [1]. Oil shale deposits are found in several horizons mainly that of upper Cretaceous Muwaqqar Chalk Marl Formation with more than 350 m in the Yarmouk area [2-4].

The upper Cretaceous OS in Jordan is not true shale but bituminous carbonate [5]. The inorganic constituents are: carbonates, clays, silica, phosphates, and sulfur. On the other hand, the organic matter fraction, which may exceeds 25% wt in some horizons, is essentially kerogen with low quantities of bitumen [6].

Oil shale is expected to play crucial role in Jordanian economy in the near future. The exploration for commercial OS deposits in Jordan was started since more than four decades (e.g. [7]), as early as 1980 and later, the research continued through many geologic and technologic studies for evaluation and utilization purposes (e.g. [8]). It is the purpose of this paper is to highlight the status of Jordanian OS deposits, especially those in central Jordan, in terms of their variability, processing technologies and utilization options.

2. Methodology

To illustrate the variability of Jordanian OS deposits, data available from the Natural Resources Authority (NRA) have been reviewed and analyzed. Certain important figures therefore constructed. Experimentally, and for the purpose of examining OS variability, three OS samples were provided from different OS deposits in central and south Jordan (namely; El-Lajjun, Sultani, and Jurf Ed-Darawish) (Fig. 1). About 2 grams of finely ground (< 150 μm) OS samples were soxhlet extracted for 24 hours using 200 ml Tetrahydrofuran. The shale oil yield data then were recorded.

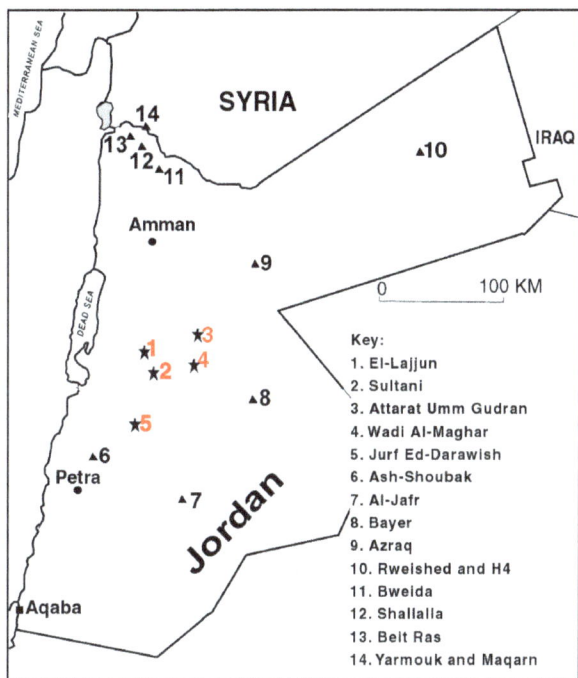

Figure 1. *Oil Shale distribution in Jordan. Central Jordan oil shale deposits are highlighted in stars.*

3. Oil Shale Variability

3.1. Geologic Variability

Figure 2. *Local oil shale outcrop in El-Lajjun area, Central Jordan.*

The geologic variability of Jordanian OS is studied in detail by [5]. The author reported the variability in terms of their rock types, chemical and physical properties, and beds thickness. In central Jordan, OS deposits are found in localized basins [9]. The upper most part of OS succession is locally exposed (e.g. Fig. 2). In selecting the future OS processing technology, this variability should be considered.

3.2. Lateral Variability

Oil shale in central Jordan has variable thickness and stripping ratio (Fig. 3). To highlight the OS variability, Fig. 3 and later figures, except Fig. 6, are constructed from the

average data of OS analyses results of various chemical and physical laboratory tests performed by the NRA and reported during the year 2006 by [10] from the NRA team.

In central Jordan, the OS deposits are quite near surface where the stripping ratio ranges between 1 to1.6. Relatively low stripping ratio indicates the suitability of surface mining techniques for OS extraction. Thickness variability is related to different geologic and formational conditions as discussed by [5] and [9].

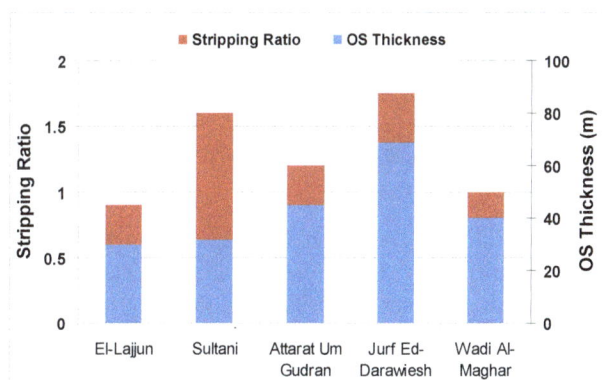

Figure 3. *Thickness and stripping ratio variation across central Jordan oil shale deposits.*

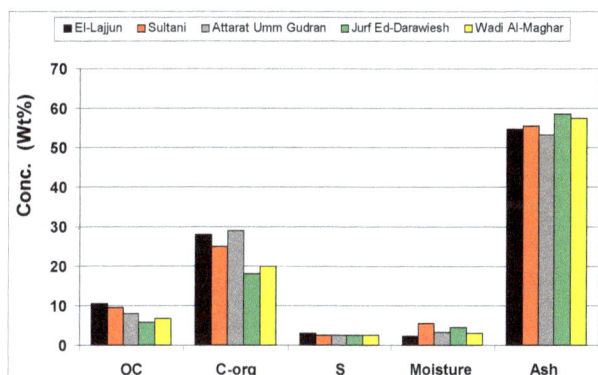

Figure 4. *Variability in some chemical and physical properties of oil shale deposits in central Jordan. OC: average oil content. C-org: average organic carbon content. S: Sulfur.*

The chemical and physical variability between the main OS deposits in central Jordan is illustrated in Fig.4 and Fig. 5 respectively. The average oil content varies between 5.7 to 10.5 % by wt. High total organic matter content can be inferred from the high organic carbon values that may reach up to 29 % by wt. at the OS deposit of Attarat Umm Gudran. The SO_3 and moisture contents rise up to 5 % and 3.8 % by wt. respectively. Such variability in the chemical and physical properties indicates different quality for different OS deposits. Therefore, careful method selection is needed when considering such deposits for utilization.

The major inorganic fraction is the carbonate as indicated from the $CaCO_3$ content that varies in the range of 47 – 69 %. Other important fractions indicated from the values of major oxides presented in Fig.5 are quartz, clay, phosphate, and sulfides.

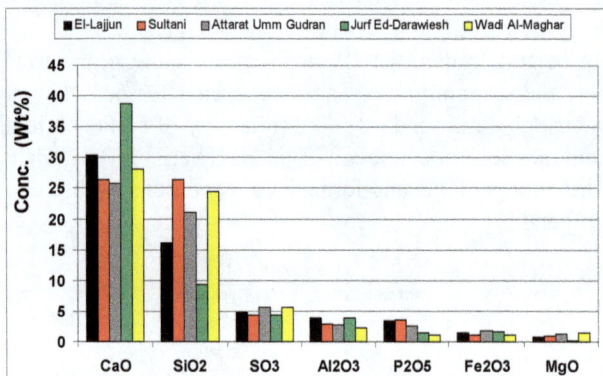

Figure 5. Variability in the chemical composition (major oxides) of oil shale deposits in central Jordan.

The variability of Jordanian OS deposit can be inferred from the results of the solvent extraction on selected oil shale samples from different deposits (Fig. 6). Such extraction variability indicates different composition of OM between OS deposits that must be taken into consideration in future utilization opportunities especially if certain extraction methodologies need to be applied.

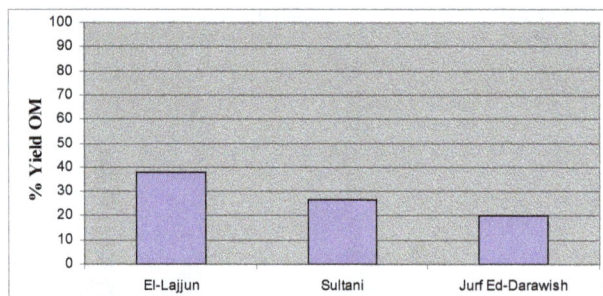

Figure 6. Variability in organic matter yield % resulted by soxhlet extraction on oil shale deposits in central Jordan.

3.3. Local Variability

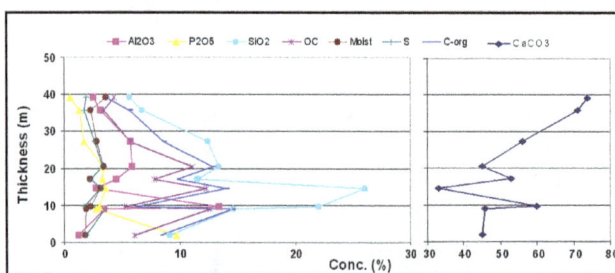

Figure 7. Vertical variability in some physical and chemical composition of El-Lajjun OS deposit, central Jordan.

To test the variability within the same deposit (e.g. El-Lajjun OS deposit), the physical and chemical properties are found to be vertically variable (Fig. 7). This variability indicates that the deposit is not uniform and such variability complicates the potential utilization methodologies. Change of formational conditions is the main reason for such variability [5]. High carbonate and sulfur contents play a

crucial role in the selection of future processing technology. Such content of these two fractions resulted in lowering the OS quality. Many issues may arise from their presence during the OS processing. Therefore, they are considered as major obstacles facing any progress in the Jordanian OS utilization.

4. Processing and Utilization Options

The interest in OS resources in Jordan developed due to the fact that Jordan energy sector is totally dependent on the imported crude oil. The dramatic increase of imported oil put extra pressure on the country economy. Therefore, the government of Jordan is seriously considering the exploitation of the country's potential OS resources. The NRA has conducted various OS assessment studies since 1980. In addition, a number of international collaborative works has been completed. Such work covered the suitability of current OS processing technologies and the application of international expertise to develop the Jordanian oil shale industry.

The main collaborative work with international companies is clearly reported in literature. For instance, Besieso [11] have summarized the previous work on the exploitation and use of the Jordanian OS resources such as the work of Klockner/Lurgi and Suncor. Since 2005, the Jordanian government has implemented a long term strategy for energy sector in Jordan including OS resources. Such strategy has aimed to attract foreign investors and to create several investment opportunities in the OS sector include OS feasibility studies, mining and exploitation of deep and near surface deposits and their processing [11]. Such strategy has succeeded and several agreements have been singed such as the agreement with Enefit and Shell companies.

Previous work conducted so far on the Jordanian OS has considered the Jordanian OS as a suitable source of energy due to its high organic content. The identified OS resources are enormous and will cover the Jordan's energy need for hundreds of years [11]. Jaber et. al. [12] summarized the most important factors in the Jordanian oil shale development. In their comprehensive study, all aspects regards the OS utilization are fully discussed. Jaber et. al. [12] stated that the costs of available extraction and processing technologies, the quality of the markets for the products and by-products, the implication of social and physical environments development, and health and environmental regulations are the major factors might affect Jordanian OS commercialization.

Technical research and exploration studies conducted so far on Jordanian oil shales proved their suitability for retorting or direct combustion without taking into account the OS variability. Retorting should take into account the high sulphur content. Direct combustion is favored if high sulphur content is overcome and combustion occurs at temperatures less than that of the deformation temperatures of carbonate. At the mean time, many technical and environmental issues would delay the OS utilization. At this point, and within the available OS processing technologies, the research on the OS resources in Jordan may be diverted to more environmentally-friendly techniques such as supercritical fluid extraction.

5. Conclusion

Jordanian OS shows local and lateral variability. Oil shale in central Jordan reveals relatively high oil content, relatively low moisture and majoring of carbonate and silicate phases with sulphur content of approximately 3 % wt. The OS variability should be taken into account in selecting the most relevant processing technology.

References

[1] Dyni, J. 2005. "Geology and Resources of Some World Oil-Shale Deposits". USGS Report 2005-5294.

[2] Abed, A.M.; Arouri, K,; Amiereh, B.S.; and Al-Hawari, Z. 2009. Characterization and Genesis of Some Jordanian Oil Shales. Dirasat, Pure Sciences, Volume 36, No. 1. Pp 7-17.

[3] Jarrar, M. and Mustafa, M. 1995. Mineralogical and geochemical study of the oil shale of Wadi Esh-Shallala (N-Jordan). Abhath Al-Yarmouk "Pure Science and Engineering Series". Vol. 4.2 .Pp 111-36.

[4] Pufahl P. K., Grimm K. A., Abed A. M., Sadaqah R. M. Y. (2003). Upper Cretaceous (Campanian) phosphorites in Jordan: implications for the formation of a south Tethyan phosphorite giant. Sedimentary Geology, vol 161. 175 - 205.

[5] Alnawafleh, H.M. 2007. Geological Factors Controlling the Variability of Maastrichtian Bituminous Rocks In Jordan." PhD Thesis. The University of Nottingham, UK.

[6] Abed, A.M. and Arouri, K. 2006. Characterization and Genesis of Oil shales from Jordan. International Conference on Oils Shale: "Recent Trends in Oil Shale", 7-9 November, Amman, Jordan. PAPER NO. rtos-A121.

[7] Speers, G. C. 1969. "El-Lajjun Oil Shale Deposit Jordan". Natural Resources Authority, Amman, and BP Research Centre.

[8] Hufnagel, H. 1980. "Investigation of the El-Lajjun Oil Shale Deposit". Bundesanstalt fur Geowissenschaften und Rohstoffe, Hannover, Technical cooperation Project NO. 78.2156.5.

[9] Abed, A. M. 2000. The geology of Jordan and its environment and water (in Arabic). Publication of the Jordanian Geologists Association, Amman - Jordan.

[10] Alali, J. and Sawaqed, S.(Editors). 2006. Oil Shale Resources Development in Jordan. Unpublished report. Natural Resources Authority, Amman, Jordan.

[11] Besieso, M. 2007. Jordan's Commercial Oil Shale Strategy. 27th Oil Shale Symposium. Colorado School of Mines, Colorado, USA.

[12] Jaber, J.O., Thomas A. Sladek, T.A., Mernitz,S., Tarawneh, T. M. 2008. Future Policies and Strategies for Oil Shale Development in Jordan .Jordan Journal of Mechanical and Industrial Engineering. Vol. 2. No. 1. Pp 31-44.

Transesterification of Palm Oil to Biodiesel and Optimization of Production Conditions i.e. Methanol, Sodium Hydroxide and Temperature

Shaila Siddiqua, Abdullah Al Mamun, Sheikh Md. Enayetul Babar

Biotechnology and Genetic Engineering Discipline, Khulna University, Khulna, Bangladesh

Email address:

siddiquashaila@gmail.com (S. Siddiqua), mamun.bge.ku@gmail.com (A. A. Mamun), babarku@yahoo.com (S. Md. E. Babar)

Email address:

Shaila Siddiqua, Abdullah Al Mamun, Sheikh Md. Enayetul Babar. Transesterification of Palm Oil to Biodiesel and Optimization of Production Conditions i.e. Methanol, Sodium Hydroxide and Temperature.   .

Abstract: Biodiesel is an alkyl ester of long chain fatty acids and considered as an alternative to lower the appalling consequence of fuel on the environment. It is produced by transesterification of a fat or oil with a short chain primary alcohol like methanol and alkali like sodium hydroxide (NaOH). Palm oil (*Elaeis guineensis*) was used as source to produce biodiesel and Box Behnken experimental design was applied to see the effect of various process parameters, i.e. methanol quantity, alkali concentration and temperature for the optimization of calorific value of biodiesel. Response surface plots and contour plot were created in order to perceive the optimum condition. Though, all the three variables significantly affected the calorific value of the palm biodiesel, but it was found that methanol was more effective variable than alkali concentration and temperature. It was observed that 12.5 ml methanol/50 ml oil and 0.4 gm NaOH/50 ml oil and 55°C temperature were optimum condition, where the calorific value of palm biodiesel is 9297.206 kcal/kg.

Keywords: Biodiesel, Palm Oil, Transesterification, Calorific Value, Optimization

1. Introduction

It is anticipated that the primary energy in every form from gasoline and diesel to the non-commercial fuels like biomass consumption of the entire world in a year is almost equivalent to ten thousands million tons of oil [1]. An increase of energy consumption will rise to forty nine percent from year 2007 to 2035 is conjectured by analyzing total vend energy of the world [2]. By considering the possibility of reduction of the oil production globally and the continuous rise of energy requirement for every day, the search for alternative fuel is must to meet the energy demand. Hence, biodiesel as a source of energy could become the alternative fuel. When the fossil fuel is burned it releases carbon dioxide (CO_2) at an amount of near about twenty one billion in one year, in which merely half of the anticipated quantity of CO_2 was captivated by means of natural process. Thus, every year an excess of 10.65 billion tons of CO_2 added to the atmosphere [3]. Whereas, biodiesel burning emits less than eighty percent carbon dioxide and emissions of sulfur dioxide tends to zero [4].

An alternative type of petroleum diesel fuel is biodiesel, generally made from different vegetable oils, waste products, animal fat or recycled restaurant greases. Biodegradability and emission of minimal amount of air pollutants make it more environment friendly fuel than petroleum fuel. Basically, biodiesels are the long-chain alkyl esters derived from organic sources, resulted from the chemical reaction of lipids (e.g., vegetable oil, animal fat) and alcohol termed as transesterification process. The use of alcohol and base or acid catalyst is extensively used in chemical conversion process to generate methyl esters from the base oil. Glycerin is formed as a by-product of the transesterification process [5, 6]. It differs from the vegetable and waste oil which directly used in fuel engine requires conversion of the diesel engine, as it is made in proper diesel forms to utilize in the regular diesel engine, whether use it in its pure form or combination with petroleum diesel [7, 8].

The demand for the production of biodiesel is increasing throughout the world because of its high quality fuel properties that makes it easy to use in almost every type of diesel engine. Industrial scale manufacturing of palm biodiesel is the way to go as it will not only bridge the energy deficit in the near future, but also it will deal with the ever increasing outcry of environmental contamination. Fossil fuel acts in the opposite because it defiles the environment in a great extend [9].

Biodiesel production process optimization refers to identify the most favorable values of raw ingredients. For optimization of chemical method of palm biodiesel production Box–Behnken designs [10] were used. The Box–Behnken design, which is the response surface methods (RSM), is a very useful statistical tool to optimize multiple variables for predicting the best performing conditions by using a minimum number of experiments [11]. In this study, we extracted biodiesel from palm oil by chemical method [12] and applied Box–Behnken designs as statistical tool to optimize the values of pre-eminent element (methanol and sodium hydroxide concentration and temperature) of biodiesel production process.

2. Materials and Methods

2.1. Materials

The research work was carried out with palm oil collected from local market (Khulna, Bangladesh). Methanol and sodium hydroxide used in the transesterification reaction were supplied by Merck, Germany.

2.2. Method of Biodiesel Production

In transesterification reaction, three moles of methanol react with one mole of triglyceride. The reaction is slowed by mass transfer limitations since at the start of the reaction the methanol is only slightly soluble in the oil and later on, the glycerin is not soluble in the methyl esters. Since the catalyst tends to concentrate in the glycerin, it can become unavailable for the reaction without agitation. The procedure of making biodiesel follows several steps.

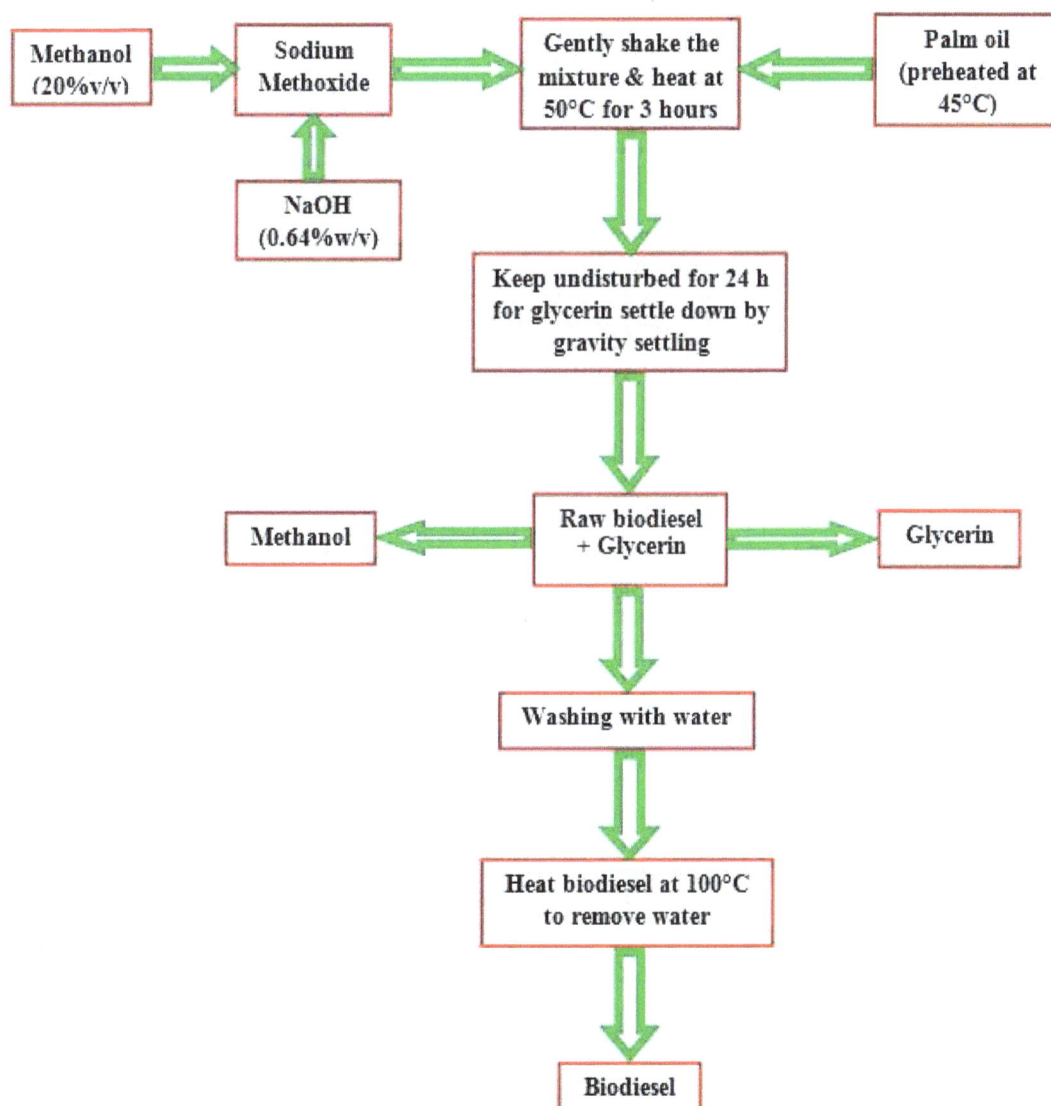

Figure 1. Flow diagram of biodiesel production from palm oil [12].

Mixing of alcohol and catalyst: The catalyst sodium hydroxide, at an amount of 0.8 % of vegetable oil was dissolved in the methanol at an amount of 25% vegetable oil by hand shaking and whirling. The methanol and catalyst sodium hydroxide were mixed properly. Water must be avoided as a solvent in this stage.

Reaction: The alcohol catalyst mix was then poured into a conical flask and the oil was added. The palm oil was heated at 45°C temperature before mixing with alcohol-catalyst mixture. After adding oil the system keeps air tight to prevent the loss of alcohol. The reaction mix was kept just below the boiling point of the alcohol to speed up the reaction and the reaction took place. The reaction was held under constant temperature around 50-60°C on the water heater. Reaction time varies from 1 to 8 hours.

Figure 2. *Different steps of biodiesel. (a) Raw biodiesel and glycerin, (b) washing of biodiesel, and (c) pure palm biodiesel.*

Separation: Once the reaction was completed, two major products exist- biodiesel and glycerin. The glycerin which was much denser than the biodiesel, separated by gravity separation and the biodiesel was ready for further processing. The glycerin separation steps were usually accomplished by gravity settling or with a centrifuge.

Alcohol Removal: Remaining methanol (3% to 6%) in the biodiesel was removed by vaporization since methanol has a propensity to work as a co-solvent for soap present in the biodiesel.

Methyl Ester Wash: After the methanol had been removed, the biodiesel needs to be washed to remove residual free glycerin, methanol, soaps and catalyst. This

was done by liquid-liquid extraction by mixing water with the biodiesel and gently agitating and the soap was separated by gravity separation. The washing process was done 3-4 times until the wash water no longer picks up soap.

Methyl Ester Drying: Remaining water present in the washed biodiesel was removed by heating at 100°C for 10 minutes. Finally usable 100% pure biodiesel was extracted.

2.3. Experimental Design

The Box-Behnken design is a type of response surface methodology (RSM) used as the experimental design. It is an independent quadratic design. These designs are rotatable (or near rotatable) and require three levels of each factor. Three variables were methanol (ml, X_1), NaOH (gm, X_2) and temperature (°C, X_3); three levels are +1, 0 and -1 respectively. Table 1 represents three variables and their coded levels. The central values (zero level) chosen for experimental design were: methanol-X_1 (12.5 ml i.e. 0.25ml methanol/ml oil), NaOH-X_2 (0.4 gm i.e. 0.008gmNaOH/ml oil) and temperature-X_3 (55°C).

Table 1. *Range of independent variables in the experimental design.*

Variables	Coded levels		
	+1	0	-1
Methanol (ml, X_1)	11	12.5	14
NaOH (gm, X_2)	0.3	0.4	0.5
Temperature (°C, X_3)	50	55	60

Where each methanol and NaOH concentration is given for 50 ml oil

Coefficients found from regression analysis were used in the second order polynomial equation to generate predicted value. Surface plots were produced using predicted value to find out the possible optimum condition.

The second order polynomial equation for three responses:

$$Y = A_0 + A_1X_1 + A_2X_2 + A_3X_3 + A_4X_1X_2 + A_5X_1X_3 + A_6X_2X_3 + A_7X_1^2 + A_8X_2^2 + A_9X_3^2 \tag{1}$$

Where Y is the response (calorific value); X_1 methanol, X_2 NaOH concentration, X_3 temperature; A_0 the regression coefficient, A_1–A_3 are the linear coefficients, A_4–A_6 the cross product coefficients, and A_7–A_9 are the quadratic coefficients. The regression analysis, statistical significance and analysis of variance (ANOVA) were carried out using Microsoft Office Excel. Surface plots and contour plot were developed using the same software along with Sigma Plot software.

3. Results and Discussion

For the optimization of chemical parameters of biodiesel production process from palm oil we used the most efficient method found in our previous work [12]. The Box–Behnken design was employed as a statistical means to optimize this

production process [13].

3.1. Process Optimization by Box-Behnken Method

In transesterification process methanol (alcohol) reacts with palm oil to form fatty acid alkyl esters (biodiesel) and glycerin. This reaction needs heat and sodium hydroxide as a strong base catalyst. Thus the important factors for the production of alkaline catalyzed transesterification reaction of palm oil (vegetable oil) are methanol, alkali (NaOH) and temperature. Hence, these factors are considered as the independent variables and their effects on calorific value of biodiesel are studied using Box- Behnken design of Response surface methodology (RSM).

Table 2. The Box-Behnken design matrixes employed for three independent variables (methanol, NaOH and temperature) with observed calorific values for palm biodiesel.

Run No.	Methanol (X_1) (ml)	NaOH (X_2) (gm)	Temperature (X_3) (°C)	Calorific value (c) (kcal/kg)
1	11	0.3	55	9120.511
2	14	0.3	55	9136.32
3	11	0.5	55	9131.531
4	14	0.5	55	9145.291
5	11	0.4	50	9156.396
6	14	0.4	50	9123.81
7	11	0.4	60	9124.825
8	14	0.4	60	9166.736
9	12.5	0.3	50	9168.34
10	12.5	0.5	50	9202.726
11	12.5	0.3	60	9245.954
12	12.5	0.5	60	9240.515
13	12.5	0.4	55	9297.221
14	12.5	0.4	55	9297.199
15	12.5	0.4	55	9297.21

The results of Box-Behnken design experiments for studying the effects of three independent variables, viz., methanol, NaOH and temperature on calorific value are presented in Table 2. These values are used for analysis of regression where Microsoft office excel tool "data analysis" is used. 95% confidence level is kept and the calculation of regression gives the coefficient values (A_0 to A_9 of Equation1). For each response different set of these coefficients were obtained. At the condition of the central values (zero level) i.e. 12.5 ml methanol, 0.4 gm NaOH and 55°C temperature the calorific value is 9297.221 kcal/kg.

From regression analysis all nine coefficients are used in making the response equation. The second order polynomial equations for each response were found as follows:

$$Y(c) = -3031.33 + 1174.911X_1 + 4893.243X_2 + 141.3478X_3 - 3.4155X_1X_2 + 2.483217X_1X_3 - 19.9124X_2X_3 - 52.2752X_1^2 - 4617.75X_2^2 - 1.46595X_3^2 \tag{2}$$

Where Y (c) is calorific value and X_1, X_2 and X_3 are coded values for methanol, NaOH and temperature respectively.

Table 3. Regression coefficient and corresponding probability values (p-values) for specific response (calorific value) for palm biodiesel.

Parameter (coefficient)	Calorific value (c)	
	coefficient	p-value
Constant (A_0)	-3031.33	0.110189
X_1(A_1)	1174.911	0.00017*
X_2(A_2)	4893.243	0.015437*
X_3(A_3)	141.3478	0.019215*
X_1X_2(A_4)	-3.4155	0.953784
X_1X_3(A_5)	2.483217	0.07766
X_2X_3(A_6)	-19.9124	0.289679
X_1X_1(A_7)	-52.2752	4.09E-05*
X_2X_2(A_8)	-4617.75	0.003257*
X_3X_3(A_9)	-1.46595	0.008598*

*$p < 0.05$

Table 3 represents the ANOVA analysis of design variables where values of coefficients and *p*-values are represented for the three responses. The *p*-values are used as a tool to check the significance of each coefficient, which also indicate the interaction strength between each independent variable. A *p*-value less than 0.05 indicate that the factor interacted significantly with the response. It is observed that all the *p* values are smaller than 0.05, except methanol-alkali concentration and alkali concentration- temperature on calorific value. The p-value 0.00004 obtained from methanol concentration (X_1 X_1) indicates that methanol has greater impact on calorific value than alkali and temperature. The calculation of regression analysis also gives the value of the determination coefficient R^2 represented in table 4.

Table 4. R^2 values for the ANOVA analysis of the three response output.

R^2 values	Calorific value (c)
Palm biodiesel	0.977974

3.2. Analysis of Calorific Value by Response Surface Plot

The relationship between coded variables and responses can be better understood by examining the series of 3D line plots. These 3D lines display the effect of variation of two factors while the third is kept constant. Due to three coded levels and the three coded variables total nine combinations are possible for each response. The plots are created with the aim to observe optimum condition from predicted values.

For palm biodiesel Figure 3 represents the effect of methanol, NaOH concentration and temperature on calorific value. When temperature is kept constant at 55°C [fig.3 (C)], maximum calorific value 9297.206 kcal/kg was observed at o.4gmNaOH and 12.5 ml methanol. At 0.3 gm NaOH and 50°C temperature the calorific value has a minimum value 9190 kcal/kg, when 12.5 ml methanol was constant. As the temperature increases the calorific value increases. But after 57°C the calorific value decreases. Similar effect has shown for other NaOH concentration. Thus combined NaOH and temperature has more effect on calorific value than alone NaOH or temperature. The maximum value obtained at 55°C temperature and 0.4 gm NaOH concentration when 12.5 ml methanol is constant [fig.3 (B)]. When methanol is kept constant [fig. 3 (A)] an increase or decrease in NaOH concentration yields calorific values with slight changes, but when NaOH is kept constant [fig.3 (B)] an increase or decrease in methanol concentration yields calorific values with greater changes. This indicates that the changes in methanol concentration have more effect on calorific value than changes in NaOH concentration.

From all nine surface plots for calorific value it is clear that the methanol, NaOH concentration and temperature at central values yield maximum calorific value. Changes in methanol, NaOH concentration and temperature from central values minimized the calorific values which support the theory. So, 12.5 ml methanol, 0.4 gm NaOH and 55°C temperature are optimum condition for palm biodiesel production with maximum calorific value.

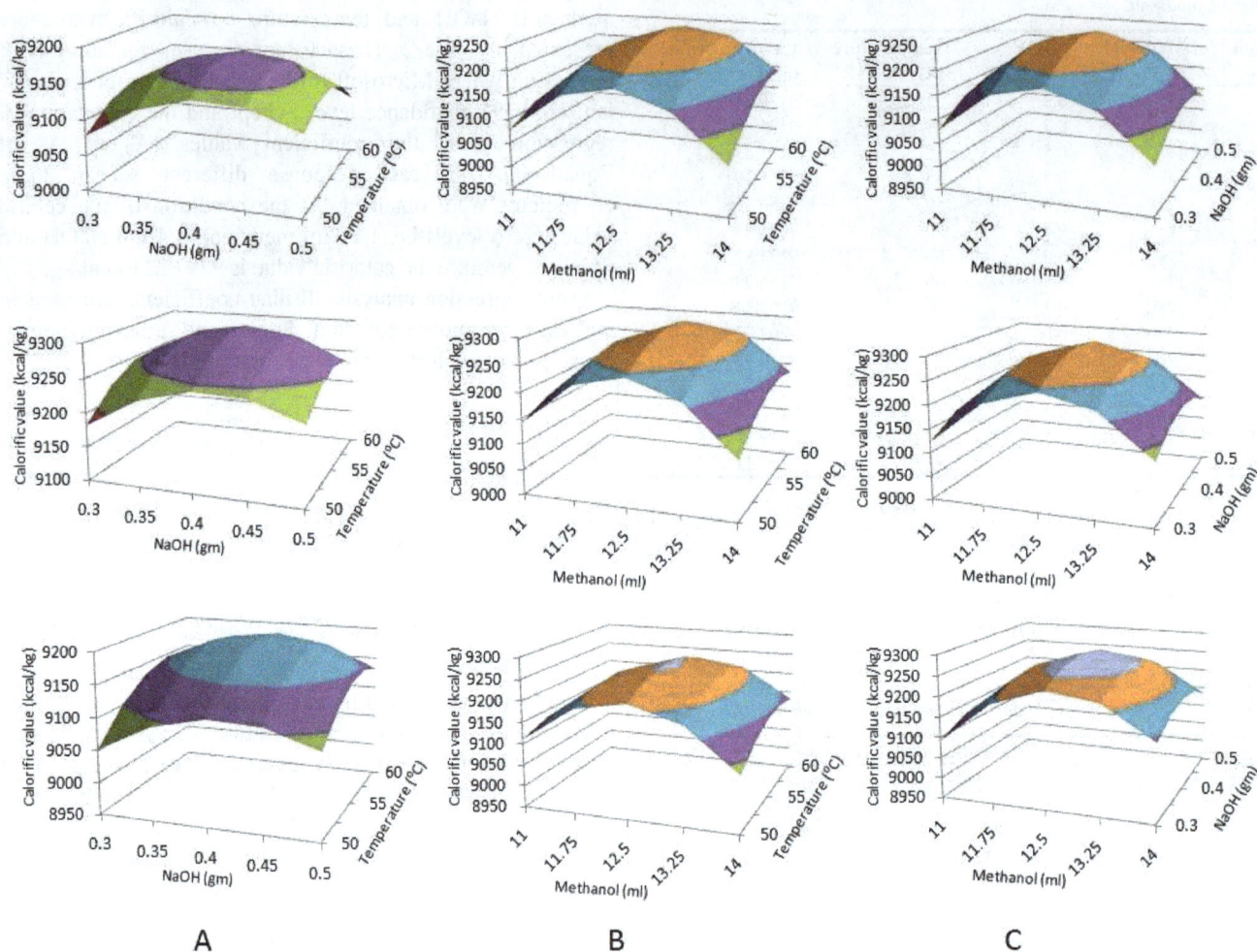

A   B   C

Figure 3. *3-D response surface plots for all design conditions.*

(A) Effects of temperature and NaOH on calorific value at constant methanol (ml), (B) Effects of temperature and methanol on calorific value at constant NaOH concentration (gm), and (C) Effects of methanol and NaOH on calorific value at constant temperature (°C).

3.3. Optimization and Method Validation

Optimal aspects for production process are confirmed with the assist of a contour plot. The plot was developed by plotting NaOH concentration (y-axis) against methanol concentration (x-axis) for a series of predicted calorific values at a constant 55°C temperature. Figure 4 represents the contour curve where different values articulated a different zone. Here, X is methanol concentration (ml/50 ml oil), Y is NaOH concentration (gm/50 ml oil) and Z is calorific value (kcal/kg).

Experimentally, only one combination produced maximum calorific value (Table 2). From the curve, it is found that several combinations of NaOH concentrations and methanol are likely to produce the same line. It can be observed that a smaller portion on the middle of the curve consists of high calorific value (9290 kcal/kg) zones and smaller lower calorific value (9150 kcal/kg) zone at the left corner.

A higher calorific value zone could be obtained using a NaOH concentrations range of 0.375-0.440 gm/50 ml oil and

methanol 12.4-12.6 ml/50 ml oil. While a lower calorific value zone resulted from a NaOH concentrations range of 0.3-0.31 gm/50 ml oil and methanol 11-11.1 ml /50 ml oil. This curve signifies that high calorific value is more responsive to the combined effect of NaOH and methanol concentrations.

Table 5. *Experimental and predicted calorific values for method validation experiment.*

Methanol (X_1) (ml)	NaOH (X_2) (gm)	Temperature (X_3) (°C)	Calorific value (c) (kcal/kg)	
			Experimental	Predicted
11	0.3	50	9111.677	9103.06
11	0.4	55	9145.384	9164.451
12.5	0.4	50	9233.589	9239.407
14	0.5	60	9198.542	9174.498

Validation of the process is done by selecting a few values randomly from the combinations. The results are presented in Table 5 shows that the experimental values are close to predicted value. NaOH concentration and temperature at central values yield maximum calorific value. Changes in methanol, NaOH concentration and temperature from central values minimized the calorific values which support the theory. As these chemical parameters produce higher calorific value, these parameters can be precise as the better production condition.

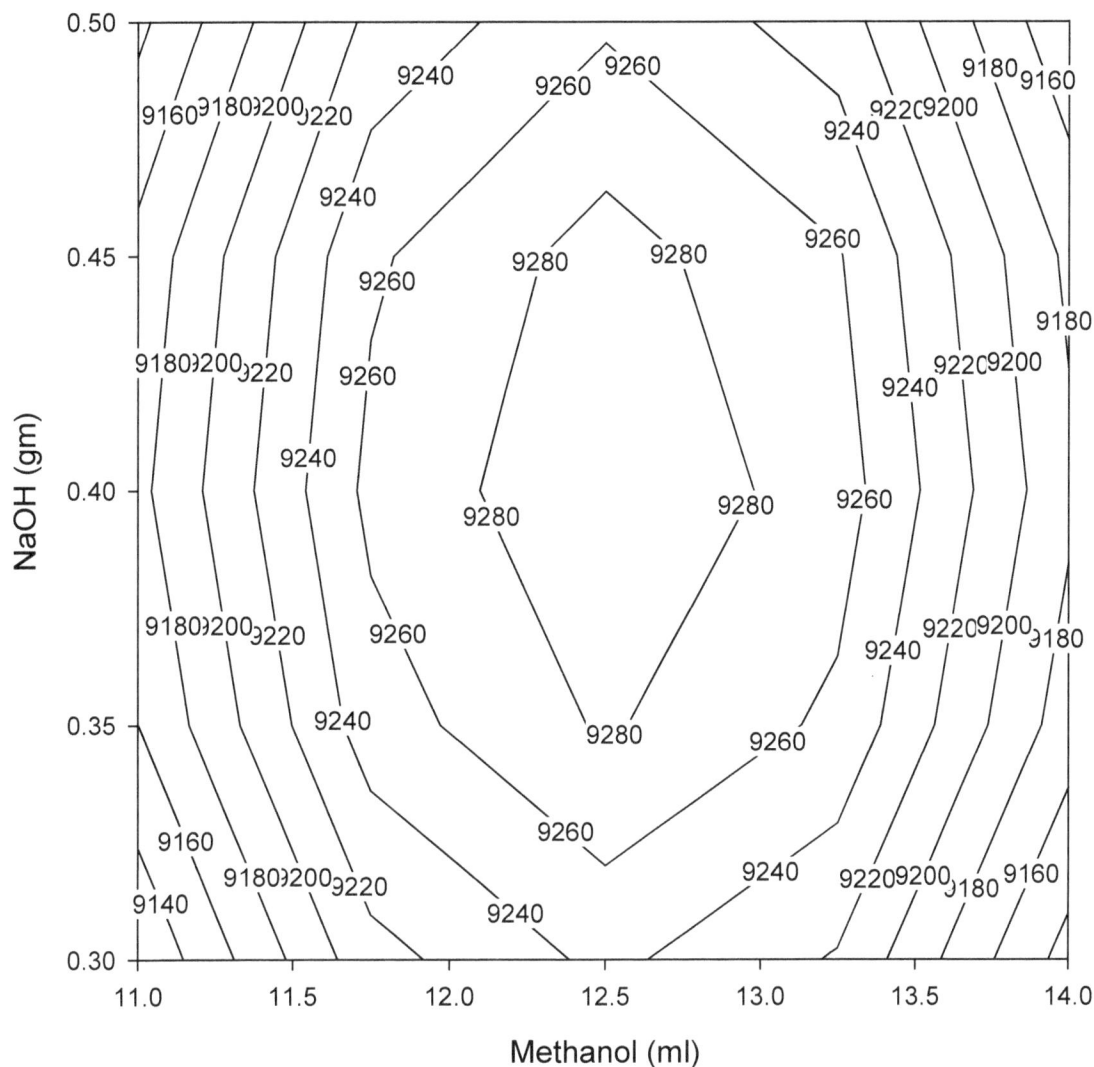

Figure 4. *Effect of methanol and NaOH concentration on calorific value of palm biodiesel at constant 55°C.*

4. Conclusion

In this experiment response surface methodology was used, and a quadratic polynomial equation was obtained for each value by multiple regression analysis. Total 15 combinations of methanol, NaOH and temperature were employed to develop Box-Behnken experimental design. Qualities of resulted products were evaluated in terms calorific value. It was found that 12.5 ml methanol/50 ml oil and 0.4gmNaOH/50 ml oil and 55°C temperature were optimum condition, where the calorific value of palm biodiesel is maximum 9297.206 kcal/kg. The optimum condition will be used to investigate the effect of methanol, alkali concentration and temperature on reaction time and on other fuel properties of biodiesel.

Acknowledgements

The authors acknowledge Khulna University, Bangladesh for the financial and technical support and Department of Chemical Engineering of Bangladesh University of Engineering and Technology for chemical analysis.

References

[1]   (2011) Commission services, Organization for Economic Co-operation and Development, [Online]. Available: http://www.inforse.dk/europe/dieret/WHY/why.html.

[2]   (2011) International Energy Outlook 2010, U.S. Energy Information Administration,[Online]. Available: http://205.254.135.24/oiaf/ieo/highlights.html.

[3]   (2011) US Department of Energy on greenhouse gases, [Online]. Available: http://en.wikipedia.org/wiki/Fossil_fuel.

[4]   (2011) U.S. Department of Energy and the U.S. Department of Agriculture, [Online]. Available: http://www.jatrophabiodiesel.org/bioDiesel.php.

[5]   A. Nag, "Biofuels Refining and Performance", New York, NY: McGraw-Hill, 2007.

[6]   S. Paweetida, J. Hiroi, K. Yoshikawa and T.Namioka, "Basic Chemical Reaction Study on Biodiesel Fuel Production from Plant Oil", Tokyo Institute of Technology, paper presented at 2nd AUN SEEDNet Regional Conference on New and Renewable Energy, Thailand, January 2010.

[7]   R. Burton and L. Forer, (2015), "Introduction to Biofuels: Biodiesel and Straight Vegetable Oil", [online]. Available at: www.biofuels.coop/pdfs/1_intro.pdf.

[8]   EG. Shay, "Diesel fuel from vegetable oil: status and opportunities", Biomass Bioenergy, 1993; 4(4):227^4-2, 1993.

[9]   L. Attanatho, S.Magmee and P. Jenvanitpanjakul, "Factors Affecting the Synthesis of Biodiesel from Crude Palm Kernel Oil", the Joint International Conference on "Sustainable Energy and Environment (SEE)" 1-3 December 2004, HuaHin, Thailand.

[10]   Box, G. E. P., Behnken, D. W., Technometrics 1960, 2, 455 – 475.

[11]   Cocharn, W. G. and Cox, G. M., Experimental Designs, 2nd Ed., Wiley, New York 1992.

[12]   Mamun, A. A., Siddiqua, S. and Babar, S. M. E, "Selection of an Efficient Method of Biodiesel Production from Vegetable Oil Based on Fuel Properties", International Journal of Trends and Technology,   2013, V4 (8):3289-3293.

[13]   Babar, S.M.E., Song, S.J., Hasan, M.N. and Yoo, Y.S., "Experimental design optimization of the capillary electrophoresis separation of leucine enkephalin and its immune complex", Wiley Inter Science, 2007.

9

Renewable Energy: A Solution to Hazardous Emissions

Ahmed Bilal Awan

Electrical Engineering Department, College of Engineering, Majmaah University, Majmaah, KSA

Email address:

a.awan@mu.edu.sa

Abstract: The problem of energy security, increasing prices of energy, the aspect of environmental pollution and depletion of the known fuel reserves in future have created a scope for utilization of renewable resources. Increasing prices of fossil fuels and costs associated with emissions may affect the economy of a country severely. Similarly, fossil fuels although produce useful energy, are responsible for production of harmful emissions like CO_2, SO_x, NO_x etc. These dangerous emissions are an acute threat to human health on our plannet. The obvious choice available is to use renewable energy, which can play a critical role to mitigate these emissions. In this article, hazardous environmental effect of fossil fuels is discussed. The status of existing renewable energy technologies especially wind and solar energy and their future growth trend is presented in this article. In this article a focused literature review on research articles discussing the environmental impact of replacement of fossil fuel energy technologies with renewable technologies, with goals to prove that if fossil fuel energy is replaced by renewable technologies can be a solution to hazardous emissions. Last part of the article provides directions for renewable energy policies of a country, which could help to increase the renewable energy mix in the traditional energy production.

Keywords: Renewable Energy, Emissions, Greenhouse Gases, Mercury Emissions, Wind Energy, Solar Energy

1. Introduction

Technological and environmental developme-nts in today's world are causing a steep rise in energy demand. World economy is increasing at a rate of 3.3% /year and energy demand is increasing at 3.6%/year since the last 30 years. International energy outlook 2009 indicates the increase of energy demand from 472 quadrillion Btu in 2006 to 552 quadrillion Btu in 2015 and to 678 quadrillion Btu in 2030. The historical increasing energy demand and the projected demand is shown in Fig. 1 [1].

Various industrial processes and conventional power generation plants are releasing hazardous gasses to contaminate the environment.

The rapid growth of human activities in the recent past has resulted in a dangerous level of greenhouse gases (GHG) in the atmosphere. Control of these GHG emissions is necessary to avoid the negative consequences on climate. Fossil fuels are the main source of energy in today's world but at the same time they are the main source of CO_2 emissions as well [2]. According to IPCC study, The level of GHG emissions has to be controlled in order to bound the temperature increase to 2^oC above pre industrial level [3].

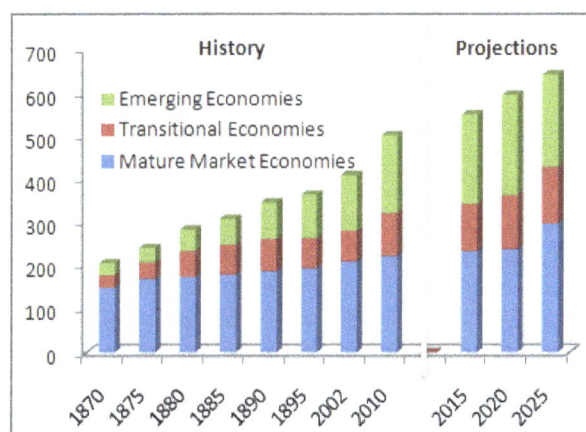

Fig. 1. World energy demand growth.

Negative impact of fossil fuel on our environment and other associated problems of fossil fuels have forced many countries to shift to environmental friendly renewable alternatives that could sustain the rapid growth in energy demand. Environmental issues has received highest attentions in many countries. One of the example is the 20-20-20 target of European Union (EU). According to that target of EU, the share of renewable energy must be increased by 20%, GHG emissions cut by 20% and the use of primary energy decreased by 20% [4].

2. Emissions from Conventional Power Plants

Carbon Dioxide Emissions: Carbon dioxide is one of the main source of climate change. About 90% of the total CO_2 emissions are coming from the energy sector and accounts for 75% of the global GHG emissions in the developed countries [5]. Power plants and refineries account for about 50% of these emissions. Carbon is emitted as CO_2 during fossil fuel combustion, some carbon is emitted as carbon monoxide (CO), methane (CH_4) or non-methane volatile organic compound (NMVOCs). For CO_2, emission factors majorly depend on the carbon contents of the fuel rather than combustion conditions. CO_2 emissions can be accurately estimated from the amount of combusted fuel in the combustion process [5]. South Korea is generating 154.7million tons of carbon per year, out of which CO_2 emission is 136.9 million tonnes which is 85% of total GHG emissions. Coal combustion is generating huge quantity of CO_2 per unit heat energy as compared to other fossil fuels [6].

Mercury Emissions from Coal Fired Power Plants: Mercury (Hg) is one of the most dangerous emission to the atmosphere, land and water. Global environmental emissions of mercury estimation in 2005 was 1930 tonnes from all anthropogenic resources. Coal combustion processes are one of the major source of mercury emission to the atmosphere [7,8]. Mercury emissions from coal combustion processes accounts for around 45% of global anthropogenic mercury emissions [9]. Mercury emissions from the coal depends upon the amount of coal combusted. Although, the quantity of mercury content in the coal is not very high, the Hg emitted from the coal combustion process is globally quite significant mainly because of the huge amount of coal used in coal fired power plants [9]. Mercury concentration in coal mainly depends on the type of coal and its origin [10].

Poland is ranked Europe's fourth highest anthropogenic mercury emitting country in 2005 [10]. According to World Coal Institute (2008), Poland was ranked the top country using coal for the generation of electricity. In 2006, 93% of the electricity was generated from the burning of brown and hard coal. In 2008, Poland's power generation mix indicates that 33% of the electricity was produced by brown coal and 62% electricity was generated by hard coal [10].

South Africa is the world's second highest mercury emitting country [11]. The primary source of mercury emissions is the coal combustion process in the coal-fired power plants. South Africa is the world's third highest coal producing country. About 64% of the primary energy supply in the country is coming from coal. Coal-fired power plants are responsible for 61% of the total consumption of coal in the country. These power plants are producing more than 90% of country's electricity. Mercury content in the South African

coal is 0.2mm and estimates of mercury emissions in the country were based on the quantity of coal combusted in these power plants, which was about 112.3Mt/year [12]. Mercury emissions in South Africa are 50 tonnes Hg/year. Mercury emissions from coal-fired power plants in various countries are presented in Table 1[12].

Table 1. *Mercury emissions from coal fired power plants.*

Country	Emission Tones Hg/years	Fraction of total electricity generation from coal (Ratio)
Canada	1.3	0.27
China	72.86	0.47
Mexico	1.6	0.78
Poland	20.6	0.96
Russia	16	0.68
USA	42.6	0.7
South Africa	50.0	0.92

Results of a recent study presented in [13] show that 49% of lakes in USA contain fish with concentration of mercury above the permitted safe limit. Conventional electricity production plants emit 50-1000 times more mercury (Hg) to the environment than solar power plants, i.e. about 15g Hg/GWh from coal as compared to around 0.1g Hg/Gwh from solar equipment [14, 15].

3. Emissions Mitigation via Renewable Energy

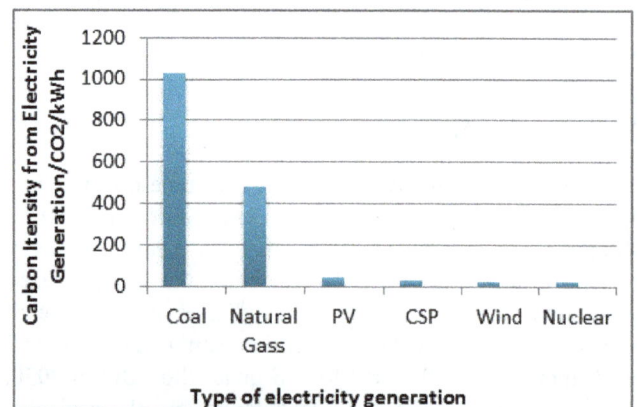

Fig. 2. *CO_2 emissions from differnct electricity generaration methods.*

The environmental impact of any technology for energy can be characterized from its hazardous carbon emissions intensity, which is the measured quantity of mercury emissions, carbon dioxide or carbon dioxide equivalent per unit of energy generation. Here carbon dioxide equivalent means any non-CO_2 greenhouse emissions like nitrous oxide, methane etc. which are the result of carbon rich fossil fuels combustion for various human activities. On the other hand, renewable technologies like solar and wind energy will produce very little or no emissions at the operation stage. These technologies can release some emissions at the manufacturing stage. CO_2 emissions per Kwh from different renewable and conventional power generation methods are shown in Fig. 2 [16]. It can be seen in the figure that renewable technologies produce very small amount of CO_2 emission as compared to carbon rich conventional fossil fuel technologies.

The amount of emissions mitigation depends on the type of energy resource displaced by the renewable technologies, the amount of convention energy generation resource replaced and the type and amount of energy used during the manufacturing stage, installation phase and during operation of these renewable energy technologies.

4. Solar and Wind Energy; A solution to Environmental Emissions

Global warming effects and other dangerous climate changes associated with fossil fuels are being consider as a serious threat to human health [17, 18]. There is a growing concern about the increasing energy demand and its environment contamination problem. In order to address these concerns, the global community is taking major steps to include alternative source of energy. Renewable resources especially, solar and wind energy, will play a significant role to meet the future energy demand and to reduce the environmental pollution caused by the conventional fossil fuel resources.

4.1. Solar Energy

In one second, some $1.73 \times 10^{17} J$ of energy falls on earth in term of solar radiations [19]. Almost 4 million Hexa-Joules $(1EJ = 10^{18})$ of energy reaches the surface of earth from sun during a period of one year. Out of which approximately $5 \times 10^4 EJ$ could be harvested. This harvested amount is much more than our existing primary energy needs of 533 EJ in 2010 and projected energy demand of 782 EJ I 2035 [20]. Inspire of this massive potential only 0.5% of our electricity need is being provided by solar energy [21]. Solar energy is very important for low carbon development in the developing countries. Developing countries, in general, enjoy a higher level of solar radiations [20].

4.1.1. Impact on Human Health and Well-being

Solar power generation is rapidly increasing day by day. Currently world's installed capacity of solar is more than 22.9GW and is escalating at about 40%/year [13]. Table 2 depicts the impact of solar PV energy in forested area on human well-being and human health. The impact is globally very beneficial due to reduced toxics emissions resulting from the use of fossil fuels. NOx, SO2 and other significant pollutants are the result of conventional fossil fuel plants. About 64% of the world's greenhouse emissions are coming from fossil fuels electricity plants [12] and bulk of the remaining emissions are the result of petroleum use which can be replaced by green energy resources. These emissions are a major health hazards being faced by the humanity in today's world.

4.1.2. Importance of Solar PV Energy In Terms of Emissions

Damon Turney and Vasilis Fthenakis calculated the CO_2 emission per KWh of solar PV electricity for a forest region in USA [13]. The calculation were made assuming a plant life of 30 years, operating under isolation of

Table 2. Impact of solar PV on environment.

Impact category	Effect relative to traditional power	Beneficial or not beneficial
Exposure to hazardous chemicals		
Emissions of mercury	Reduces emissions	Beneficial
Emissions of cadmium	Reduces emissions	Beneficial
Emissions of other toxics	Reduces emissions	Beneficial
Emissions of particulates	Reduces emissions	Beneficial
Other Impacts		
Noise	Reduce noise	Beneficial
Recreational resources	Reduces pollution	Beneficial
Visual aesthetics	Similar to fossils	Neutral
Climate change	Reduces change	Beneficial
Land occupation	Similar to fossils	Neutral

$1700 kWh/m^2$ per day, having a module conversion efficiency of 13% and a 0.5% per year degradation rate in the module performance. The results are presented in Table 3 [13] that shows the following: (i) 0 to 9 g emissions of CO_2/kWh resulted from the loss of forest sequestration (ii) 0 to 2 g emissions of CO_2/kWh in the 10 years following deforestation (iii) 0 to 36 g emissions of CO_2/kWh due to removal of initial vegetation (iv) 16 to 40 g emissions of CO_2/kWh due to life cycle of solar system (v) 650 g CO_2/kWh avoidance from the conventional electricity generation. The results in Table 3 depicts that solar is very beneficial in terms of carbon emission and could be a very useful alternative to conventional power plants. The results presented in Table 3 are calculated for a forest region. In true deserted area, the solar power environmental impact would be much more beneficial.

Table 3. Carbon dioxide emissions saving.

	Carbon dioxide emissions (g Co^2/Kwh)	
	Best case	Worst case
Loss of forest sequestration	+ 0.0	+ 8.6
Respiration of soil biomass	+ 0.0	+ 1.9
Oxidation of cut biomass	+ 0.0	+ 35.8
Other phases of the life cycle	+ 16.0	+ 40.0
Total emissions of solar	+ 16.0	+ 86.3
Fossil fuel emissions avoidance	- 850.0	- 650.0
Total including avoidance	- 834.0	- 563.7

4.2. Concentrated Solar Thermal Power

This technology concentrate the sunlight to heat up a fluid to very high temperature. This hot fluid can derive a heat engine or a steam turbine to produce electricity. Different reflecting and concentrating efficient methods exists which could concentrate the sunlight by a multiplying factor of 70 times. These concentrating methods include, solar trough, solar parabolic dish, solar tower system, linear feresnal reflector. The worldwide installed capacity of concentrated solar power was 2.5GW and most of that was in USA and Spain [16]. The major advantage of CSP over PV technology is the easy storage of thermal energy as compared to PV. By stored heated fluid, the power can be generated during the off hours when no Sun energy is available. This storage advantage leads to less intermittency during the cloudy weather and enables the power system to match the energy demand. About 60% of the installed capacity of CSP in Spain can store the thermal energy in molten salt for six hours, which means the plants can generate full power for six hours using the stored solar heat. The other method of heat storage is steam storage but it is less efficient and heat storage can last under one hour [23].

Carbon Mitigation Using CSP: Different life cycle analyses show that carbon CO_2 quantity for each unit of electricity (Kwh) is 20-50 g [24]. If CSP system is integrated into the power grid than the thermal heat storage system would help to reduce the other storage components like PV.

4.3. Wind Power Generation

A rapid growth of wind power generation has been observed in the last decade. About 2% of the world consumption is being provided by wind generation [18]. China is the world leading country with 44GW of installed wind capacity followed by United States with 40GW and Germany with 27GW of installed wind power generation [25]. More than 50GW of wind power generation plants are installed in the European Union region [26]. The cost of wind power generation per MWh are declining day by day thanks to the technology advancements in wind turbines, their control and wind atlases. This decreasing trend in the cost will help to increase the wind generated power share in the energy mix of the planet.

4.4. Positive Impacts of Wind Power Generation

Unlike other conventional resources (gas, coal and other petroleum based fuel), wind energy is not environment pollutant. It can help to mitigate environmental pollution by replacing conventional resources. It is available in abundance and be harvested on land or oceans.

4.4.1. Savings in Water Consumption

Water consumption is vital in water stressed countries like Saudi Arabia, Singapore and UAE where clean water resources are scarce. Convention power plants use huge quantity of water for cooling and condensing purposes. Water is also used for cleaning and fuel processing on coal power plant. Use renewable energy generation processes can save millions of liters water per day. The water consumption per KWh for various conventional power plants and renewable power generation methods is shown in Table 5 [18].

Table 4. *Water consumption from various power generation technologies.*

Technology	gal/kWh	l/kWh
Nuclear	0.62	2.30
Coal	0.49	1.90
Oil	0.43	1.60
Combined cycle gas	0.25	0.95
Solar	0.030	0.110
Wind	0.001	0.004

4.4.2. Reduction in CO2 Emissions

Normally wind turbines has no adverse effect on the environment only some amount of carbon dioxide emits during the construction and maintenance. Every electricity unit generated by wind will replace an electricity unit by conventional power plants and would save CO_2 emissions in the environment because it does not produce hazardous emissions like fossil fuels [27]. According to German Federal Ministry for the environment, in 2006, about 67 million tons of carbon dioxide were saved by producing electricity from renewable resources.

Wind power has the great capacity to mitigate the hazardous emissions on our planet. In [26], the authors calculated the emissions savings from the wind power plants in Texas. Electricity production in USA mainly depends on fossil fuel with 42% share of coal, 25% share from natural gas, 19% from nuclear, compared to 8% from hydropower, less than 1% from each of geothermal, solar and biomass, and 3% from the wind power generation.

Average emissions from a coal power plant in USA are 6 lbs/MWh of NO_x, 13 lbs/MWh of SO_2 and 1.1 tons/MWh of CO_2. Average emissions in USA resulting from natural gas plants are are 1.71 lbs/MWh of NO_x, 0.10 lbs/MWh of SO_2 and 0.57 tons/MWh of CO_2. If wind power replace one MWh of USA energy mix, an emission savings of 0.79 lbs/MWh of NO_x, 1.3 lbs/MWh of SO_2 and 0.52 tons/MWh of CO_2 can be achieved at ERCOT plant [26].

4.4.3. Comparison of Wind Power with Other Power Generations

Wind power generation has less adverse impacts as compares to other sources of energy. A comparison of wind power with other resources is shown in Table 4 [18].

4.4.4. Increasing Trend of Wind Power Generation

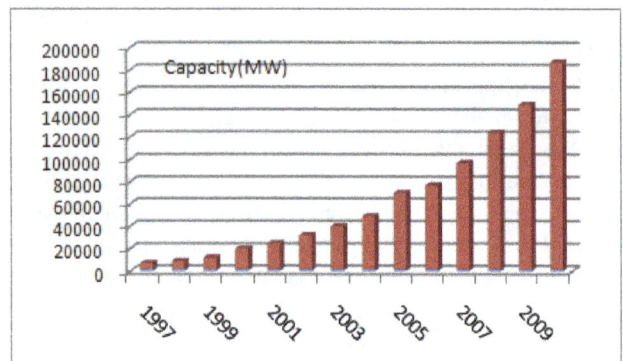

Fig. 3. *Wind Capacity of the Planet.*

Wind power generation does not cause air pollution like thermal power generation that depends on fossil fuel combustion (coal or natural gas). Wind turbines do not emit GHG emissions or acid rain. Due to its real tiny effect on the environment, this form of energy is considered as the true green energy. These environmental benefits are pushing the word to accelerate the installation of wind energy. The increasing trend in the wind power generation capacity is shown in Fig. 3 [28].

5. Policy Suggestions

Renewable energy is a true green energy, which can mitigate our environmental pollution but still it is facing some problems to become an integral part of our national grids. Electricity production from renewable resources needs to be critically analyzed. Some suggestion for renewable energy production to mitigate the negative environmental impact are presented in this section:

a) One of the hurdle to solar PV is the high generation cost. The prices of PV technologies have coming down at a great pace in the last decade or so, yet the cost of production is having issues to achieve grid parity in most of the countries. It has been seen that the developing and poor countries enjoy very good solar irradiance. The solar PV production factories can be located in these high solar irradiant countries near the load centers. The cost of Labor in these countries is very low which would help to lower the prices of PV panels.

b) One big issue with solar energy is the availability of flat land area. Normally the solar technologies PV and Concentrated Solar Power (CSP) need a lot of flat land. In the countries where flat land is not available like European countries, this problem can be addressed by installing solar PV and CSP plants in the high solar irradiant countries like Middle East with lot of flat deserts and transmitting the produced power to Europe by High Voltage DC (HVDC) transmission lines.

c) There is a need to conduct an extensive study by the countries to know the wind potential. Based on the availability of land, both On-shore and Off-shore wind forms can deployed.

d) The government can set renewable energy share targets and provide a proper financial support to achieve the set targets. In this regard, the government can provide tax incentives, relief in the duty of renewable energy technologies import and feed in tariff.

In this article, the catastrophic effect of fossil fuels on over planet are discussed. Fossil fuels although produce useful energy but they are also responsible for hazardous emissions, which are contaminating our environment. This article focused on the poisonous emissions of conventional fossil fuel power generation methods. Coal combustion is generating huge quantity of emissions per unit heat energy as compared to other fossil fuels. It has been seen that the leading countries using coal fired power plants are releasing huge amount of emissions to the environment. It has been seen that these emissions can be avoided/decreased by increasing the share of renewable energy in the energy mix. The emissions from renewable energy resources and conventional energy resources are compared and the comparison shows that renewable technologies especially wind and photovoltaic produce very little emissions over their entire life span. Today's world is taking aggressive steps to increase the renewable energy production. Further steps are required to be taken by the governments to include the alternative energy resources to replace conventional energy resources. Renewable energy policy suggestion are proposed in the last part of the article. These suggestions could help to increase the renewable energy share on our planet. The facts presented in this article by a survey of recent research of fossil fuels impact and the suggestions included regarding the policy making could be helpful for the authorities to mitigate the catastrophic impact these fossil fuel emissions.

Table 5. Impact of wind power generation vs other power generation methods.

Habitat impacts	Coal	Natural gas	Oil	Nuclear	Hydropower	Wind
Air and water pollution	Yes	Yes	Yes			
Global warning	yes	yes	yes			
Thermal pollution of water				yes		
Flooding of land						
Waste disposal	yes			yes	yes	
Mining and drilling	yes	yes	yes	yes		
Construction of plants	yes	yes	yes	yes	yes	yes

References

[1] International Energy Agency (IEA). World energy outlook. Medium term oil and gas market report.

[2] International Energy Agency (IEA), CO2 emissions from fossil fuel. Combustion

[3] Mitigation of climate change. IPCC Fourth Assessment report by Working Group III of the International Panel on Climate Change. (hppt:// www.ipcc.ch)

[4] Second strategic energy review. An EU energy and solidarity action plan COM (2008) 781 final

[5] IPCC. 2006 IPCC guidelines for national greenhouse gas inventories; 2006.

[6] Energy Information Administration. Emissions of greenhouse gases in the United States 1985-1990. DOE/EIA-1573; 1993. P. 16.

[7] Pacyna, J.M., Pacyna , E.G., Steenhussein, F., Wilson, S.,2003. Mapping 1995 global athropogenic emissions of mercury. Atmospheric Environment 37, 109-117.

[8] Pacyna, J.M., Pacyba, E,G., 2006. Mercury Strategy Development in the EU and UN; Current global emissions and their scenarios, MEC3 Third International Expert's Workshop, Katowice, Poland.

[9] Pacyna, J.M., Munthe, J., Wilson, S., Maxson, P., Sundseth, K., Pacyna, E.G., Harper, E., Kindbom, K., Wangberg, I., Panasiuk, D., Glodek, A., Leaner, J., Dabrowski, J., 2008. Technical Background Report to th Global tmospheric Mercury assessment. Arctic Monitoring and Assessment Programme/ UNEP Chemical Branch. www.chem.unep.ch/mercury.

[10] Anna Glodek, Jozef M. Pacyna, Mercury emission from coal-fired power plants in Poland, Atmospheric Environment, Volume 43, Issue 35, November 2009, Pages 5668-5673, ISSN 1352-2310

[11] Pacyna, E.G., Pacyna, J.M., Steenhuisen, F., Wilson, S., 2006. Global anthropogenic mercury emissions inventory for 2000. Atmospheric Environment 40, 4048-4063.

[12] James M. Dabrowski, Peter J. Ashton, Kevin Murray, Joy J. Leaner, Robert P. Mason, Anthropogenic mercury emissions in South Africa: Coal combustion in power plants, Atmospheric Environment, Volume 42, Issue 27, September 2008, Pages 6620-6626, ISSN 1352-2310.

[13] Damon Turney, Vasilis Fthenakis, Environmental impacts from the installation and operation of large-scale solar power plants, Renewable and Sustainable Energy Reviews, Volume 15, Issue 6, August 2011, Pages 3261-3270, ISSN 1364-0321.

[14] Vasilis M. Fthenakis, Hyung Chul Kim, Erik Alsema., Emissions from Photovoltaic Life Cycles. Environmental Science and Technology 2008; 42: 2168-2174.

[15] Ruud Meij, Henk te Winkel, The emissions of heavy metals and persistent organic pollutants from modern coal-fired power stations, Atmospheric Environment, Volume 41, Issue 40, December 2007, Pages 9262-9272, ISSN 1352-2310.

[16] Jenny Nelson, Ajay Gambhir and Ned Ekins-Daukes., Solar Power for CO2 Mitigation. Grantham Institute for Climate Change Imperial College London Briefing Paper no. 11.

[17] Zhong Xiang Zhang, Asian energy and environmental policy: Promoting growth while preserving the environment, Energy Policy, Volume 36, Issue 10, October 2008, Pages 3905-3924, ISSN 0301-4215.

[18] R. Saidur, N.A. Rahim, M.R. Islam, K.H. Solangi, Environmental impact of wind energy, Renewable and Sustainable Energy Reviews, Volume 15, Issue 5, June 2011, Pages 2423-2430, ISSN 1364-0321.

[19] Kopp, G. and Lean, J.L. A new, Lower Value od Total Solar Irradiance: Evidence and Climate Signicance, Geophy. Res Letters Frontier article, 38, L01706, doi: 10.1029/2010GL045777, 2011.

[20] World Energy Outlook, International Energy Agency (IEA) 2012.

[21] Greenpeace International, European Renewable Energy Council GWE. 2012 Energy Revolution: a sustainable world energy outlook (http://www. Energyblueprint.info).

[22] IPCC. Fourth assessment report of the International Panel on climate change: mitigation of climate change; 2007.

[23] Sarada Kuravi et al. Thermal energy technologies and systems for concentrating solar power plants, Progress in Energy and Combustion Science, Volume 39, Issue 4, Pages 285-319.

[24] John J. Burkhardt III, Garvin Health, and Elliot Cohen, Life Cycle Greenhouse Gas Emissions of Trough and Tower Concentrating Solar Power Electricity Generation Systematic Review and Harmonization, Journal of Industrial Ecology 16, S1; DOI: 10.1111/j.1530-9290.2012.00474.x, 2012.

[25] Ahmad Bilal Awan, Zeeshan Ali Khan, Recent progress in renewable energy – Remedy of energy crisis in Pakistan, Renewable and Sustainable Energy Reviews, Volume 33, May 2014, Pages 236-253, ISSN 1364-0321.

[26] Kaffine Daniel T, McBee, Brannin J, Lieskovsky, Jozef ., Emissions Savings from Wind Power Generation in Texas, The Energy Journal, Volume 34, Issue 1, January 2013, Page 155.

[27] R.H. Crawford, Life cycle energy and greenhouse emissions analysis of wind turbines and the effect of size on energy yield, Renewable and Sustainable Energy Reviews, Volume 13, Issue 9, December 2009, Pages 2653-2660, ISSN 1364-0321.

[28] Ali Mostafaeipour, Productivity and development issues of global wind turbine industry, Renewable and Sustainable Energy Reviews, Volume 14, Issue 3, April 2010, Pages 1048-1058, ISSN 1364-0321.

10

Foraging Patterns of Birds in Resource Partitioning in Tropical Mixed Dry Deciduous Forest, India

Nirmala Thivyanathan

Principal, Research Centre of Zoology, Jayaraj Annapackiam College for Women, Periyakulam, Theni District, Tamilnadu, India

Email address:

principal@annejac.com

Abstract: A study was undertaken in the tropical mixed dry deciduous forest of India. Direct observation on foraging of birds was made on twelve days in a month within four hours after sunrise with direct observation. For each foraging attempt microhabitat details such as the foraging height, substrate, method, canopy and the plant species were recorded. Vegetation profile consisted of tree species from 2 to 6m and shrubs from 0 to 1m height. In total, 3982 foraging observations were made on 36 bird species. A higher percentage of foraging manoeuvre was recorded at 3-6m height. 29 bird species were gleaner. Majority of the canopy layers used for foraging of bird species were edge edge (23%) followed by ground (18%) and middle lower (17%). Grey Jungle Fowl, Vernal Hanging Parrot and Red-rumped Swallow are specialists. The higest mean niche overlap among the species was found in method followed by canopy and height. The two major guilds are gleaner and sallier.

Keywords: Foraging Method, Foraging Substrate, Foraging Canopy, Foraging Height, Guild, Niche Overlap, Resource Partitioning, Tropical Mixed Dry Deciduous Forest

1. Introduction

Birds prefer some specific habitats and coexist as guilds with the available pattern of food resources [1]. Guild segregates themselves into specific ecological niches by adopting foraging behaviour and differs in microhabitat use and foraging tactics [2]. The foraging tactics include various methods to exploit the resources. Insectivore birds exhibit different methods of exploiting resources such as gleaning, sallying, probing, pouncing and hawking [3], [4], [5].

Although resource partitioning has been well documented for bird species from temperate forests [3], [6], [7], no such studies are available in India except the study of Gokula and Vijayan [5] in the dry deciduous forest of Mudumalai Wildlife Sanctuary. Moreover, knowledge of the ways in which birds exploit resources within a forest will increase the understanding of their habitat use and the essential requirements for their survival. The following objectives were set to analyse the patterns of feeding behavior, method of feeding and microhabitat use by birds in the mixed dry deciduous forest.

2. Study Area

The study was undertaken in the tropical mixed dry deciduous forest of Anaikatty hills [8], the foothills of the Nilgiri in the Nilgiri Biosphere Reserve, Western Ghats, India situated at an elevation of about 610-1200m above MSL between 76° 39' and 76° 47'E and from 11° 5' to 11° 31'N in Coimbatore, TamilNadu, Southern India. The climate is moderate and pleasant for most part of the year except summer which is relatively hot and dry.

Based on the climate, four different seasons were observed as follows. *Southwest monsoon (June, July and August):* The study area received 5% of the total annual rainfall during this season. The mean rainfall received was around 40 mm. *Northeast monsoon (September, October and November):* The study area received more than half (69%) of the total annual rainfall during this season. The mean rainfall received was around 500 mm.

Winter (December, January and February): It was the least rainy period of the year with the annual rainfall of 34 mm. This season was the colder period with the minimum temperature falling to 18°C. *Summer (March, April and May):* This area received 21% of the annual rainfall in this season from the pre-monsoon showers. This was the period of maximum temperature, which leaped up to 37°C with low relative humidity.

Temperature varied between 18°C and 37°C and Relative humidity showed fluctuation in different seasons between 31% - 75% at 08:30 hrs. and 72% - 89% at 17:30 hrs. Monthly windspeed varied between 3 and 14 km/h. The tropical mixed dry deciduous forest, India has the major tree community of *Acacia leucophloea, Ziziphus mauritiana, Chloroxylon swietenia, Albizia amara, Tamarindus indicus, Albizia lebbeck, Acacia polyacantha, Diospyros ferrea, Cassia fistula* and *Commiphora caudata.* Major shrubs are *Chromolaena odorata, Elaeodendron glaucum, Pavetta indica, Lantana camara, Randia dumetorum, Premna tomentosa, Flacourtia indica* and *Mundulea sericea.*

3. Materials and Methods

Foraging records of birds were made during May 1999 to May 2001 on twelve days in a month from the tropical mixed dry deciduous forest, India. Most of the observations were done within four hours after sunrise. This is the most active foraging time for birds [9]. Only initial record was taken from any individual encountered as done by MacNally [7] to provide precise estimate of foraging location rather than that of the subsequent ones [10].

Table 1. *Definition of foraging activities used to assess guild structure of avifauna.*

Foraging method	Sub categories
Sally	Above canopy-sally
	Below canopy-sally
	Herb-sally
	Shrub-sally
	Sally (sally to the ground)
Glean	Flower-glean
	Fruit-gleaning
	Ground-gleaning
	Litter-gleaning
	Main trunk-gleaning
	Secondary branch-gleaning
	Twig-glean
	Leaf-glean
Pounce	Ground-pounce
Probing	Ground-probing
	Litter-probing
	Main trunk-probing
	Secondary branch-probing
Tear	Leaf-tear
Hover	Hovering/aerial capture

For each foraging attempt microhabitat details such as the foraging height, substrate, method, canopy and the plant species at which the prey was found were recorded. *Foraging attempts* were assigned to 12 height categories. A *substrate* is the place from where food is taken by birds in 7 different areas. *Foraging methods* of birds were categorized as, *Glean, Probe, Sally* or *fly catching* and *Pounce,* To cluster the species

on a micro level, these methods were classified further into finer levels based on the substrate, which is given in Table 1 and described by Crome [3] and expanded by Holmes *et al.,* [4], Ramsen and Robinson [11] and MacNally [12].

The *canopy layers* used by the bird species were classified into ten layers and were possibly distinguished from three layers namely lower canopy, middle canopy and upper/edge canopy (Figure 1). Lower canopy was further distinguished as lower center, lower middle and lower edge. Middle canopy was classified further into middle center, middle middle and middle edge. c). Upper/edge canopy was classified as edge center, edge middle and edge edge. d). Birds, which do not use plant at all for its prey was grouped under ground/air/under canopy.

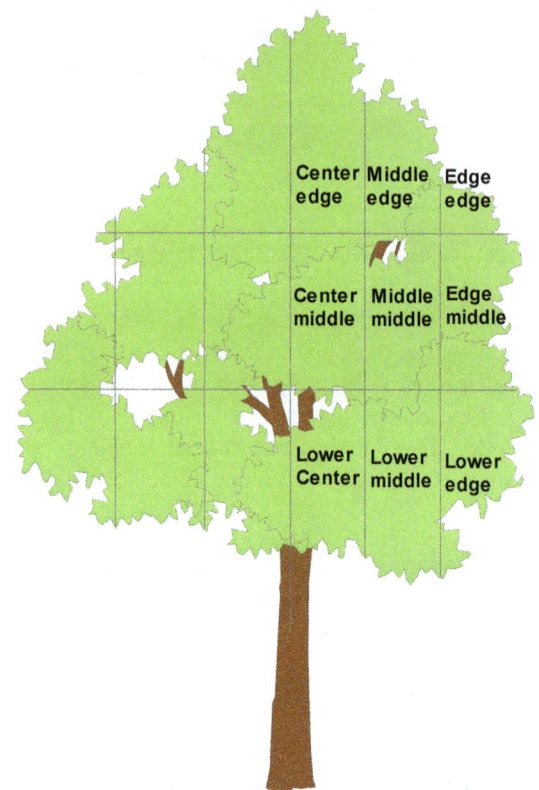

Figure 1. *Diagrammatic Representation of the Canopy Layers of a Plant.*

As thirty independent observations are recommended to represent the behavior of a bird accurately [13], species with more than 30 observations were taken for analysis.

4. Statistical Analysis

4.1. Specialist-Index J'

The foraging specialization of each foraging parameter (method, substrate, height and canopy) was analyzed using the Shannon-Weaver [14] index. These values were then converted to a standardized range using the formula $J' = H'/H_{max}$ (Where J' = specialization and H_{max} = the maximum H' value) following Crome [3] and Recher *et al.* [6]. J' value ranges between one and zero, with foraging specialization increases as J' decreases.

4.2. Niche Overlap

The extent to which resource use overlaps between species pairs is niche overlap. The degree of species overlap in niche utilization for the different categories recorded (foraging method, substrate, canopy and foraging height) has been quantitatively expressed using Horn's index [15].

4.3. Cluster Analysis

To compare foraging behavior (substrate, height, canopy use and method adopted) by various species, cluster analyses were performed on a data matrix (species * characteristics), following Holmes *et al.* [4]. This analysis used the

unweighted pair group clustering method with arithmetic averages (UPGMA) and Squared Euclidean Distance [16], [17]. The SPSS statistical software [18] was used for the data analyses.

5. Results

In total, 3982 foraging observations were made on 36 species in the tropical mixed dry deciduous forest, India (Table 2).

Table 2. Number of foraging records on each bird species observed in the tropical mixed dry deciduous forest during 1999-2001.

S. No	Common name of the species	Scientific name	Family	Number of foraging observations
1	GREY JUNGLEFOWL	Gallus sonneratii	Phasianidae	94
2	INDIAN PEAFOWL	Pavo cristatus	Phasianidae	36
3	BLOSSOM-HEADED PARAKEET	Psittacula roseate	Psittacidae	45
4	MALABAR PARAKEET	Psittacula columboides	Psittacidae	225
5	VERNAL HANGING PARROT	Loriculus vernalis	Psittacidae	53
6	BLUE-FACED MALKOHA	Phaenicophaeus viridirostris	Cuculidae	74
7	GREEN BEE-EATER	Merops orientalis	Meropidae	42
8	CHESTNUT-HEADED BEE-EATER	Merops leschenaultia	Meropidae	56
9	COMMON HOOPOE	Upupa epops	Upupidae	35
10	BROWN-HEADED BARBET	Megalaima zylanica	Capitonidae	31
11	STREAK-THROATED WOODPECKER	Picus xanthopygaeus	Picidae	28
12	PALE-BILLED FLOWERPECKER	Dicaeum erythrorynchos	Picidae	124
13	RED-RUMPED SWALLOW	Hirundo daurica	Hirundinidae	49
14	RED-WHISKERED BULBUL	Pycnonotus jocosus	Pycnonotidae	84
15	RED-VENTED BULBUL	Pycnonotus cafer	Pycnonotidae	102
16	WHITE-BROWED BULBUL	Pycnonotus luteolus	Pycnonotidae	333
17	BLACK BULBUL	Hypsipetes leucocephalus	Pycnonotidae	74
18	COMMON IORA	Aegithina typhia	Irenidae	368
19	BLUE-WINGED LEAFBIRD	Chloropsis cochinchinensis	Irenidae	115
20	TAWNY-BELLIED BABBLER	Dumetia hyperythra	Muscicapidae	125
21	JUNGLE BABBLER	Turdoides striatus	Muscicapidae	240
22	YELLOW-BILLED BABBLER	Turdoides affinis	Muscicapidae	189
23	BLYTH'S REED WARBLER	Phylloscopus reguloides	Muscicapidae	128
24	COMMON TAILORBIRD	Orthotomus sutorius	Muscicapidae	40
25	GREENISH WARBLER	Phylloscopus trochiloides	Muscicapidae	159
26	LARGE-BILLED LEAF WARBLER	Phylloscopus magnirostris	Muscicapidae	147
27	ASIAN PARADISE FLYCATCHER	Terpsiphone paradise	Muscicapidae	82
28	PLAIN FLOWERPECKER	Dicaeum concolor	Dicaeidae	74
29	PURPLE-RUMPED SUNBIRD	Nectarinia zeylonica	Nectariniidae	361
30	LOTEN'S SUNBIRD	Nectarinia lotenia	Nectariniidae	90
31	PURPLE SUNBIRD	Nectarinia asiatica	Nectariniidae	57
32	JUNGLE MYNA	Acridotheres fuscus	Sturnidae	72
33	BLACK-HOODED ORIOLE	Oriolus xanthornus	Oriolidae	41
34	BLACK DRONGO	Dicrurus macrocercus	Dicruridae	73
35	ASHY DRONGO	Dicrurus leucophaeus	Dicruridae	33
36	WHITE-BELLIED DRONGO	Dicrurus caerulescens	Dicruridae	103
	Total			3982

Nomenclature following Grimmette *et al.* (1998)

5.1. Foraging Height

All the 12 height categories were utilized by 36 bird species in the tropical mixed dry deciduous forest, India (Table 3). Although most species fed over a broad range of heights, they were grouped according to the layer of vegetation in which the majority of their foraging was recorded. Foliage was partitioned as three layers of strata; ground (0m), shrub/short trees (0.1-3), and tree layers (>3).

In the community as a whole, a higher percentage of foraging manoeuvre were recorded in the layers of 3-6m height.

Six species foraged mainly at ground level. Among them, Grey Junglefowl absolutely used the ground layer while Jungle Myna, Yellow-billed Babbler, Indian Peafowl, Common Hoopoe, and Jungle Babbler showed variety in their height preference.

The 0.1-3m height category of shrub and short tree layers

were utilized by Blossom-headed Parakeet, Tawny-bellied Babbler, Loten's Sunbird, Common Tailorbird, Red-vented Bulbul, White-browed Bulbul, Blyth's Reed Warbler, Purple-rumped Sunbird and Asian Paradise Flycatcher.

The tree layer (>3m) was used by 21 bird species. Within the tree layers, higher percentage of foraging manoeuvre was recorded in the 3-6m height category. All the foraging attacks of the Ashy Drongo, Large-billed Leaf Warbler, Black Drongo, Vernal Hanging Parrot and Red-rumped Swallow were at >6m height. For the foraging community as a whole in the tropical mixed dry deciduous forest, a higher number of foraging manoeuvres was recorded in the tree layers (>3m height).

5.2. Foraging Substrate

Majority of the bird species used foliage (24 bird species) followed by twigs (22 bird species) as their substrate (Table 4). Only 11 species used ground and flower to find their food.

The ground-foraging guild was with five species viz. Jungle Babbler, Grey Jungle fowl, Indian Peafowl, Common Hoopoe and Yellow-billed Babbler.

Streak-throated Woodpecker and Black-Hooded Oriole largely obtained their prey from the trunk. In addition to this substrate, these birds also used twigs and fruits. Five bird species such as Blyth's Reed Warbler, Bluewinged Leafbird, Plain Flowerpecker, Black Bulbul and Large-billed Leaf Warbler used this substrate. Blue-faced Malkoha, Greenish Warbler, Tawny-bellied Babbler, Common Tailorbird and Common Iora used exclusively twigs as substrate (Table 4). Only Vernal Hanging Parrot alone used flower as its substrate while Parakeets used fruit predominantly with a little overlap of flower. Other species such as Bulbuls and Large Green Barbet used this substrate and also other substrates for their prey. Red-rumped Swallow obtained its prey exclusively from air. Drongos predominantly used air for their prey and in addition, they also used foliage to a lesser extent.

Table 3. Percentage of prey attacks by different species of birds at various height categories in the tropical mixed dry deciduous forest, India.

Name of the bird species	Foraging heights (m)												
	0	0.1-1	1-2	2-3	3-4	4-5	5-6	6-7	7-8	8-9	9-10	>10	H'
Grey junglefowl	100	0	0	0	0	0	0	0	0	0	0	0	0.0
Common hoopoe	91	0	0	3	0	0	3	3	0	0	0	0	1.9
Yellow-billed babbler	97	2	0	0	0	0	1	0	0	0	0	0	0.4
Jungle babbler	81	6	4	4	1	1	2	0	0	0	0	0	0.7
Indian peafowl	77	23	0	0	0	0	0	0	0	0	0	0	1.5
Jungle myna	39	19	0	0	0	0	21	3	18	0	0	0	1.5
Blossom-headed parakeet	0	88	0	0	0	0	0	0	0	0	2	9	0.3
Tawny-bellied babbler	0	72	24	4	0	0	0	0	0	0	0	0	0.7
Loten's sunbird	0	44	11	2	8	19	8	6	2	0	0	0	1.7
Common tailorbird	0	48	30	10	8	5	0	0	0	0	0	0	0.9
Red-vented bulbul	0	40	15	6	14	1	4	5	1	3	0	12	1.8
White-browed bulbul	2	16	30	22	17	12	1	1	1	0	0	0	0.9
Blyth's reed warbler	0	27	30	18	15	8	2	0	0	0	1	0	1.5
Purple-rumped sunbird	0	22	18	9	14	16	11	6	3	1	1	1	0.9
Asian paradise flycatcher	0	18	26	45	5	5	1	0	0	0	0	0	1.4
Brown-headed barbet	0	0	6	29	10	6	19	16	10	0	0	3	2.3
Black bulbul	0	0	0	26	0	3	38	24	0	0	9	0	1.5
Red-whiskered bulbul	0	14	19	24	8	12	12	7	1	1	1	0	2.2
Streak-throated woodpecker	0	0	7	21	57	0	11	0	4	0	0	0	0.3
Green bee-eater	0	19	0	0	52	7	0	0	0	0	0	21	0.7
Malabar parakeet	0	3	2	3	32	28	12	4	7	3	4	3	1.2
Blue-winged leafbird	0	0	0	5	30	39	17	5	2	1	1	0	1.4
Purple sunbird	0	5	23	2	19	7	28	0	11	0	5	0	2.2
Pale-billed flowerpecker	0	4	9	10	27	16	21	4	3	3	3	1	1.9
Common iora	0	3	13	17	22	17	15	7	2	0	2	2	0.9
Blue-faced malkoha	19	14	3	5	20	23	8	4	0	3	0	1	2.3
Greenish warbler	0	0	1	8	16	25	18	6	16	3	2	7	1.5
Plain flowerpecker	0	0	8	14	15	42	13	0	8	0	0	0	1.8
Black-hooded oriole	0	0	0	0	10	51	17	7	2	2	0	10	1.4
White-bellied drongo	4	1	0	1	13	22	14	17	12	10	4	4	2.1
Chestnut-headed bee-eater	2	5	7	2	2	16	16	13	13	0	2	23	2.7
Ashy drongo	0	0	0	0	3	15	30	15	3	0	18	15	1.9
Large-billed leaf warbler	0	1	6	12	19	28	22	7	3	2	1	0	1.5
Black drongo	8	3	1	3	4	0	21	10	22	3	7	19	2.4
Vernal hanging parrot	0	0	0	0	2	0	2	2	34	8	53	0	0.9
Red-rumped swallow	0	0	0	2	0	29	0	0	0	0	59	10	0.6

Table 4. Percentage use of various foraging substrates by different species of birds in the tropical mixed dry deciduous forest, India.

Name of the bird species	Ground	Trunk	Foliage	Twigs	Flower	Fruit	Air
ASHY DRONGO	0	0	0	0	24	0	76
BLACK BULBUL	0	1	45	49	0	5	0
BLACK DRONGO	0	0	1	0	0	0	97
BLACK-HOODED ORIOLE	0	66	0	32	0	2	0
BLOSSOM-HEADED PARAKEET	0	0	0	0	7	93	0
BLYTH'S REED WARBLER	0	0	56	44	0	0	0
MALABAR PARAKEET	0	0	0	0	2	98	0
BLUE-FACED MALKOHA	24	8	9	35	0	23	0
GREY JUNGLEFOWL	97	0	3	0	0	0	0
BLUE-WINGED LEAFBIRD	0	0	56	17	20	8	0
COMMON HOOPOE	91	6	0	3	0	0	0
COMMON IORA	0	0	27	72	0	0	0
JUNGLE BABBLER	81	2	0	17	0	0	0
JUNGLE MYNA	39	3	0	0	53	6	0
LARGE-BILLED LEAF WARBLER	0	0	90	10	0	0	0
BROWN-HEADED BARBET	0	6	6	0	0	87	0
VERNAL HANGING PARROT	0	0	0	0	100	0	0
LOTEN'S SUNBIRD	0	0	0	6	93	1	0
STREAK-THROATED WOODPECKER	0	93	0	7	0	0	0
PLAIN FLOWERPECKER	0	0	46	40	14	0	0
INDIAN PEAFOWL	74	0	3	0	0	23	0
ASIAN PARADISE FLYCATCHER	0	0	12	0	0	0	88
GREENISH WARBLER	0	9	27	64	0	0	0
PURPLE-RUMPED SUNBIRD	0	0	5	7	87	1	0
PURPLE SUNBIRD	0	7	16	4	74	0	0
TAWNY-BELLIED BABBLER	0	0	14	85	0	1	0
RED-RUMPED SWALLOW	0	0	0	0	0	0	100
RED-VENTED BULBUL	2	12	4	12	0	71	0
RED-WHISKERED BULBUL	1	4	1	24	0	70	0
GREEN BEE-EATER	0	0	21	0	0	0	79
COMMON TAILORBIRD	3	0	35	58	0	5	0
PALE-BILLED FLOWERPECKER	0	0	15	21	64	0	0
WHITE-BROWED BULBUL	2	0	1	15	0	77	5
WHITE-BELLIED DRONGO	0	0	2	0	0	0	98
YELLOW-BILLED BABBLER	97	0	1	2	0	0	0
CHESTNUT-HEADED BEE-EATER	0	0	13	0	0	0	88
Substrate preference by Number of Bird Species	11	12	24	22	11	16	8

Table 5. Percentage of prey attack manoeuvres by different bird species in the tropical mixed dry deciduous forest, India.

Name of the Bird Species	Glean	hover	Pounce	Probe	Sally
BLACK BULBUL	100	0	0	0	0
BLACK-HOODED ORIOLE	83	0	17	0	0
BLOSSOM-HEADED PARAKEET	100	0	0	0	0
BLYTH'S REED WARBLER	100	0	0	0	0
MALABAR PARAKEET	100	0	0	0	0
BLUE-FACED MALKOHA	99	0	0	1	0
GREY JUNGLEFOWL	95	0	0	5	0
BLUE-WINGED LEAFBIRD	100	0	0	0	0
COMMON HOOPOE	74	0	0	26	0
COMMON IORA	100	0	0	0	0
JUNGLE BABBLER	100	0	0	0	0
JUNGLE MYNA	97	0	0	3	0
LARGE-BILLED LEAF WARBLER	100	0	0	0	0
BROWN-HEADED BARBET	100	0	0	0	0
VERNAL HANGING PARROT	100	0	0	0	0
LOTEN'S SUNBIRD	100	0	0	0	0
STREAK-THROATED WOODPECKER	71	0	0	29	0
PLAIN FLOWERPECKER	100	0	0	0	0
INDIAN PEAFOWL	67	0	0	33	0
GREENISH WARBLER	100	0	0	0	0
PURPLE-RUMPED SUNBIRD	100	0	0	0	0
PURPLE SUNBIRD	100	0	0	0	0
TAWNY-BELLIED BABBLER	100	0	0	0	0
RED-VENTED BULBUL	99	0	1	0	0
RED-WHISKERED BULBUL	100	0	0	0	0

Name of the Bird Species	Glean	hover	Pounce	Probe	Sally
COMMON TAILORBIRD	100	0	0	0	0
PALE-BILLED FLOWERPECKER	100	0	0	0	0
WHITE-BROWED BULBUL	95	0	0	0	5
YELLOW-BILLED BABBLER	100	0	0	0	0
RED-RUMPED SWALLOW	0	100	0	0	0
ASHY DRONGO	24	0	0	0	76
BLACK DRONGO	0	0	0	0	100
ASIAN PARADISE FLYCATCHER	0	0	0	0	100
GREEN BEE-EATER	0	0	0	0	100
WHITE-BELLIED DRONGO	0	0	0	0	100
CHESTNUT-HEADED BEE-EATER	0	0	0	0	100

5.3. Foraging Methods

Birds such as gleaner (88%), sallier (10%), prober (1%), pouncer and hoverer (1%) were recorded from this forest (Table 5). Twenty-nine species were recorded as gleaner, of which 24 species predominantly used (100%) gleaning. Six species such as Asian Paradise Flycatcher, White-bellied Drongo, Green Bee-eater, Black Drongo, Ashy Drongo and Chestnut-headed Bee-eater used sallying to obtain their prey. Except Ashy Drongo, all other birds of this guild used sally as the only prey attacking manoeuvre. Red-rumped Swallow was recognized as hoverer or aerial capture, which used this method alone as the prey-attacking manoeuvre.

Prey attack manoeuvre by gleaners: Since gleaning formed the major method adopted by the birds of tropical mixed dry deciduous forest, India, it's usage was further bifurcated into eight types (Table 6). In total, gleaning of flower (21%), fruit (21%) and twig (21%) formed 63% of gleaning. Gleaning on

ground (18%) and leaf (11%) was comparatively less, while on trunk (4%) and stem (4%) it was very little.

Flower Gleaning: Six species exploited the flowers by gleaning for nectar. Vernal Hanging Parrot alone used only this method for feeding. Loten's Sunbird, Purple-rumped Sunbird and Purple Sunbird used this method predominantly while Pale-Billed Flowerpecker and Jungle Myna used this method frequently (Table 6). *Fruit Gleaning:* Malabar Parakeet, Blossom-headed Parakeet frequently used this method along with flower gleaning. Brown-headed Barbet, White-browed Bulbul, Red-vented Bulbul and Red-whiskered Bulbul also used this method along with other methods. *Ground Gleaning:* Yellow-billed (White-headed) Babbler, Grey Junglefowl, Jungle Babbler, Indian Peafowl and Common Hoopoe formed the groud gleaner to get their prey from ground and overlap with litter gleaning.

Table 6. Percentage of Prey Attack Manoeuvres by different types of Gleaner Bird Species in the Tropical Mixed Dry Deciduous Forest, India.

Name of the bird species	Flower Gleaner	Fruit Gleaner	Ground Gleaner	Leaf Gleaner	Litter Gleaner	Trunk Gleaner	Stem Gleaner	Twig Gleaner
LARGE-BILLED LEAF WARBLER	0	0	0	90	0	0	0	10
BLYTH'S REED WARBLER	0	0	0	56	0	0	0	44
STREAK-THROATED WOODPECKER	0	0	0	0	0	65	25	10
COMMON IORA	0	0	0	28	0	0	0	72
GREENISH WARBLER	0	0	0	27	0	0	9	64
COMMON HOOPOE	0	0	46	0	46	4	0	4
JUNGLE BABBLER	0	0	55	0	26	0	2	17
GREY JUNGLEFOWL	0	0	74	3	22	0	0	0
TAWNY-BELLIED BABBLER	0	1	0	14	0	0	0	85
BLACK-HOODED ORIOLE	0	3	0	0	0	9	50	38
COMMON TAILORBIRD	0	5	3	35	0	0	0	58
BLUE-FACED MALKOHA	0	23	7	10	16	0	8	36
BLACK BULBUL	0	32	0	0	0	0	2	66
INDIAN PEAFOWL	0	33	54	4	8	0	0	0
RED-WHISKERED BULBUL	0	70	1	1	0	0	4	24
RED-VENTED BULBUL	0	71	1	4	0	0	12	12
BROWN-HEADED BARBET	0	87	0	6	0	0	6	0
YELLOW-BILLED BABBLER	1	0	93	1	4	0	0	2
WHITE-BROWED BULBUL	1	81	2	1	0	0	0	16
MALABAR PARAKEET	2	98	0	0	0	0	0	0
BLOSSOM-HEADED PARAKEET	11	89	0	0	0	0	0	0
PLAIN FLOWERPECKER	16	0	0	45	0	0	0	39
JUNGLE MYNA	54	6	40	0	0	0	0	0
PALE-BILLED FLOWERPECKER	67	0	0	14	0	0	0	19
PURPLE SUNBIRD	74	0	0	16	0	0	7	4
PURPLE-RUMPED SUNBIRD	88	0	0	5	0	0	0	7
LOTEN'S SUNBIRD	93	1	0	0	0	0	0	6
VERNAL HANGING PARROT	100	0	0	0	0	0	0	0

Leaf Gleaning: Large-billed Leaf Warbler, Blyth's Reed Warbler and Plain Flowerpecker used this method with twig and flower gleaning.

Trunk Gleaning: Streak-throated Woodpecker alone used this type of feeding along with gleaning on stem and twig.

Stem Gleaning: Black-hooded Oriole was the only bird species, which used this method. This species also used fruit, trunk and twig as substrate for collecting food.

Twig Gleaning: Common Iora, Greenish Warbler, Tawnybellied Babbler, Black Bulbul, Common Tailorbird and Blue-faced Malkoha were recognized as twig gleaners (Table 6).

5.4. Position in the Canopy

Majority of the canopy layers used for foraging of bird species were edge edge (23%) followed by ground (18%) and middle lower (17%). Five major canopy layers out of 10 categories were distinctly used by 36 bird species in the tropical mixed dry deciduous forest, India. The major canopy positions foraged were Edge edge, center middle, center edge, middle edge and the birds used ground or air also (Table 7).

5.4.1. Ground/Air (Under/over Canopy)

Bird species such as Grey Junglefowl, Red-rumped Swallow, Yellow-billed Babbler, Common Hoopoe, Jungle Babbler, Indian Peafowl, Green Bee-eater, Jungle Myna, Chestnut-headed Bee-Eater and Blue-faced Malkoha occupied this for its prey. Grey Junglefowl and Red-rumped Swallow depend only on these strata and the other bird species extends overlap with other layers in the canopy (Table 7).

5.4.2. Center Center (Lower Canopy)

Bird species perched on the middle main axis of the plant canopy were Streak-throated Woodpecker, Black-hooded Oriole and Tawny-bellied Babbler. They also feed on the edge edge and middle middle canopy. No species was restricted to any particular layer alone.

5.4.3. Middle Edge (Middle Canopy)

Bird species feeding on the upper canopy was White-bellied Drongo which feeds on the upper and middle canopy. *Center edge (upper canopy):* Birds perched for preying over the upper canopy was Common Tailorbird. *Edge edge (upper canopy):* Twenty-one bird species (Table 7) were feeding on the upper canopy of the plant. Asian Paradise flycatcher and Common Tailorbird exploited food from other canopies too.

5.5. Specialists

Among the four dimensions, number of specialists (J'=0) was more in the substrates (2) method (2) and canopy (1) followed by height (Table 8). Grey Jungle Fowl, Vernal Hanging Parrot and Red-rumped Swallow are specialists as their J' values were zero. On the contrary, generalists were Blue-faced Malkoha, Common Tailorbird and Chestnut-headed Bee-eater (Table 8).

5.6. Niche Overlap

Niche overlap was calculated with foraging height (12 categories), foraging manoeuvre (20 categories), canopy (10 categories) and foraging substrate (7 categories). Among the foraging dimensions the higest mean niche overlap among the species was found in method (White-browed Bulbul) followed by canopy, height and the lowest in foraging method (Redrumped Swallow).

Height: Blue-faced Malkoha and Purple-rumped Sunbird had the highest mean niche overlap (0.75) while the lowest (0.36) was found in Yellow-billed Babbler (Table 9). *Method*: The mean niche overlap in feeding method was highest in the White-browed Bulbul (0.83) and lowest (0.14) in the Red-rumped Swallow (Table 9). *Canopy*: The highest mean niche overlap was found in Brown-headed Barbet (0.82) and lowest was in Yellow-billed Babbler (0.42). *Substrate*: The highest mean niche overlap was in Common Tailorbird (0.63) and the lowest was in Yellow-billed Babbler (Table 9).

Table 7. *Percentage of Ten Foraging Canopy Layers preferred by different bird species in the Tropical Mixed Dry Deciduous Forest, India.*

Name of the bird species	Ground/ Air	Centre lower	Centre middle	Centre edge	Middle lower	Middle middle	Middle edge	Edge lower	Edge middle	Edge edge
GREY JUNGLEFOWL	100	0	0	0	0	0	0	0	0	0
STREAK-THROATED WOODPECKER	0	7	75	0	0	11	0	0	7	0
RED-RUMPED SWALLOW	100	0	0	0	0	0	0	0	0	0
JUNGLE BABBLER	81	2	4	0	0	10	0	0	2	1
INDIAN PEAFOWL	77	0	0	0	0	0	0	6	6	11
GREEN BEE-EATER	71	0	0	0	0	2	0	0	0	26
YELLOW-BILLED BABBLER	97	0	2	0	0	0	1	1	0	1
ASHY DRONGO	18	0	0	0	0	15	3	9	21	33
ASIAN PARADISE FLYCATCHER	9	2	9	1	10	10	4	16	18	22
BLUE-WINGED LEAFBIRD	0	0	0	1	0	3	10	0	40	46
BLACK-HOODED ORIOLE	0	2	56	2	0	27	0	2	0	10
COMMON HOOPOE	91	0	0	3	0	0	0	0	6	0
LARGE-BILLED LEAF WARBLER	0	0	2	3	0	16	15	6	29	29
BLACK DRONGO	10	1	5	4	0	3	5	8	23	40
TAWNY-BELLIED BABBLER	0	3	32	4	1	14	7	0	10	30
WHITE-BELLIED DRONGO	11	0	1	4	0	10	17	4	29	25
BLUE-FACED MALKOHA	24	0	20	5	0	23	5	3	7	12

Name of the bird species	Ground/ Air	Centre lower	Centre middle	Centre edge	Middle lower	Middle middle	Middle edge	Edge lower	Edge middle	Edge edge
JUNGLE MYNA	39	0	0	7	1	8	11	0	28	6
BLYTH'S REED WARBLER	0	0	5	7	2	16	16	5	17	31
GREENISH WARBLER	0	0	8	8	0	17	10	3	31	23
PALE-BILLED FLOWERPECKER	0	0	2	10	0	9	21	3	11	45
RED-WHISKERED BULBUL	1	0	7	11	1	2	19	6	13	39
BLOSSOM-HEADED PARAKEET	0	0	0	12	0	0	21	0	0	67
COMMON IORA	0	0	4	13	1	18	10	3	20	31
PLAIN FLOWERPECKER	0	0	4	14	0	7	6	1	19	49
WHITE-BROWED BULBUL	2	0	5	17	0	10	8	1	16	40
MALABAR PARAKEET	0	0	4	18	0	0	9	0	9	59
LOTEN'S SUNBIRD	0	0	1	18	0	4	11	0	10	56
PURPLE SUNBIRD	0	0	5	18	2	11	12	4	23	26
BROWN-HEADED BARBET	0	0	3	19	0	10	16	3	13	35
PURPLE-RUMPED SUNBIRD	0	0	1	22	1	2	11	4	11	49
COMMON TAILORBIRD	3	5	15	23	5	5	8	8	13	18
CHESTNUT-HEADED BEE-EATER	29	0	0	25	0	0	11	5	25	5
RED-VENTED BULBUL	2	0	4	26	0	1	10	1	9	47
VERNAL HANGING PARROT	0	0	2	26	0	0	32	2	4	34
BLACK BULBUL	0	0	0	35	0	19	9	1	0	35
Total	765	22	276	356	712	283	318	105	470	981
Total in %	*18*	*1*	*6*	*8*	*17*	*7*	*7*	*2*	*11*	*23*

Table 8. *Extent of specialization (J') by different bird species in foraging substrate, foraging canopy, foraging method and foraging height in the tropical mixed dry deciduous forest, India (J' values range from 001 and specialization increases as J' decreases; Specialists are indicated in bold numbers).*

Name of the bird species	Foraging Substrate J'	Foraging Canopy J'	Foraging Method J'	Foraging Height J'
GREY JUNGLEFOWL	0.09	0.00	0.52	0.00
INDIAN PEAFOWL	0.45	0.37	0.89	0.56
BLOSSOM-HEADED PARAKEET	0.17	0.40	0.15	0.11
MALABAR PARAKEET	0.06	0.58	0.05	0.44
VERNAL HANGING PARROT	0.00	0.64	0.00	0.33
BLUE-FACED MALKOHA	*1.00*	0.88	*1.00*	0.85
CHESTNUT-HEADED BEE-EATER	0.26	0.75	0.65	*1.00*
GREEN BEE-EATER	0.35	0.32	0.41	0.26
COMMON HOOPOE	0.24	0.17	0.92	0.70
STREAK-THROATED WOODPECKER	0.18	0.39	0.22	0.11
REDRUMPED SWALLOW	0.00	0.00	0.00	0.22
BLACK-HOODED ORIOLE	0.49	0.56	0.78	0.52
BLACK DRONGO	0.05	0.82	0.72	0.89
ASHY DRONGO	0.37	0.76	0.68	0.71
WHITE-BELLIED DRONGO	0.06	0.83	0.74	0.78
JUNGLE MYNA	0.66	0.75	0.57	0.56
COMMON IORA	0.41	0.84	0.37	0.33
BLUE-WINGED LEAFBIRD	0.78	0.52	0.68	0.52
RED-WHISKERED BULBUL	0.55	0.82	0.49	0.81
RED-VENTED BULBUL	0.64	0.68	0.58	0.67
WHITE-BROWED BULBUL	0.51	0.79	0.45	0.33
BLACK BULBUL	0.63	0.63	0.56	0.54
TAWNY-BELLIED BABBLER	0.31	0.80	0.28	0.26
JUNGLE BABBLER	0.39	0.35	0.64	0.26
YELLOW-BILLED BABBLER	0.09	0.08	0.17	0.15
ASIAN PARADISE FLYCATCHER	0.25	0.98	0.62	0.52
COMMON TAILORBIRD	0.63	*1.00*	0.56	0.33
BLYTH'S REED WARBLER	0.47	0.21	0.41	0.54
LARGE-BILLED LEAF WARBLER	0.22	0.78	0.20	0.56
GREENISH WARBLER	0.59	0.82	0.51	0.56
BROWN-HEADED BARBET	0.32	0.80	0.28	0.85
PALE-BILLED FLOWERPECKER	0.61	0.73	0.54	0.70
PLAIN FLOWERPECKER	0.67	0.70	0.60	0.67
PURPLE-RUMPED SUNBIRD	0.33	0.68	0.28	0.33
LOTEN'S SUNBIRD	0.19	0.60	0.16	0.63
PURPLE SUNBIRD	0.55	0.86	0.49	0.81
Number of specialists	2	2	2	1
Number of generalist	1	1	1	1

Table 9. *Mean niche overlap for different bird species in foraging height, foraging substrate, foraging canopy and foraging method in the tropical mixed dry deciduous forest, India (niche overlap ranges from 001 and high niche overlap is indicated in bold numbers).*

Name of the Bird species	Niche overlap				
	Foraging Height	Foraging Substrate	Foraging Canopy	Foraging Method	All dimensions
ASHY DRONGO	0.68	0.54	0.80	0.56	0.67
BLACK BULBUL	0.59	0.58	0.71	0.82	0.63
BLACK DRONGO	0.66	0.43	0.81	0.26	0.63
BLACK-HOODED ORIOLE	0.69	0.54	0.61	0.77	0.61
BLOSSOM-HEADED PARAKEET	0.54	0.51	0.73	0.82	0.59
BLYTH'S REED WARBLER	0.66	0.51	0.79	0.82	0.65
MALABAR PARAKEET	0.71	0.36	0.72	0.82	0.59
BLUE-FACED MALKOHA	0.75	0.61	0.76	0.82	0.70
GREY JUNGLEFOWL	0.38	0.42	0.43	0.82	0.41
BLUE-WINGED LEAFBIRD	0.66	0.57	0.71	0.81	0.64
COMMON HOOPOE	0.53	0.52	0.58	0.75	0.54
COMMON IORA	0.73	0.45	0.78	0.82	0.65
JUNGLE BABBLER	0.51	0.42	0.53	0.82	0.48
JUNGLE MYNA	0.61	0.50	0.73	0.82	0.61
LARGE-BILLED LEAF WARBLER	0.71	0.44	0.76	0.82	0.63
BROWN-HEADED BARBET	0.73	0.55	0.82	0.82	0.69
VERNAL HANGING PARROT	0.49	0.46	0.74	0.82	0.56
LOTEN'S SUNBIRD	0.70	0.46	0.76	0.82	0.64
STREAK-THROATED WOODPECKER	0.70	0.51	0.57	0.74	0.59
PLAIN FLOWERPECKER	0.70	0.59	0.79	0.82	0.69
INDIAN PEAFOWL	0.56	0.55	0.65	0.73	0.59
ASIAN PARADISE FLYCATHER	0.63	0.43	0.77	0.26	0.60
GREENISH WARBLER	0.69	0.51	0.77	0.82	0.65
PURPLE-RUMPED SUNBIRD	0.75	0.42	0.74	0.82	0.63
PURPLE SUNBIRD	0.72	0.54	0.80	0.82	0.68
TAWNY-BELLIED BABBLER	0.49	0.51	0.73	0.82	0.57
REDRUMPED SWALLOW	0.53	0.36	0.49	0.14	0.46
RED-VENTED BULBUL	0.68	0.52	0.77	0.82	0.65
RED-WHISKERED BULBUL	0.74	0.54	0.80	0.82	0.69
GREEN BEE0EATER	0.64	0.47	0.66	0.26	0.59
COMMON TAILORBIRD	0.68	0.63	0.78	0.82	0.69
PALE-BILLED FLOWERPECKER	0.74	0.54	0.78	0.82	0.68
WHITE-BROWED BULBUL	0.66	0.47	0.80	0.83	0.64
WHITE-BELLIED DRONGO	0.69	0.40	0.80	0.27	0.63
YELLOW-BILLED BABBLER	0.36	0.35	0.42	0.82	0.38
CHESTNUT-HEADED BEE-EATER	0.71	0.47	0.73	0.26	0.63

All dimensions: All the dimensions together when combined, Yellow-billed Babbler showed the lowest overlap (0.38) and Blue-faced Malkoha (0.70) showed the highest overlap among the 36 species (Table 9).

5.7. Foraging Guilds

Species were separated into a number of distinct groups whose members exploit food resources from similar substrates or height using similar methods and thereby considered as guilds. The guild formed in the tropical mixed dry deciduous forest, India based on the use of substrates, methods, canopy and height, their relationships among the 36 bird species are summarized in the cluster diagram (Figure 2). Two distinct major guilds (gleaner and sallier) were

arbitrarily recognized from the cluster diagram (Figure 2). The gleaner was further consisted of three distinct guilds based on the substrates of gleaning, namely 1. Fruit, 2. Flower, 3. Ground and 4. Stem (trunk and twigs).

Guild I consisted of birds that glean their prey on fruit (Frugivore). Guild II consisted of birds that glean their food from the flower (Nectarivore). The guild III consisted of birds that largely obtained their food mainly insects or other invertebrates from all strata (ground, plant and air) (Figure 2). Within this guild, two major groups were obvious such as purely insectivore and omnivore. This was bifurcated again into five groups based on the substrates: ground, twigs and leaf, main trunk and air.

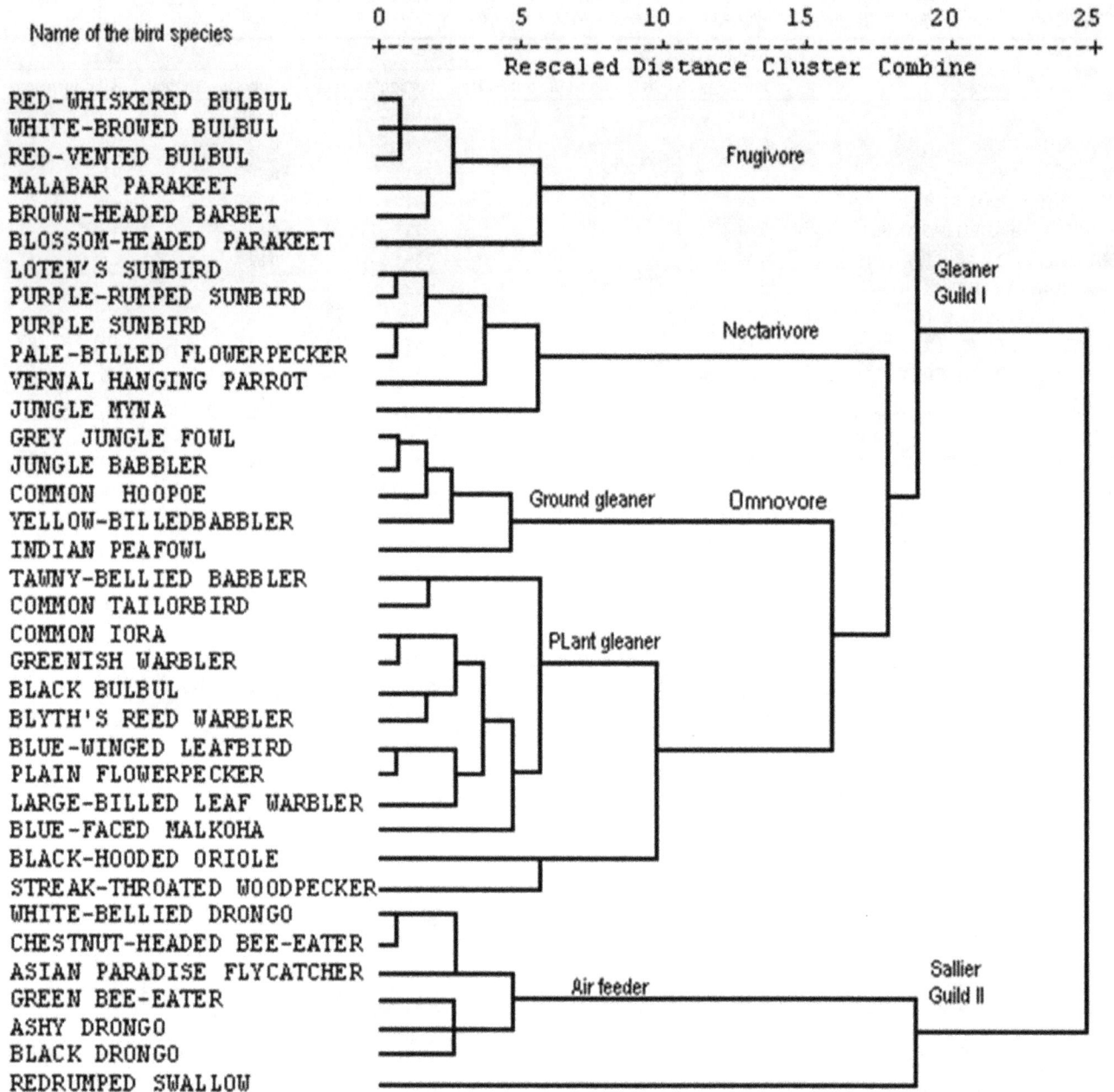

Figure 2. *Dendrogram showing interspecific relationships of 36 bird species based on multivariate analyses of foraging method, substrate and height use in the tropical mixed dry deciduous forest, India.*

Based on the observational data, birds foraged in similar ways or exploited the same resources for foods were grouped in a schematic representation (Figure 3). The schematic portrayal of the groupings relies on the foraging behavior, foraging height, canopy and foraging substrate differences to associate species. Of the 36 species, major group of birds was of insectivores, which comprised of 24 bird species followed by nectarivores such as Vernal Hanging Parrot, Loten's Sunbird, Purple-rumped Sunbird, Pale-billed Flowerpecker and Jungle Myna. Frugivore guild comprised of (fruit, flower, insect and grain feeder) Red-whiskered Bulbul, Red-vented Bulbul, White-browed Bulbul, Malabar Parakeet, Brown-headed Barbet and Blossom-headed Parakeet. Insectivores largely obtain their food from plants or from air

by sally (Red-rumped Swallow). Among the plant forms, the number of species, which obtained their food from twig and leaf were more than that depending on other substrates such as main trunk and secondary branches. Six bird species sallying from four different positions in the canopy were distinguished as insectivore's viz. Chestnut-headed Bee-eater, White-bellied Drongo, Ashy Drongo, Paradise Flycatcher and Green Bee-eater (Figure 3). Other insectivores guild, feeding from plants were Streak-throated Woodpecker, Black-hooded Oriole, Tawny-bellied Babbler, Common Tailorbird, Common Iora, Greenish Warbler, Blyth's Reed Warbler, Blue-winged Leafbird, Plain Flowerpecker, Large-billed Leaf Warbler and Blue-faced Malkoha.

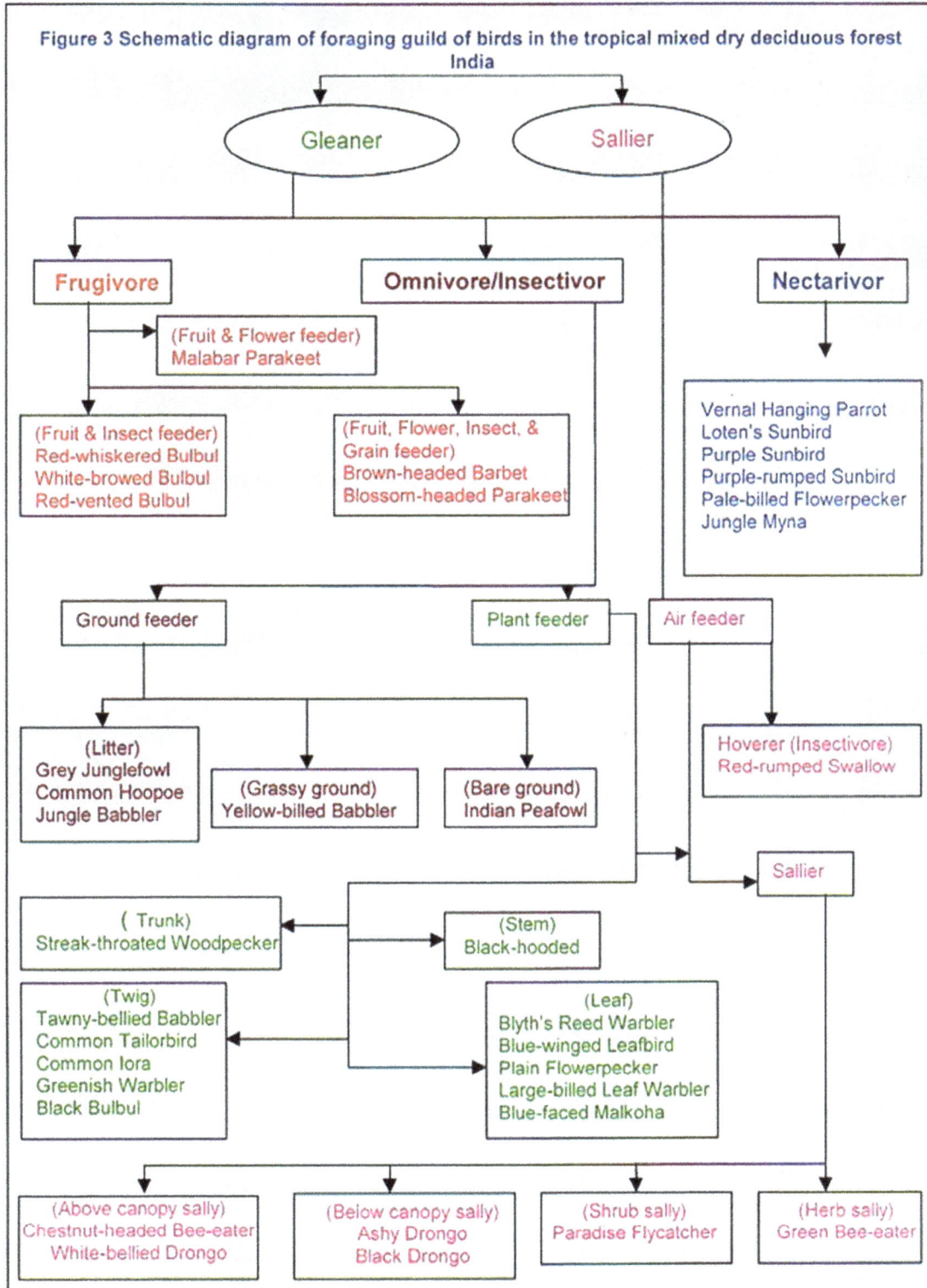

Figure 3. Schematic diagram of foraging guild of birds in the trophical mixed dry desiduous forest india.

5.8. Plant Community in the Tropical Mixed Dry Deciduous Forest

Vegetation profile of mixed dry deciduous forest consisted mostly of tree species of 2-6m height (Figure 4) and the upper stratum was thinned out with a few tall trees such as *Ficus* sp., *Tamarindus indica, Acacia polyacantha, Albizia* *amara, Canthium dicoccum, Celtis philippensis* and *Commiphora caudata*. Shrubs formed the lower stratum at 0-2m. and it occupied a predominant place from ground to 1m height in mixed dry deciduous forest. Moreover the number of shrub species are higher (45) than the tree species (27). Higher foliage profile layers harbour more bird species [19] was true in this habitat as studied by [20].

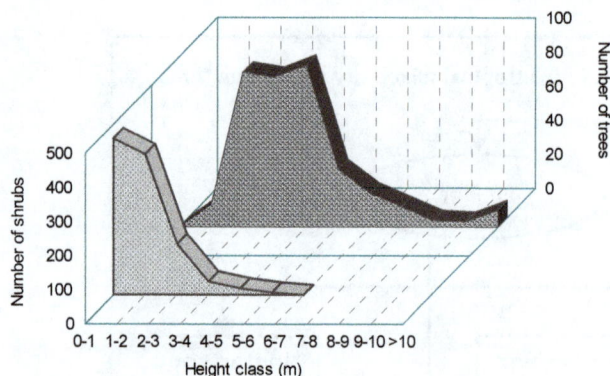

Figure 4. Vegetation profile of mixed dry deciduous forest.

6. Discussion

Tree layers found to be a distinctive foraging environment for birds in the tropical mixed dry deciduous forest, India followed by shrubs/short tree layers due to the availability of high foliage layer in the trees and more foliage overlap between short trees and shrubs. Sucessful foraging by avian predators is influenced largely by prey availabilty, which encompasses not only the density of prey but also its vulnerabilty to capture [21]. An intersting observation was this forest comprised of two guilds namely gleaners and salliers. Feeding methods are more specialised in each species. Species generalised in feeding tend to vary in feeding technique, substrate choice, canopy and height when the type of food varies. Yet another intersting observation was large scale utilization of layers at different height such as 0.1-2m and 3-6m. This might perhaps be due to the foliages of majority of the trees in the tropical mixed dry deciduous forest of India are spread between 3-6m height which formed the upper stratum and shrubs of 0-2m height formed the lower stratum and that gives more opportunity to birds for exploitation. Moreover the number of shrub species are higher than the tree species. The availability of various plant forms such as shrubs, short trees and trees in these habitats not only increases the vertical and horizontal foliage layering and complexity, but also provides many supporting substrates. So majority of birds in this habitat used these strata for foraging. Foraging birds require a large number of small preys to maintain resting metabolic rates [22]. Information on the foraging height, attack maneuvers; substrate and foliage density was collected independently for each foraging bird [23].

Three major substrates namely ground, plant and air were recognized. Of which, more bird species fell under the plant guild because plant offers a great variety of microhabitats (trunk, branches, twigs, foliage, flower and fruit) to find their suitable and favourable food. Foliage and twigs were utilized by more number of birds because branches with leaves offer a great variety of places to find food along with concealment. Moreover most of the trees in this habitat withered their dryleaves and emerging of new leaves tookplace during winter, thus increasing the opportunity of searching and finding their prey or vicinity of the prey

becomes more. In total, bird species used 12 methods to obtain food from the tropical mixed dry deciduous forest of India. Searching patterns are largely a function of the morphological and perceptual traits of each species, which allow the birds to move through the foliage to locate, detect and capture prey in specific ways. Similar study was reported in thorn forest [5] of Mudumalai Wildlife Sanctuary, India. Information on the foraging height, attack maneuvers; substrate and foliage density was collected independently for each foraging bird [23].

The availability of diverse food items may vary between habitats [20] and hence birds that feed on variety of foods (e.g., insects, seed, nectar and fruit) may change their manoeuvre according to the habitat. Moreover, changes in the foraging manoeuvres may be a strategy to avoid competition. Hence it is likely that the combination of factors such as availability of food, habitat structure and interspecific competition are responsible for the variations in the foraging behaviour of birds observed in this forest. Predation of two adult birds was recorded during the study period. Also, predation of fledglings of almost all the breeding birds was observed. Interspecific competition also can alter foraging behavior of Warblers and Babblers [24], [25], [26]. Thus, changes in the foraging manoeuvre may be a strategy to avoid competition. Hence, it is likely that the combination of factors such as availability of food, habitat structure and interspecific competition are responsible for the changes. Foraging behavior and foraging success of the reddish egret were studied by [27] focusing on whether their foraging behavior or success varied with age, color morph, group size and habitat measures.

In this study, closely related species used the same basic foraging method indicating the importance of phylogeny in determining the feeding patterns of birds [28], [4]. Resource partitioning reduces the effect of competition by decreasing the amount of overlap between the competing species [2]. Partitioning of foraging dimensions among birds could occur in this habitat as reported earlier for bird communities of various places and habitats [6], [29], [5], [20] Foraging behavior and foraging success of the reddish egret were studied by [27] focusing on whether their foraging behavior or success varied with age, color morph, group size and habitat measures.

Many species fed from different strata and positions in the canopy overlapping with others where specialists such as the Yellow-billed Babbler fed by only gleaning and that too from ground thus sharing high specialization or preference and thus having very little overlap with other species. When food availability is high they feed on the outer part of tree canopies in this study as found by Diaz *et al.* [30] in Tits. Birds selected foraging sites with a higher mean prey density than at random sites [31].

Some species of water birds have been found to forage at the interface of open water and vegetation [32], [33], [34].

Bird species evolved with specialization for a particular type of habitat or substrate or prey that resulted in a specialist for a particular habitat. Greenberg [35], [36] investigated

Warblers' response to different substrate and inferred that the species that had a diverse foraging behavior were conservative in their use of substrates. Thus it can be inferred that niche overlap can be attributed to the availability of food resources, morphology of species and competition as suggested by Alatalo [24], Rolando and Robotti [37], Szekely [38] and Gokula and Vijayan [5]. Successful foraging by avian predators is influenced largely by prey availability, which encompasses not only the density of prey but also its vulnerability to capture [21].

7. Conclusion

Foraging data were collected early in the morning during the study period. In total, 36 species were observed from the mixed dry deciduous forest. Various foraging dimensions such as method, substrate, height and position in the canopy were analyzed. Foraging attempts were assigned to 12 height categories, seven substrate categories, 9 positions in the canopy and 20 foraging methods. Thirteen species shared change in the use of substrate while only five species changed the method used. Five bird species were considered as specialists as their J' values were zero. In four dimensions highest mean niche overlap is found in the use of foraging height. There are two major guilds, namely gleaners and salliers and gleaners are grouped into four major guilds. There are four major groupings among the bird species based on the food eaten such as insectivores, nectarivores, frugivores and omnivores. The plant (shrubs and trees) surface provides microhabitats such as foliage, twig, flower, fruit, secondary branches and trunk and the proportion of foliage use at different heights is higher. Specialization of species and their niche overlap with others are analysed. Foraging method is specialized being constrained by morphology in many species while substrates and strata are used opportunistically depending on the environment.

References

[1] N. B. Davies, "Ecological questions about territorial behaviour. In Behavioural ecology- an evolutionary approach" (Eds.) J. R. Krebs and N. B. Davies. Blackwell scientific publications. Oxford, London Edinburgh and Melbourne. 1978, pp. 317-350.

[2] J. A. Wiens, "Ecology of bird communities". Cambridge University Press, Cambridge, 1989 Vol. 1 & 2.

[3] F. H. J Crome, Foraging ecology of an assemblage of birds in lowland rainforest in Northern Queensland. *Aust. J. Ecol.* 3: 1978, pp. 195-212.

[4] R. T. Holmes, R. E. Jr. Bonney and S. W Dalala, "Guild structure of the Hubbard Brook bird community: a multivariate approach" *Ecology.* 60: 1979, pp. 512-520.

[5] V. Gokula and L. Vijayan, "Foraging patterns of birds in the thorn forest of Mudumalai Wildlife Sanctuary, Southern India" *J. South Asian Nat. Hist.* 5 (2): 2001, pp. 143–152.

[6] H. F. Recher, R. T. Holmes, M. Schulz, J. Shields and R.

Kavanagh, "Foraging patterns of breeding birds in eucalyptus forest and woodland of Southern Australia". *Aust. J. Ecol.* 10: 1985, pp. 399-419.

[7] R. Mac Nally, "Habitat specific guild structure of forest birds in southeastern Australia: a regional scales perspective", *J. Anim. Ecol.* 63: 1994, pp. 988-1001.

[8] H. G. Champion and S. K. Seth, "A revised survey of the forest types of India", Government of India publications". New Delhi, India, 1968.

[9] C. J. Bibby, N. D Burgess and D. A. Hill, "Bird census techniques. British Trust for Ornithology and the Royal Society for the Protection of Birds" Academic Press, London. 1993, pp. 66-84.

[10] G. P. Bell, G. A. Bartholomew and K. A. Nagy, "The roles of energetics, water economy, foraging behavior, and geothermal refugia in the distribution of the bat, *Macrotus californicus".* *J. Comp. Psych. B.* 156: 1986, pp. 441-450.

[11] J. V. Ramsen and S. K. Robinson, - A classification scheme for foraging behaviour of birds in terrestrial habitats. *Studies in Avian Biology* 123: 1990, 144-160.

[12] R. C. Mac Nally, "On characterizing foraging versatalization illustrated by using birds" *Oikos.* 69: 1994, pp. 95-106.

[13] M. L. Morrison, "Influence of sample size and sampling design on analysis of avian foraging behavior" *Condor.* 86: 1984, 146-150.

[14] C. E. Shannon and W. Weaver, "The mathematical theory of communication. University of Illinois Press", Urbana, 1949.

[15] H. S. Horn, - The measurement of 'overlap' in comparative ecological studies. *Am. Nat.* 100: 1966, pp. 419-424.

[16] L. Legendre and P. Legendre, - Numerical ecology. Developments in environmental modelling, Elsevier Sci. Publ. Co., Amsterdem, 1983.

[17] F. L Rohlf., "Numerical Taxonomy and Multivariate Analysis System". Exeter publishing, Setauket, New York 1987.

[18] M. J. Norusis, -"SPSS Inc. SPSS release 6.0 for Unisys 6000", Chicago, Illinois, USA, 1994.

[19] R. H. Mac Arthur and J. W. MacArthur, - On bird species diversity. *Ecology.* 42 (3): 1961, pp. 594-598.

[20] T. Nirmala, L. Vijayan, "Breeding behaviour of the Indian Robin Saxicoloides fulicata in the Anaikatty hills, Coimbatore", pp. 43–46. In: Proceedings of 28th ESI Conference, (2003), pp. 7–8.

[21] M. Samantha, Lantz, E. Dale, Gawlie, I. Marce and Cook, "The effect of water depth and submerged Aquatic vegetation on the success of wading birds Condor 112 (3): 2010, pp. 460-469.

[22] T. Piersma, "Energetic bottles and other de-sing constrains in avian annual cycles" Integrative and comparative Biology, 42: 2002, pp. 51-67.

[23] S. Mohammad, Mansor and A. Shahrul Mohamed Sah, "Foraging Patterns reveal niche separation in tropical insectivorous birds", Acta Ornithologic Vol. 47, 2012.

[24] R. V. Alatalo, "Interspecific competition in tits *Parus* spp. and the Goldcrest *Regulus regulus*: foraging shifts in multispecific flocks" *Oikos.* 37: 1981, pp. 335-344.

[25] L. Vijayan, "Comparative biology of Drongos (Family: Dicruridae, Class: Aves) with special reference to ecological Isolation" *Ph. D. Thesis,* Bombay University, Bombay, 1984.

[26] J. H. Carothers, -"Behavioral and ecological correlates of interference competition among some Hawaiian drepanidinae" *Auk.* 103: 1986, pp. 564–574.

[27] M. Elizabeth Baters and Bart M. Ballard "Factors influencing behavior and success of foraging Reddish Egrets (Egretta rufescents)", water birds 37 (2): 2014, pp. 191-202.

[28] S. K. Robinson and R. T. Holmes, -"Foraging behaviour of forest birds: the relationships among search tactics, diet, and habitat structure", *Ecology.* 63: 1982, pp. 1918-1931.

[29] A. G. Wheeler and M. C. Calver, "Resource partitioning in an Island community of insectivores birds during winter" *Emu.* 96: 1996, pp. 23-31.

[30] M. Diaz, J. C Illera. and J. C. Atienza, "Food resource matching by foraging tits parus spp. during spring – summer in a mediterranean mixed forest; evidence for an ideal free distribution" *Ibis.* 140: 1998, pp. 654–660.

[31] L. Racheal, Pierce and E. Dale Gawlik "Wadding birds foraging habitats selection in the florida Everglader" Water birds 33 (4): 2010, pp. 494-503.

[32] R. J. Saffron, M. A. Colwell, C. R. Isola and O. E. Taft, "Foraging site selection by nonbreeding White-faced Ibis" Condor 102: 2000, pp. 221-225.

[33] R. E. Bennetts, P. C. Darby and L. B. Karunaratne, "Foraging patch selection by snail kites in response to vegetation structure and prey abundance and availability" Water birds 29: 2006, pp. 88-94.

[34] E. D. Stolen, J. A. Collazo and H. F Percivan, "Vegetation effects on fish distribution in impounded salt marshes" South Eastern Naturalist 8: 2010, pp. 503-514.

[35] R. Greenberg, - The Winter exploitation systems of Bay-breasted and Chestnut-sided warblers in Panama. University of California. *Publ. Zool.* 116: 1984a, pp. 1-107.

[36] R. Greenberg, "Neophobia in the foraging site selection of a neotropical migrant bird: an experimental study. *Proc. Nat. Acad. Sci.* 81: 1984b pp. 3778-3780.

[37] A. Rolando and C. A Robotti., "Foraging niches of tits and associated species in north-western Italy" *Boll. Zool.* 52: 1985, pp. 281-297.

[38] T. Szekely, "Foraging structure of the foliage-gleaning and bark-foraging guild in winter and spring. *Proc. Fifth Nordic Ornithological Congress*: 1985, pp. 140-146.

11

Preparation of Charcoal Pellets from Eucalyptus Wood with Different Binders

Alejandro Amaya, Mariana Corengia, Andrés Cuña, Jorge De Vivo, Andrés Sarachik, Nestor Tancredi*

DETEMA, Facultad de Química, Universidad de la República, Montevideo, Uruguay

Email address:

nestor@fq.edu.uy (N. Tancredi)

Abstract: At present, there is great interest in using biomass as an alternative energetic source, as it is renewable and environmentally friendly. In the case of solid fuels, biomass has low energetic density, although it can be increased by charring and pelletizing. These methods also allow the improvement of physical properties, such as hydrophobicity and resistance to microbiological attack. In this work, the agglomeration of charcoal dust produced from sawmill waste with three different binders (wood tar, molasses and starch) was studied. The procedure included agglomeration and curing by heating in air atmosphere. The prepared charcoal pellets showed appropriate mechanical resistance, higher heating value than the original wood residues and higher energetic density than charcoal. Molasses and tar used as binders in the preparation of fuel pellets allow energy densification and an adequate durability of the products.

Keywords: Charcoal, Pellet, Binders, Eucalyptus Wood, Renewable Energy

1. Introduction

In the last decades high interest has been raised worldwide in using biomass materials as an alternative to fossil fuels [1-4]. The main advantages of this substitution are that biomass is a renewable and environmental friendly energetic source since it leads to lower emissions of greenhouse and acid gases.

Some problems associated to biomass materials in their original form are high moisture content, irregular shape and sizes and low bulk and energetic densities [5]. These factors increase storage, handling and transportation costs. Moreover, biomass is subject to microbiological attack during storage [6, 7] and may cause plague infestation [8].

Carbonization is a way to increase energetic density as char heating value is about 25-30 MJ kg-1 compared to 15 MJ kg-1 for raw biomass [9-11]. Also, carbonization increases hydrophobicity, resulting in a decrease of moisture content and microbiological growing [12]. Nevertheless carbonization products are highly friable and lead to the generation of carbon dust, a difficult product to handle and that may cause explosions [13]. These problems may be overcome by milling the charcoal and pelletizing carbon dust by means of adequate binders. Pelletization would contribute not only to the durability of the product but also to an increase of its energetic density. In addition, carbon pellets have the same advantages as charcoal compared to the direct use of biomass as fuel (such as higher heating value and lower moisture content) [14-18].

Different binders have been used in order to achieve particle agglomeration and good cohesion properties [19-20]. In some cases, binding has been explained as due to adherence between surfaces enhanced by bonds with the binder [21].

At present, forest biomass in Uruguay, as in other developing countries [22-24], appears as an important source of renewable energy [25]. In addition to this, due to the activity of sawmills, huge quantities of residues are produced and accumulated as sawdust or other irregular-shaped pieces of wood. These residues could be used as a raw material in the production of energy [26].

In this work, the agglomeration of charcoal produced from sawmill waste with three different binders (wood tar, molasses and starch) was studied. The procedure included agglomeration and curing by heating in air atmosphere. Physico-chemical properties of raw material and charcoal pellets were determined.

2. Materials and Methods

2.1. Raw Materials

Sawmill residues (Eucalyptus grandis wood, pre-dried), with prismatic shapes, about 2 cm thickness, 5 cm width, 10-30 cm length, were carbonized in a pyrolysis kiln in a previous experiment [27], obtaining charcoal pieces of similar shape and size (yield 28.3%, dry basis) and wood tar as a by-product. Charcoal was grinded in a mortar, sieved and 30-50 mesh size fractions were selected. Sugar cane molasses were obtained from ANCAP, the national company dedicated to oil refinery and ethanol production. Aqueous starch solutions were prepared from potato starch.

2.2. Pellet Preparation

Pellets were prepared by mixing grinded charcoal with the binder, pressing the mixture in a manual Parr press, at 15 MPa during 1 min and cured as indicated below. When tar was used as a binder, a mass ratio tar/charcoal of 1.1 was used to prepare the mixture. Then cylindrical pellets of mass 0.55-0.65 g, diameter 1.1 cm and height 0.7 cm were made by cold compression. They were cured in an oven at 105 °C during 24 h, in order to obtain uniform pellets with adequate durability. After curing, pellets were stored in a desiccator and then mechanical resistance tests were performed. Other charcoal/tar ratios were tested, but those pellets showed very low mechanical resistance.

When molasses was used as a binder, the best molasses/charcoal weight ratio for achieving a good agglomeration was 1.3; cylindrical pellets of mass 0.5 g, diameter 1.1 cm and height 0.7 cm were prepared. Curing was carried out in an oven at 162-179 °C for 24 h. This temperature range was chosen from TG data in order to avoid molasses combustion and allow caramelization.

When starch was used as a binder, a mixture of 85 cm3 of water and 3 g of starch was heated up to the boiling point; at this time 50 g of 30-50 mesh preheated charcoal were added while stirring slowly. The obtained mixture was cooled, shaped into cylindrical pellets as described above .and cured at 105 °C for 24 h. Other preparation procedures, including the change of water/starch/charcoal ratio or the use of cold water were discarded as the obtained pellets had inadequate mechanical strength.

2.3. Analysis of Raw Materials and Products

For raw materials and products proximate analysis, elemental analysis, determination of heating value, apparent and bulk density and TGA were carried out. For pellets mechanical properties were also tested.

Proximate analysis included: moisture determination (ASTM D 2867-70 for wood, charcoal and starch), ashes determination (ASTM 2866-70) and volatile matter determination (ISO 5621-1981). As molasses has a high volatile content and caramelization occurs at about 118-129 °C [28], its moisture content was estimated from TGA in air. For wood tar, also with a high volatile content, moisture content was determined by distillation with toluene.

Elemental analysis was carried out in a Carlo Erba model EA 1108 CHNS – OR equipment. For the determinations of heating value a Parr 1341 Plain Oxygen Bomb Calorimeter was used. Apparent densities were determined by mercury immersion. Bulk densities were determined by pouring the material into a graduated container and measuring its mass. Thermogravimetric analyses were carried out in a Shimadzu TG-50 equipment, in air atmosphere (50 cm^3 min^{-1} STP, dried by carbon molecular sieves), at a heating rate of 2 °C min^{-1} up to 850 °C. Pellets mechanical properties (Impact Resistance Index, IRI, and friability) were tested by dropping the pellets from 1 m (IRI) and by rolling 6 g of pellets for 4 min at 25 rpm (friability, Erweka equipment) [19, 29]. SEM images of selected samples were obtained with a JEOL JSM 5900 L Scanning Electron Microscope (High Technology Service, School of Sciences).

3. Results and Discussion

3.1. Raw Materials

In Tables 1 and 2 raw materials and products characterization are shown.

Differences among wood residue and charcoal values can be explained by the devolatilization during carbonization, which produces an increase in ashes content, fixed carbon and carbon content, as well as a decrease in oxygen content as a consequence of CO and CO_2 formation. Molasses and starch showed low fixed carbon content as expected. For charcoal and charcoal dust, their heating values are the highest for all the samples and their apparent and bulk densities are the lowest. The low value of charcoal bulk density is the justification for a densification attempt.

Fig. 1 and 2 show the TG and DTG of the different raw materials in air atmosphere. In air, wood residues combustion shows a maximum rate at 300 °C; the autoignition temperature is 250 °C, in accordance with reported values in literature [30].

Autoignition temperatures of about 300 °C, 270 °C and 380 °C are found for wood tar, starch and charcoal, respectively. For all the graphs but that of charcoal two main peaks are shown: one at low temperatures, corresponding to volatile ignition, and a second one close to 500 °C, corresponding to char combustion. In the case of molasses a peak beginning at 120 °C with maximum at 140 °C appears; it is attributable to molasses caramelization [28]. At 177 °C volatile matter combustion begins, reaching a maximum at 200 °C. At 440 °C char combustion starts.

Table 1. *Proximate and Elemental analysis of raw materials (%, dry basis).*

Sample	Proximate Analysis				Elemental Analysis (ash free)			
	Moisture	Ash	Volatile Matter	Fixed Carbon	C	H	N	O^a
Wood residue	9.4	0.2	87.8	12.0	48.5	5.9	< 0.1	45.6
Charcoal	6.1	1.1	19.3	79.6	77.2	2.5	< 0.1	20.3
Wood tar	12.6	4.1	67.9	28.0	40.0	6.3	< 0.1	53.7
Molasses	19.0	5.6	88.4	6.1	42.7	3.8	0.9	52.6
Starch	12.0	0.3	98.2	1.5	46.9	5.6	0.3	47.2
Tar pellet	9.5	2.8	46.1	51.1	74.4	3.7	0.3	21.6
Molasses pellet	1.8	5.0	33.4	61.6	67.8	3.1	0.4	28.7
Starch pellet	3.0	1.3	25.1	73.6	80.1	2.9	< 0.1	17.0

[a]by difference

3.2. Pellets

Pellets properties are shown in Tables 1 and 2. All the pellet properties are close to that of charcoal, as the curing process affects mainly the binder. This is especially remarkable for starch pellet, as charcoal is largely the major component. The only difference is observed in bulk density, larger for pellets, as expected. Regarding to mechanical resistance, all the pellets passed the IRI test and the losses after friability test were lower than 3% in all cases. These results are considered acceptable.

The SEM of the pellets and the charcoal are shown in Fig. 3. The completely separated charcoals particles are shown in Fig. 3a. The binding properties of starch are observed in Fig. 3b, as white connections between the charcoal particles. The union of different particles is also clearly observed for the tar pellet in Fig. 3c and for the molasses pellet in Fig. 3d.

In Fig. 4 and 5 TG and DTG in air for the prepared pellets and charcoal are compared. The main charcoal combustion peak is also seen for the pellets agglomerated with tar and starch, but for the pellets agglomerated with molasses the peak shifts to lower values. For molasses and tar pellets the curing and pelletization process cause the lowering of their autoignition temperatures compared to charcoal, though none of them were lower than 250 °C. This is due to the high volatile content of the binders. TG and DTG for starch pellets are very close to that of charcoal: in this case the binder content is very low and curing is almost a drying process.

Table 2. *Heating values (dry basis) and apparent and bulk densities of raw materials (P: pellets).*

Sample	High heating value (MJ kg^{-1})	Low heating value (MJ kg^{-1})	Apparent density (g cm^{-3})	Bulk density (g cm^{-3})
Wood residue	18.9	17.6	0.72	0.21- 0.28
Sawdust	18.9	17.6	-	0.28
Charcoal	29.6	29.2	0.33	0.12
Charcoal dust	29.6	29.2	-	0.22
Wood tar	21.6	20.4	1.17	0.31
Molasses	15.0	11.7	1.48	0.30
Starch	17.1	16.1	0.55	0.19
Tar P.	30.1	29.5	0.58	0.31
Molasses P.	25.4	24.8	0.56	0.30
Starch P.	30.8	30.2	0.36	0.19

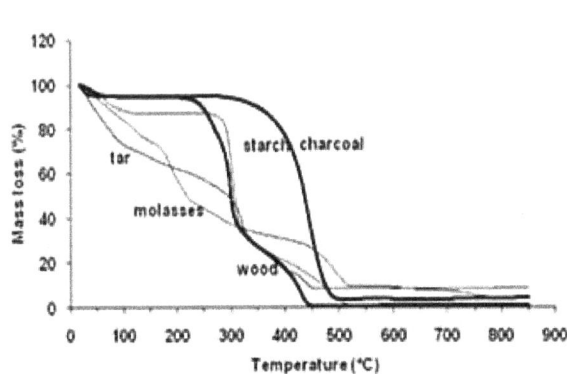

Figure 1. *TG curves for raw materials in air atmosphere.*

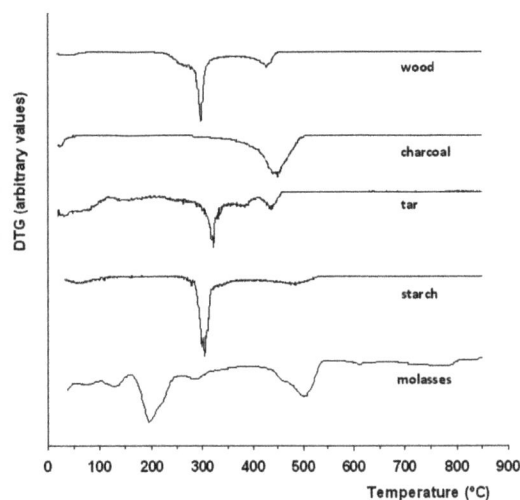

Figure 2. *DTG curves for raw materials in air atmosphere.*

a) charcoal

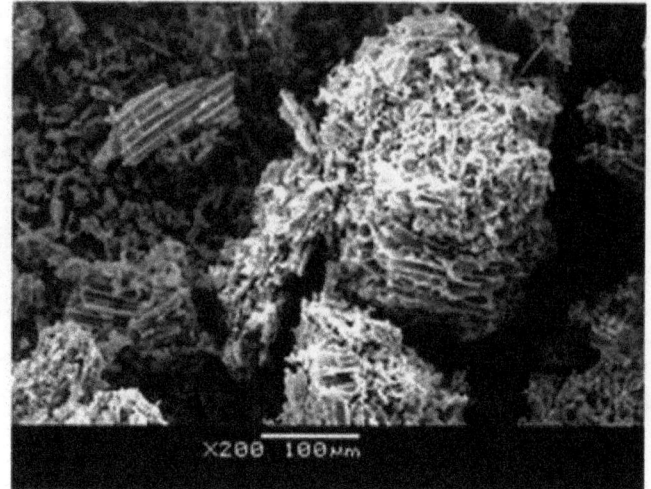

d) molasses pellet

Figure 3. *SEM images.*

3.3. Energetic Densities

As fuels are usually transported by volume, energetic density expressed as energy by volume is a better indicator of the fuel energy content than heating value.

An attempt to compare energetic densities of raw materials and products was made by defining the Energetic Density (ED) as:

ED = Low heating value (w.b.) x apparent density(1)

The calculated EDs are shown in Table 3. For pellets, EDs are higher than that for charcoal, and in the case of tar pellets ED is higher than that for wood also. These results indicate that pelletization worked as an energy densification method. Despite that ED is a property independent of pellet shape (useful for comparison among different fuels), for transportation purposes the space between the pieces of each material should be taken into account. Then, a new parameter, Packed Energetic Density (PED) was defined:

PED = Low heating value (w.b.) x bulk density(2)

Calculated values for PED are shown in Table 5, including the values for sawdust and charcoal dust. Wood and charcoal as dust show higher PED values than the materials in pieces, because of the lower void volumes.

PED increases in the order charcoal < wood < wood sawdust < charcoal dust < pellets with the exception of starch pellets, with a PED lower than charcoal dust. In order to evaluate these results, the difficulties for dust transport and storage and health risks involved in its handling should be considered. With the exception of starch pellets, energetic densification was achieved by the pelletization method described in this work.

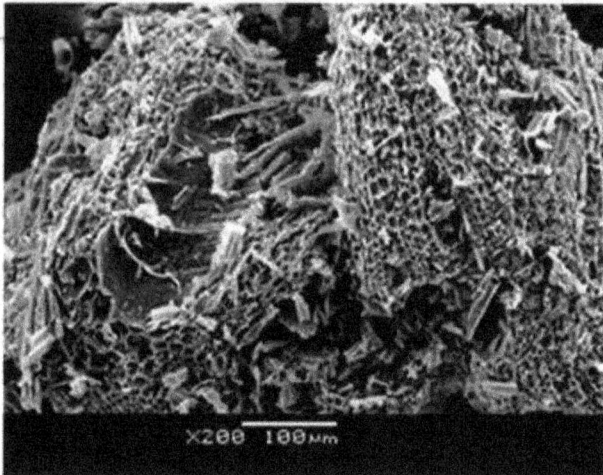

b) starch pellet

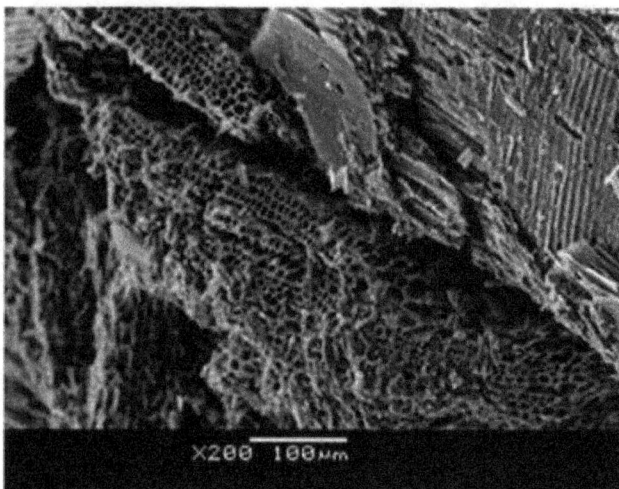

c) tar pellet

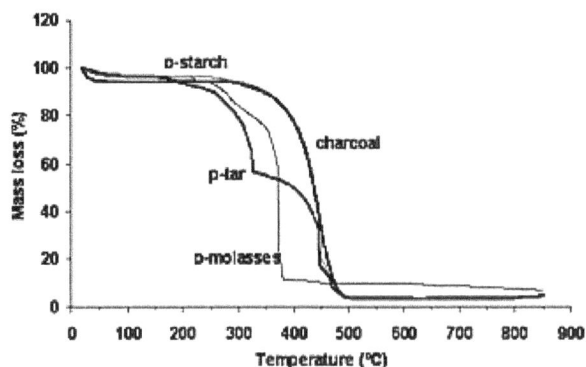

Figure 4. TG in air atmosphere of pellets and charcoal.

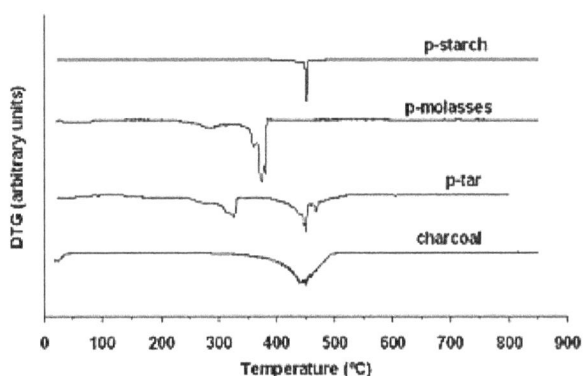

Figure 5. DTG in air atmosphere of pellets and charcoal.

Table 3. Energetic densities.

Sample	Energetic Density ED (GJ m^{-3})	Packed Energetic Density PED (GJ m^{-3})
Wood Residue	13.6	3.7 – 5.3
Sawdust	-	5.3
Charcoal	9.6	3.5
Charcoal Dust	-	6.4
Tar Pellet	17.1	9.1
Molasses Pellet	13.9	7.4
Starch Pellet	10.9	5.7

4. Conclusions

Charcoal pellets prepared from charcoal dust and three different binders showed appropriate mechanical resistance. They also exhibited higher heating value than the original wood residues and higher energetic density than charcoal. For molasses and tar pellets, packed energetic densities were higher than those for wood and charcoal, either as pieces or dust. Molasses and tar used as binders in the preparation of fuel pellets allow energy densification and an adequate durability of the products.

Acknowledgements

This work was financed by Project PDT 47/08, DINACYT, Ministerio de Educación y Cultura, Uruguay. Elemental analysis was performed at "Departamento Estrella Campos", Facultad de Química, Universidad de la República, Uruguay.

References

[1] I. Hannula. "Co-production of synthetic fuels and district heat from biomass residues, carbon dioxide and electricity: Performance and cost analysis. Biomass and Bioenerg., vol. 74, pp. 26-46, March 2015.

[2] J. Bisquert. "Materials for production and storage of renewable energy". J. Phys. Chem. Lett., vol 2, pp. 270-271, February 2011.

[3] P. Moriarty and D. Honnery. "The transition to renewable energy: make haste slowly". Environ. Sci. Technol., vol. 45, pp. 2527-2528, March 2011.

[4] N. Kaliyan and V. Morey. "Factors affecting strength and durability of densified biomass products". Biomass Bioenerg., vol. 33, pp. 337-359, March 2009.

[5] Z. Miao, T. Grift, A. Hansen, and K. C. Ting. "Energy requirement for lignocellulosic feedstock densifications in relation to particle physical properties, preheating, and binding agents". Energ. Fuel vol. 27, pp. 588-595, January 2013.

[6] R. Wakeling and P. Morris. "Wood deterioration: Ground contact hazards". ACS Sym. Ser., vol. 1158, pp. 131-146, June 2014.

[7] M. Barontini, S. Crognale, A. Scarfone, P. Gallo, F. Gallucci, M.Petruccioli, L. Pesciaroli and L. Pari "Airborne fungi in biofuel wood chip storage sites". Int. Biodeter. Biodegr., vol. 90, pp. 17-22, May 2014.

[8] M. Overbeck and M. Schmidt. "Modelling infestation risk of Norway spruce by Ips typographus (L.) in the Lower Saxon Harz Mountains (Germany)". Forest Ecol. Manag., vol. 266, pp. 115-125, February 2012.

[9] I. Niedziółka, M. Szpryngiel, M. Kachel-Jakubowska, A. Kraszkiewicz, K. Zawiślak, P. Sobczak and R. Nadulski. "Assessment of the energetic and mechanical properties of pellets produced from agricultural biomass" Renew. Energ., vol. 76, pp. 312-317, April 2015.

[10] V. Bustamante-García, A. Carrillo-Parra, H. González-Rodríguez, R. Ramírez-Lozano, J. J. Corral-Rivas, and F. Garza-Ocañas. "Evaluation of a charcoal production process from forest residues of Quercus sideroxyla and Humb., & Bonpl. in a Brazilian beehive kiln". Ind. Crop Prod., vol. 42, pp. 169-174, March 2013.

[11] M. Horio, A. Suri, J. Asahara, S. Sagawa and C. Aida. "Development of biomass charcoal combustion heater for household utilization". Ind. Eng. Chem. Res., vol. 48, pp. 361-372, January 2009.

[12] D. Medic, M. Darr, A. Shah and S. Rahn. "Effect of torrefaction on water vapor adsorption properties and resistance to microbial degradation of corn stover". Energ. Fuel, vol. 26, pp. 2386-2393, April 2012.

[13] R. K. Eckhoff, Dust Explosions in the Process Industries, 3rd ed. Boston: Gulf Professional Publishing/Elsevier, 2003, pp. 256-263.

[14] H. Li, L. Jiang, C. Li, J. Liang, X. Yuan, Z. Xiao, Z. Xiao and H. Wang "Co-pelletization of sewage sludge and biomass: The energy input and properties of pellets" Fuel Process. Technol.,vol.132, pp. 55-61, April 2015.

[15] F. Fonseca, C. A. Luengo, J. A. Suárez and P. A. Beatón. "Wood briquette torrefaction". Energ Sust. Dev., vol. 9, pp. 19-22, March 2005.

[16] M. Katzer, S. Pirl, S. Esser, J. Kopietz, T. Rickmann, J. Behnisch and C. J. Klasen. "Residence time distribution in granulation drums, on the example of industrial carbon black". Chem. Eng. Technol., vol. 27, pp. 578-582, May 2004.

[17] T. H. Mwampamba, M. Owen and M. Pigaht. "Opportunities, challenges and way forward for the charcoal briquette industry in Sub-Saharan Africa". Energ Sust. Dev., vol 17 pp. 158–170, April 2013.

[18] S. R. Teixeira, A. F. V. Pena and A. G. Migue. "Briquetting of charcoal from sugar-cane bagasse fly ash (scbfa) as an alternative fuel". Waste Manage., vol. 30, pp. 804-807, May 2010.

[19] A. Amaya, N. Medero, N. Tancredi, H. Silva, F. Sardella and C. Deiana. "Activated carbon briquettes from biomass materials". Bioresource Technol., vol. 98, pp. 1635-1641, May 2007.

[20] A. Amaya, J. Píriz, N. Tancredi and T. Cordero. "Activated carbon pellets from eucalyptus char and tar TG studies". J. Therm. Anal. Calorim., vol. 89, pp. 987-991, September 2007.

[21] N. Kaliyan and V. Morey. "Natural binders and solid bridge type binding mechanisms in briquettes and pellets made from corn stover and switchgrass". Bioresource Technol., vol. 101, pp. 1082-1090, February 2010.

[22] A. Kumar, N. Kumar, P. Baredar and A. Shukla. "A review on biomass energy resources, potential, conversion and policy in India". Renew. Sust. Energy Reviews, vol. 45, pp. 530–539, February 2015.

[23] P. K. Halder, N. Paul and M.R.A. Beg. "Assessment of biomass energy resources and related technologies practice in Bangladesh". Renew. Sust. Energy Reviews, vol. 39, pp. 444–460, August 2014.

[24] O. A. Sotannde, A. O. Oluyege and G. B. Abah. "Physical and combustion properties of charcoal briquettes from neem wood residues". Int. Agrophys., vol. 24, pp. 189-194, June 2010.

[25] DINACYT. "El enorme potencial de la madera uruguaya". Noticias DINACYT Nr. 253, March 2005.

[26] C. Faroppa, Evaluación de la disponibilidad de residuos o subproductos de biomasa a nivel nacional. Montevideo: Ministerio de Industria, Energía y Minería, Dirección Nacional de Energía y Tecnología Nuclear; ONUDI; Uruguay, 2010, pp. 5-18.

[27] N. Tancredi, A. Cuña, J. P., Luizzi, M. Corengia, A. Sarachik, A. Amaya."Obtention of charcoal from eucalyptus wood in a steel pilot scale kiln". In Charcoal: Chemical Properties, Production Methods and Applications. New York: Nova Science Inc. Publishers, pp 61-74, 2013.

[28] E. Purlis. "Browning development in bakery products – A review". J. Food Eng., vol. 99, pp. 239–249, August 2010.

[29] K. J. Zhang and Y. Guo "Physical properties of solid fuel briquettes made from Caragana korshinskii". Powder Technol., vol. 256, pp. 293-299, April 2014

[30] M. van Blijderveen, E. A. Bramer and G. Brem. "Modelling spontaneous ignition of wood, char and RDF in a lab-scale packed bed". Fuel, vol. 108, pp. 190–196, June 2013.

12

Pricing for Natural Gas

Valentyna Novosad

Scientific Company "MAE", 33 Horyva st., Kyiv, Ukraine

Email address:

mae2010@meta.ua

Abstract: In this article, I showed the existing problems in the pricing for natural gas and tried to explore the ways to overcome them. Market research of natural gas consumers can determine the conditions of natural gas supply acceptable for each consumer. Giving for consumers the choice from many different conditions of natural gas supplies and corresponding amendments to the basic price, you can solve many problems in the pricing of the natural gas.

Keywords: Economy, Marketing, Natural Gas Pricing, Natural Gas Market

1. Introduction

Natural gas is a valued source of energy because it is versatile and burns cleanly. As result natural gas is common place in applications including cooking, residential and commercial heating, industrial process feed stocks and electrical generations. [7] Currently, the share of natural gas in the global structure of energy consumption is about 24%. According to the date of International Energy Agency the consumption of natural gas is expected to increase in the overall energy balance. [1]. Main factors contributing to the growth of natural gas consumption in the world are:

-growth of number of power stations that used the natural gas,

- decrease in attractiveness of nuclear energy and environmental problems.

Natural gas production in the United States has increase substantially due technological advancements in natural gas extraction methods [10].The United States have the largest amount of natural gas consumption. there are over 6300 producers of natural gas [2]. Over 530 natural gas processing plants are responsible for processing almost 15 trillion cubic feet of natural gas and extracting over 630 million barrels of natural gas liquids [2].

160 pipeline companies in the United States operate over 300000 miles of pipe.[2]. About 123 natural gas storage operations control approximately 400 underground storage facilities with storage capacity of 4059 billion cubic feet (Bcf) of natural gas and average daily deliverability of 85Bcf per day[2]. About 1200 natural gas distribution companies have 1, 2 million miles of distribution pipe. Many of these companies maintain monopoly status in their distribution region. [2].

This huge natural gas industry requires special attention. The introduction of the market in this important industry creates competition among the companies that are part of a system for providing of consumers with natural gas.

At the same time with introduction of the natural gas market, a huge amount of new relationships between producers, processing plants, pipeline companies, storage operations, distribution companies and their consumers was created. Also a lot of the different natural gas prices appeared with the development of the natural gas market.

The use of main principles of marketing in researches of relationship in this industry, opportunities and needs of producers, processing plants, storage operations, pipeline companies, distribution companies and consumers will make the process of pricing in this industry beneficial to the whole society.

2. Main Part

Existing business relationships in the natural gas industry can be represented by the scheme as the stages of natural gas sale. (Fig.1)

At each stage, indicated on the scheme of business relationships in natural gas industry, there is a large amount of transactions, relationships, and, accordingly, a large amount of prices. The system of prices in this scheme is based on the needs and opportunities to maneuver of producers, marketers, LDC, and transporters.

However this scheme of business in natural gas industry is

built in order to satisfy the needs of natural gas consumers. In the same time modern issues to improve the climate on the Earth and to keep useful fossil fuels impose some conditions on the consumption of natural gas by consumers.

So, we have two sides that play different roles in the

process of natural gas pricing. On one side there are the companies that are involved in the process of ensuring with natural gas of end users. Their price level is based on the desire to cover the costs, to ensure the development of production and to receive a profit.

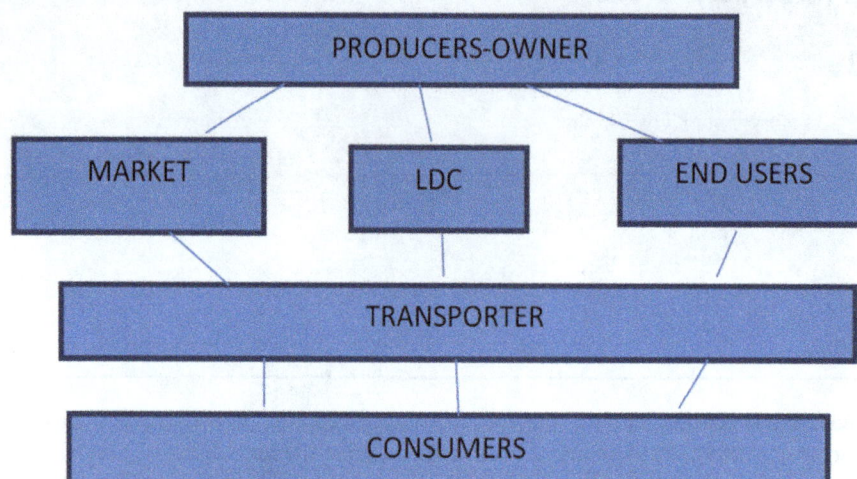

Fig. 1. *Scheme of business relationships in natural gas industry.*

On the other side are the consumers which want to have a price that corresponds to their abilities to pay accounts. The introduction of the natural gas market will adjust relationships between these two sides. Such markets of natural gas were established in North America and Europe.

The main principles on which is based their work are:

1). Ensuring a high level of protection of the rights and interests of consumers, and security of supply through diversification of sources of natural gas supply.

2). Free trade of natural gas and equality of the market subjects regardless of the state.

3). Free choice of natural gas supplier.

4). Equality of rights to import and export of natural gas to other areas.

5). Non-interference of the state in the functioning of the natural gas market, except when it is necessary to ensure the common interests.

6). Ensuring equal rights to access to the gas transmission and gas distribution systems, and gas storages, and local distribution companies (LDC).

7) The inadmissibility of restriction on competition.

8). Compliance with the established norms and safety standards.

9). Protection of the environment and rational using of energy resources.

10). The responsibility of the market participants for violation of rules of the natural gas market and the terms of contracts [8,9,10,11].

Active natural gas market naturally leads to lower prices for natural gas. However there are several economic drivers that provide an incentive for producers to continue producing even in conditions of declining prices:

1). If production from a natural gas well is halted; it may not be possible to restore the well's production due to

reservoir and wellbore characteristics.

2). If net present value of production in the future may be negative relative to productions on the gas today, it may be better to produce gas today than to wait until the future to produce gas. If producer chooses not to operate a well, production can't be recovered the next month. In this time there are no guarantees that the prices for gas in the future are going to be higher than prices today.

3). Some natural gas is produced in association with oil, and in order to stop the flow of natural gas, the oil production must be stopped as well, which may be economic.

4). A producer may be financially or contractually bound to produce specific volumes of natural gas.

On the other hand an increase in prices for natural gas does not always correspond with financial opportunities of consumers' .However they are forced to accept the current price. Producers and consumers react rationally to changes in prices. Fluctuations in the price of the natural gas market provide the signals to both supplies and consumers to ensure a constant move towards supply and demand equally.

Considering that many factors can have a negative impact on the price fluctuations and on the financial conditions of the both producers and consumers, it is important to have knowledge about these factors. Because consumers are those for whom this process of production and transportation of natural gas is organized, it is necessary to research these consumers in terms of the basic principles of marketing.

Determination of the target consumers has been a major step in the marketing research. The large industrial enterprises are these target consumers. They prefer direct contracts with producers for a long period of delivery and more stable prices than on the market. Providing such opportunities for this category of consumers gives them to more stable operation and the ability to plan over the long time their activity.

The next large category of consumers includes power plants used natural gas as a fuel. Taking into account that the markets of electrical energy are in many countries and that prices for natural gas have significant effect on the price of electricity from these power plants, activity of both the natural gas and electricity markets should be organized in such way that their prices are accurate.

In United States in August, 19, 2014 there was a discussion about concept of the trading platform for natural gas [5]. At this meeting, the participants of the natural gas markets discussed about the time of the transactions on the markets. The meeting invited all interested parties to establish the rules of coordination of processes of interstate natural gas pipeline and public utilities. On April, 16, 2015 the Federal Energy Regularly Commission published these final Rules. [5].This document amended the current NAESBWGO standards [6]. The changes allow coordination of the actions of the both electricity and natural gas markets which are related by pricing system. However it's only in North America.

So, two major categories of consumers (the target consumers and power plants used natural gas as fuel) constitute a basic level of total consumption of natural gas.

Another important category of consumers consists of those consumers whose needs must be satisfied, regardless of fluctuations in natural gas production. They're called «protected consumers". This category of consumers includes:

- Domestic consumers.

- Enterprises and organizations that produce goods or provide services which are important for the whole society.

- Consumers which due to technical reasons can't change of suppliers.

Within this category of consumers there are a lot of consumers groups with different abilities to pay bills. The state may have regulate pricing of some groups of such consumers through the reducing of the payments of producers to the budget for the amount of natural gas supplied for these groups of consumers.

All other categories of consumers should work according with the general rules of the natural gas market.

Despite the fact that the price of natural gas for consumers is formed on the market, terms of delivery, quality of natural gas are different in different consumers. Therefore, the price of the natural gas market can be only as a base price for all consumer groups.

Given today's possibilities of computer calculations we can create a system of pricing for natural gas that will to able to take into account all conditions of the supply of natural gas for each consumer. It would be desirable that the provider could offer several forms of tariffs, conditions of supply and payments to stimulate the consumers to the rational use of natural gas.

The individual price for natural gas consumers can be formed on the base market price, taking into account such factors:

1. The quality of natural gas, which is supplied, it's physical and-chemical properties and other specifications.

2. Terms of the ensuring the safe supply of natural gas: for category C-minimum safety standards, for category B- average safety standards, for category A -maximum safety standards.

3. Use technical means of measuring of the amount of consumed natural gas.

4. Conditions of the physical and commercial balancing.

5. Rules of distribution and regulation capacities in the critical moments of overloads.

6. Terms of the exchange of information.

7. The form and terms of payment for the consumed natural gas.

8. The type of tariff stimulating to the rational use of natural gas.

The use of such a large number of additions and amendments to the base natural gas price requires a detailed consideration of the conditions of supply of all consumers and the formation of small groups with the same conditions of natural gas supply.

Marketing research of natural gas consumers will allow to correctly determine the value of rebates and allowances to the price so that their using will have a stimulating role for the consumers and will bring benefit for suppliers. It is important that the provider will be to able to give for consumers opportunity to select the conditions that will be correspond with their needs and financial resources. In such circumstances, the consumer receives an additional opportunity to influence on the level of natural gas price, which is supplied for him personally.

3. Conclusions

The current pricing system for natural gas gives a good effect where there are the natural gas markets. At the same time, even in conditions of activity of the natural gas market there are conflicts of interests between suppliers and consumers. Such conflicts arise as a result of the equal prices for the supply of natural gas under different conditions. Such different conditions can be:

1. The quality of natural gas, which is supplied, its physical and-chemical properties and other specifications.

2. Terms of the ensuring of security supply of natural gas.

3. Using technical means of measuring of the amount of consumed natural gas.

4. Conditions of the physical and commercial balancing.

5. Rules of distribution and regulation capacities in the critical moments of overloads.

6. Terms of the exchange of information.

7. The form and terms of payment for the consumed natural gas.

8. The type of tariff stimulating to rational use of natural gas.

Marketing research of consumers once every 3 years and establishment of the variety discount and bonuses for the market price will be to able taking into account all possible terms of delivery and settlement. The consumer receives an additional opportunity to influence on the level of the natural gas price, which is supplied for him personally. So, giving for consumers the choice from many different conditions of natural gas supplies and corresponding amendments to the basic price, you can solve many problems in the pricing of the natural gas.

References

[1]   International Energy Agency http://www.iea.org/

[2]   www.naturalgas.org/business/industry

[3]   National   Energy   Marketing   Association www.energymarketers.com

[4]   Natural   Gas   Market   Reports   2015, www.reportlinker.com/Natural_Gas

[5]   Federal Energy Regulatory Commission, http://www.ferc.gov/

[6]   NAESBWGO standards, www.naesb.org

[7]   Charles   Augustin,Bob   Broxson,   Steven   Peterson "Understanding Natural Gas Markers", Overview The North American Gas Marketplace", www.spectraenergy.com

[8]   Natural Gas Market Law(Law on the Natural Gas Market and amending the Law on Electricity Market) LawN4646 Adoption Date 18.04.2001,ERRA

[9]   www.erranet.org

[10] Overview about natural gas markets Www.api.org/oil-and-naturalgas-overview

[11] Regulation   of   natural   gas companies.www.Law.cornell.edu/usecode/text15/chapter-158

13

Effect of the shape surface of absorber plate on performance of built-in-storage solar water heater

Omer Khalil Ahmad, Ahmed Hassan Ahmed, Obiad Majeed Ali

Technical institute of Hawija, Foundation of technical education, Baghdad, Iraq

Email address:

omerkalil@yahoo.com (O. K. Ahmed), ahmadhassan992000@yahoo.com (A. H. Ahmed), obaid_majeed@yahoo.com (O. M. Ali)

Abstract: An experimental and numerical study was carried out on a storage solar collectors to verify its suitability for domestic use. These storage collectors can be used as storage water tanks to replace the ordinary cubical or cylindrical tank commonly used in Iraqi houses. The paper includes study the effect of the shape surface of absorber plate on performance of storage solar collector by construction of three-box type, built –in-storage water heaters with three different shape of front absorber plat, flat, wavy, and zigzag shapes. Experiments were conducted in summer and autumn seasons, and the results were comparable to the theoretical calculation. The results indicated clearly that the storage collector can be used for providing hot water for domestic uses, the zigzag storage collector was the best to obtain a high temperature than the other two designs, also the finite difference model proved to be useful for prediction of water temperatures under variable operating conditions.

Keywords: Solar Energy, Storage Solar Collector, Absorber Plate

1. Introduction

Solar domestic hot water systems are probably, to date, the most frequently used devices which utilize solar energy. All the classical solar water heater systems contain two main components; the collector and the storage tank[1].

A compact solar water heater, incorporating the storage tank and collector in a single unit is an attractive alternative to the conventional solar water heating system. The elimination of the storage tank reduces the cost of the solar water heating system and should improve performance[2].

The built-in-storage water heater possesses several advantages over other types. First, it has higher efficiency during the daytime owing to the fact that primarily there are no heat losses during the circulation of water. Second, the superior contact of water with the absorber plate results in better heat transfer in comparison with poor bond conductance, as is the case in the thermosyphon and forced circulation type flat plate solar water heaters. Also, it is of low cost and easily manufactured from common materials without the need for high technology. This storage collector can be used as a water tank to replace the ordinary cubical or cylindrical tank commonly used in Iraqi houses.

Experimental and theoretical investigations on such systems have been carried out by a large numbers of investigators to study a large numbers of design and operation parameters [3-10]. From reconsideration this literature review, we find the effect of the surface shape wasn't study, therefore, the purpose of this paper, is to study the performance of built-in-storage heaters for different three types of shape surface, and various modes of operation.

2. Description of the Experimental Apparatus

The construction of a "built-in-storage" heater involves a rectangular box-like structure with the top face painted black and enclosed behind a single sheet of glass. The back surface and sides are properly insulated and the entire assembly is tilted at a suitable inclination and oriented due south. This work involved the construction of three-box type, built –in-storage water heaters with three different shape of front absorber plat, flat, wavy, and zigzag shapes as shown in

Fig.(1). A galvanized steel sheet of 2 mm thickness painted with ordinary blackboard paint was used as the absorber plate. The dimensions were (1 * 1 m), resulting in an absorber area of one m^2. The storage tanks were constructing by bending and welding of the steel sheets, which formed the top, bottom, and sides. The each of three storage tanks was wrap with 5 cm of fiber glass wool insulation on all sides and bottom and housed in an outer woody box. The assembly of each heater was mounted on a steel stand to face south at an angle 45° to the horizontal. This inclination is nearly 10° above, the local latitude of 35.33^o N for Kirkuk, where the experimental tests were carried out, to provide mean maximum collection of solar energy incident on the collector during the winter months [1]. The capacity of the each tank was 80 liter. Ordinary window glass of 4 mm thickness was used as the top transparent cover for tilted surface facing the sun. The distance between the absorber plate and the bottom surface of the glass was kept at 45 mm, which is within the recommended value for solar collectors [11, [12]. According to these workers, such a distance was supposed to provide a good insulating gap for the conduction- convection heat transfer from the hot absorber plate to the cooler glass cover. The glass cover edges were sealed with silicon tape to prevent the leakage of the hot air from the gap between the absorbing surface and the glass cover. The refractive index and extinction coefficient of window glass were taken as 1.526 m^{-1} and 0.02 mm^{-1} [13]. All construction work was performed at the technical institute of Hawija, Iraq (34 N^o, 44 E^o).

The hourly temperature of heated water was measured, using thermistor probes, at 7 point along the length and breadth of each heater as shown in the Fig (2). Two separate thermistors were used to measure the inlet and outlet temperatures of the storage water. In addition, the temperature of absorber plate measured by two additional thermistors and one to measure the glass temperature. Calibration of the thermistor was carried out by gradually varying the temperature and recording the corresponding resistance change. A bath of boiling water and a bowl of crushed ice were taken as the two limits of measurement. Calibration of thermistor was carried out against a standard mercury thermometer Data obtained regarding the temperature variation and the corresponding resistances are shown in Fig. (3).

Davis Vantage Pro with weather station package used to collect weather data. The standard version of the weather station package contains a rain collector, temperature sensor, humidity sensor, solar radiation sensor, ultra-violet (UV) sensor and anemometer. Temperature and humidity sensors were mounted in a passive radiation shield to minimize the impact of solar radiation on sensor readings. In our study, solar radiation and wind speed data are gotten from this station.

Fig (1). Three types of storage solar collector .

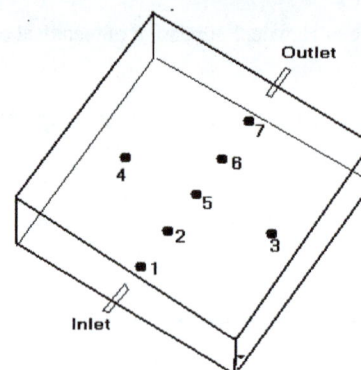

Fig (2). Positions of thermostat inside the storage solar collector

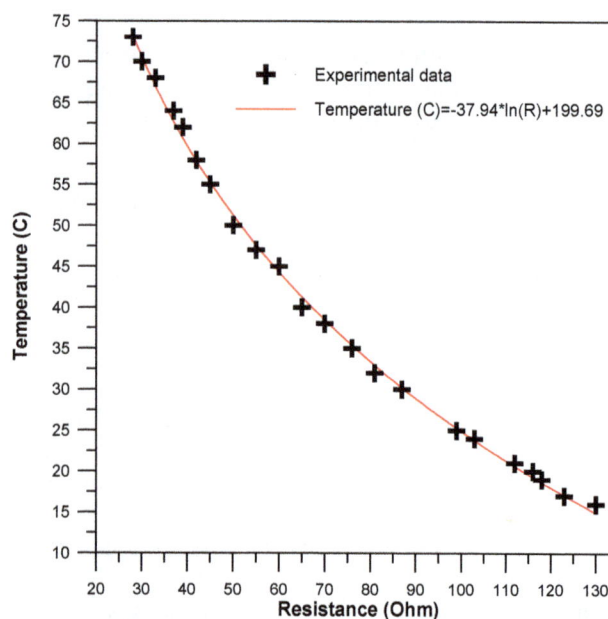

Fig (3). The calibration curve of the thermistors.

3. The Experimental Procedure

The collectors were tested under steady state conditions in which the solar intensity, wind speed, and the ambient temperature were considered constant for a period of time. Also, the inlet and outlet temperature difference and hence the useful energy delivered do not change appreciably during such a period [1]. At the onset of each experiment the collector was filled with fresh water, the glass cover was cleaned thoroughly, thermistors were checked, and the system faced south. The measuring instruments were switched on. Each test starts at sunrise and ends at sunset for the day of experimentation. The data recorded at the end of each hour included the various temperatures storage, glass cover, absorber plate and the tank inlet and outlet as well as the flow rate of the withdrawn hot water during the test run for each load pattern. The test run was repeated with and without for each load pattern. Some experiments were carried out with a load condition.

The load was connected to the storage collector in such a way that the cold inlet water entered at the bottom of collector and the hot water outlet was taken from the top of collector. A valve at the inlet of the collector regulated the mass flow rate of load water. The load mass flow rate was measured by timing the water collected in a graduated vessel of 1 liters capacity. The temperature of the inlet and outlet of the load water were record at the same time as the other data. A few experiments were conducted by imposing a large load in a short period. Load water, which 12-liter was withdrawn quickly from the collector at a period of approximately 5 minutes. In this, the average temperature of water withdrawn from the collector and the temperature of mains water entering the collector were record.

A few experiments were conducted by imposing a load on the system. In this case, the average temperature of water withdrawn from the tank and the temperature of mains water entering the tank were recorded in addition to the other measurements. A few experiments were conducted by imposing a load on the system. In this case, the average temperature of water withdrawn from the tank and the temperature of mains water entering the tank were recorded in addition to the other measurements.

4. Mathematical Model

A lumped parameter thermal balance on a storage solar collector, which is irradiated on one surface by solar energy as:

(Rate of heat absorber by surface absorber) + (rate of heat carried by water inter the collector) = (rate of heat absorbed by water) + (rate of heat absorber by metal of tank) + (rate of heat carried by water leaving the collector) + (heat lost from the collector).

Mathematically,

$$I_T(\tau_g\alpha_p)e \times F_{Sh} \times F_d \times A_a + \dot{m}_{fw} \times C_W \times T_{wi}$$
$$= m_w \times C_W \times \left(\frac{dT_w}{dt}\right) + m_c \times C_C \times \left(\frac{dT_C}{dt}\right) \qquad (1)$$
$$+ \dot{m}_{fw} \times C_W \times T_{wo} + U_L \times A_a (T_p - T_a)$$

We can re-write eq. (1) using the finite difference method as below:

$$I_T(\tau_g\alpha_p)e \times F_{Sh} \times F_d \times A_a + \dot{m}_{fw} \times C_W \times T_{wi}$$
$$= m_w \times C_w \times \left(\frac{\Delta T_w}{\Delta t}\right) + m_c \times C_c \times \left(\frac{\Delta T_C}{\Delta t}\right) \qquad (2)$$
$$+ \dot{m}_{fw} \times C_W \times T_{wo} + U_L \times A_a (T_p - T_a)$$

In other form:

$$I_T(\tau_g\alpha_p)e \times F_{sh} \times F_d \times A_a = J = \dot{m}_{fw} \times C_w (T_{wo} - T_{wi})$$
$$+ m_w \times C_w (T_{swf} - T_{swi}) \times 1/\Delta\tau \qquad (3)$$
$$+ m_c \times C_C (T_{CF} - T_{Ci}) \times 1/\Delta\tau + U_L \times A_a (T_p - Ta)$$

By simplify the eq.(3) we find the outlet temperature of load water from collector as :

$$T_{wo} = T_{wi} + \frac{J}{\dot{m}_{fw}.C_w} - \frac{m_w}{\dot{m}_{fw}.\Delta\tau}(T_{swf} - T_{swi})$$
$$- \frac{m_c.C_c}{\dot{m}_{fw}.C_w.\Delta\tau}(T_{CF} - T_{Ci}) - \frac{U_L.A_a}{\dot{m}_{fw}.C_w}(T_p - T_a) \qquad (4)$$

A program using Fortran-90 depending on eq. (4) using to find the Outlet temperature of load water from collector.

For no load condition we can used the following equation to estimate the temperature of water inside tank[1]:

$$T_w^+ = T_a + \frac{(\tau_g\alpha_p)_e I_t}{U_L} - \left[\frac{(\tau_g\alpha_p)_e I_t}{U_L} - (T_w - T_a)\right] e^{\left[-\frac{A_a U_L t}{(MC)_e}\right]} \qquad (5)$$

Eq.(5) was solved by trial and error method using a simple program achieved for this purpose.

5. Performance of the Collector

The maximum useful efficiency (MUE) was used in the paper to determine the efficiency of storage solar collector. Specifically [14],

$$\mu = KF_E\eta_o - \frac{F_E U_L (\bar{T} - \bar{T}_a)}{\bar{I}} \qquad (6)$$

In eq. (1), T is the temperature of water in the storage, T_a is the ambient temperature and I is the irradiance on the aperture plane of the collector. η_o is the optical efficiency and U_L the heat loss coefficient of the collector storage unit. K is the incidence angle modifier and F_E is an enthalpy retrieval defined as:

$$F_E = \frac{M_W C_W}{M_W C_W + M_C C_C} \qquad (7)$$

Where M_W and C_W are, respectively, the mass and the heat capacity of the water and M_C and C_C are the respective mass and heat capacity of the material from which the collector-storage unit is fabricated. Most important of all, the bars indicate time average over the daily heating period, from sunrise until the time the water reaches its maximum daily temperature typically about 8 h. That is to say, we use daily average, rather than instantaneous, values of the variables in Eq. (1). The bar over K indicates a daily energy average, which is fairly constant on a monthly basis. MUE is defined as:

$$\mu = \frac{M_W C_W (T_{max} - T_{sunrise})}{A_a \int I(t)dt} \qquad (8)$$

where A_a is the collector aperture area, and the integral is taken over the time from sunrise until the water reaches its maximum temperature.

6. Results and Discussions

The main purpose of a storage solar collector is the conversion of solar energy into thermal energy, many parameters was studied as below:

6.1. Parametric Storage Temperatures

The mean storage temperature is an important parameter. It is defined as a mass weighted average temperature, which is calculated as:

$$T_{av} = \frac{\sum_{i=1}^{n} M_i * T_i}{M_{total}} \quad (9)$$

Fig. (4) shows the variation of mean storage temperature during a typical clear summer day. It is observed that the mean storage temperature increased with time, the mean storage temperature reaches its maximum value and then decreases after 4 p.m. This is because the net energy absorbed becomes just lower than the heat losses, which means that the useful energy transferred to the water is insufficient to cause any further increase in the mean storage temperature.

The maximum temperature was recorded for the zigzag storage collector with a front area of 1.414 m2. The minimum temperature recorded for the flat storage collector with the front area of 1 m2. The maximum value of mean storage temperature was 63 °C for this particular day in the zigzag storage collector. The maximum value of the mean storage temperature of the collector and the time of its occurrence is different for the autumn days as can be observed experimentally from Fig. 5. The maximum value depends on the solar radiation intensity, the prevailing weather conditions, the starting inlet temperature and the heat losses, which are different during the days. The maximum value was 45 °C for this particular autumn day, occurred at 3 p.m., and then decreased in the late afternoon.

The agreement between the numerical predictions and the experimental data is good as shown in fig. (6).

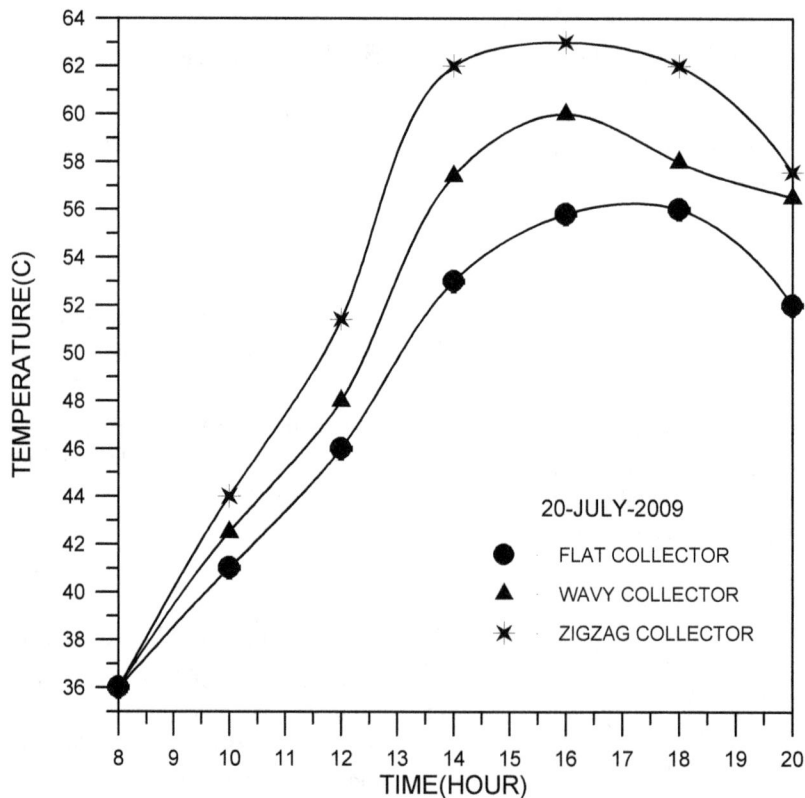

Fig (4). The variation of mean storage temperature summer day.

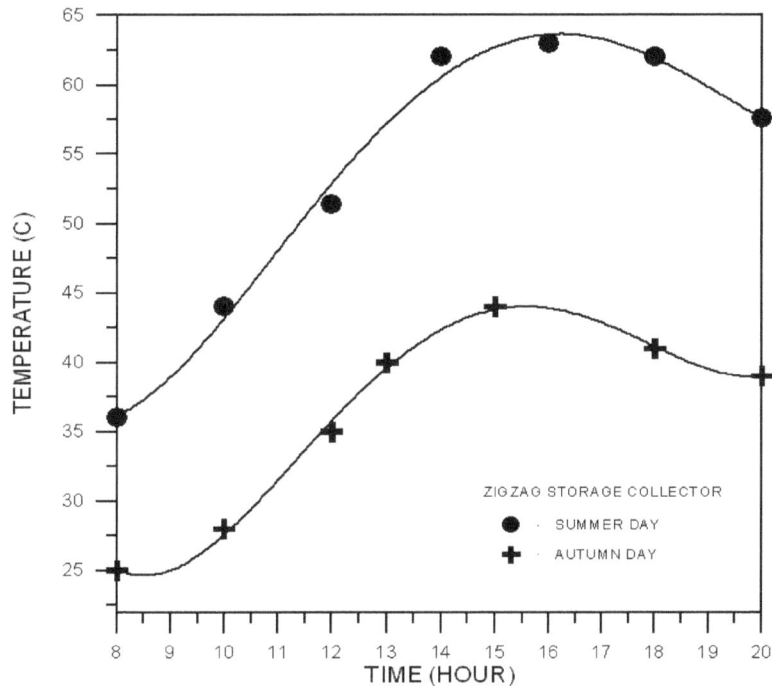

Fig (5). The variation of mean storage temperature of the zigzag collector through the autumn and summer seasons.

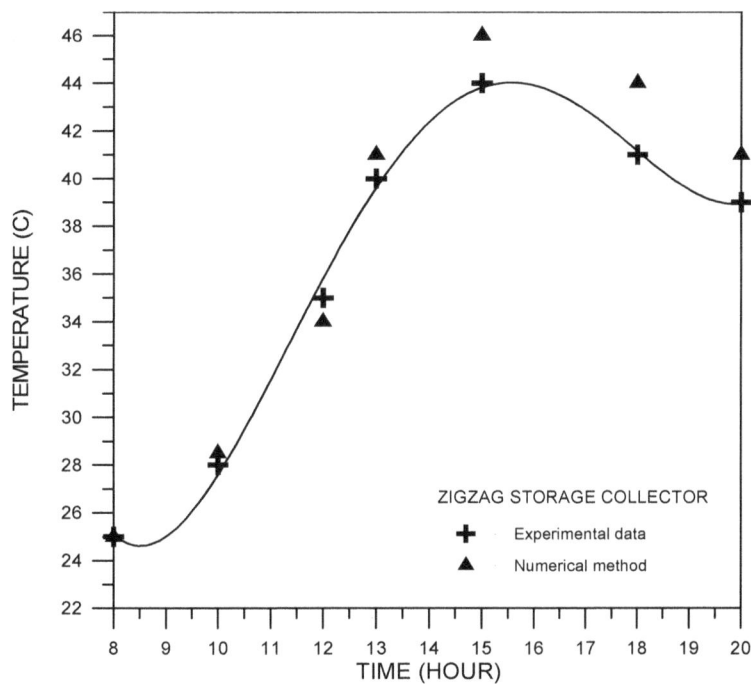

Fig(6). The variation of mean storage temperature of the zigzag collector in autumn day.

7. Effect of Loading

In order to show the effect of loading conditions on the system performance, some experiments were carried out with hot water withdrawal from the collector during the operating period in October and July 2009. The hot water withdrawn was taking continuously and intermittently. This was done by permitting cold mains water to enter the collector at its bottom, and the hot water was withdrawn from the top of the collector. The mass flow rate and temperature of the water were both measured. Fig.(7) show the variation of system temperatures for a clear day with continuous load condition. The mass flow rate was 0.2 liter/min where the total hot water removed during the whole working period was 96 and 168 liters in winter and summer seasons respectively. Fig. (7) shows the variation of inlet, outlet and mean storage temperature during autumn day. It is observed that the outlet temperatures are increasing throughout the period between 8 a.m. and 1 p.m., which indicates that the useful energy (q_u) is higher than that carried out by the load water. After 1 p.m., these temperatures start decreasing because the value of useful energy becomes lower than the energy carried out by the load water. The numerical data showed a higher value than experimental result but the agreement between the numerical and experimental data is acceptable as shown in Fig. (8).

Fig. (9) shows the variation of system temperature for a clear day with intermittent load conditions on a autumn day. This load condition was imposed by withdrawing hot water from the storage at the end of each hour. The amount of water withdrawal was 10.5 liter/hour, starting at 6 a.m. and ending at 6 p.m. The total hot water withdrawal during the day would be 126 liters. Temperature behavior is similar to the continuous load conditions of Fig. 6. The temperature difference between the outlet and inlet temperature is higher than the continuous load conditions, and the zigzag collector have the greatest difference between the other designs.

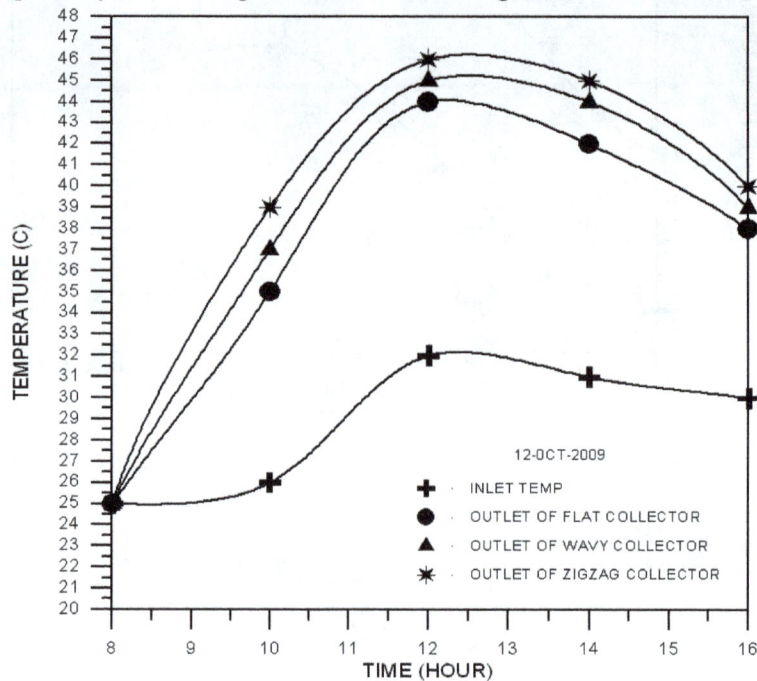

Fig (7). Variation of system temperatures of storage collector with continuous load in autumn day.

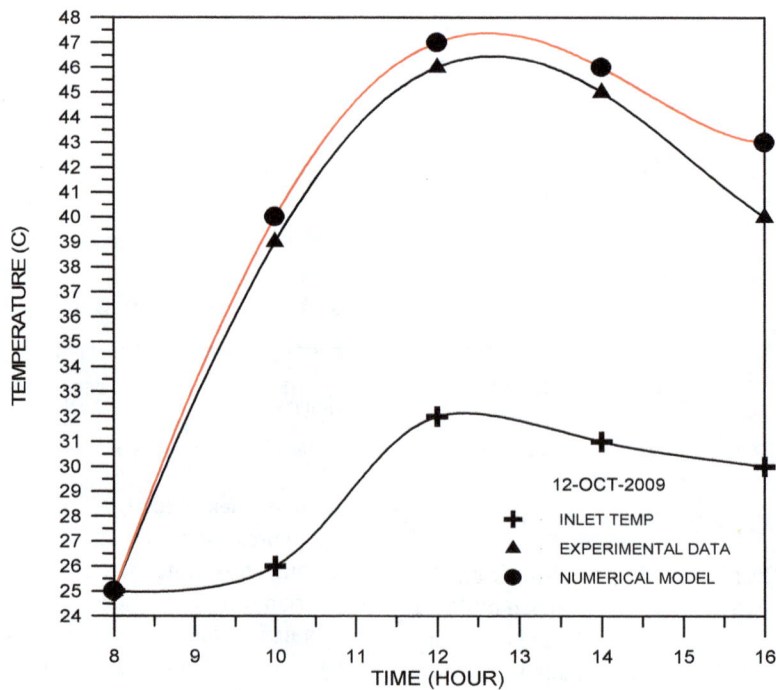

Fig (8). Comparison of experimental and numerical outlet temperature of triangular collector with continuous load.

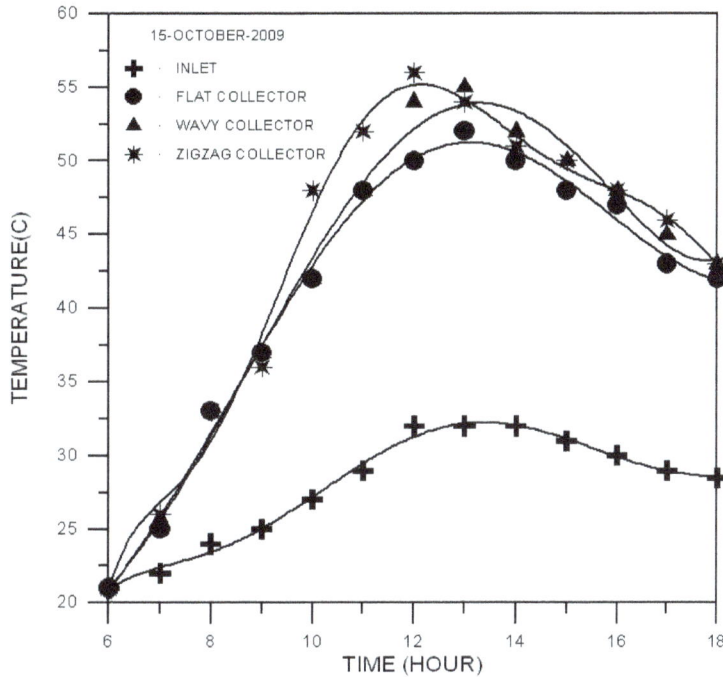

Fig (9). *Variation of system temperatures of storage collector with intermittent load in autumn day*

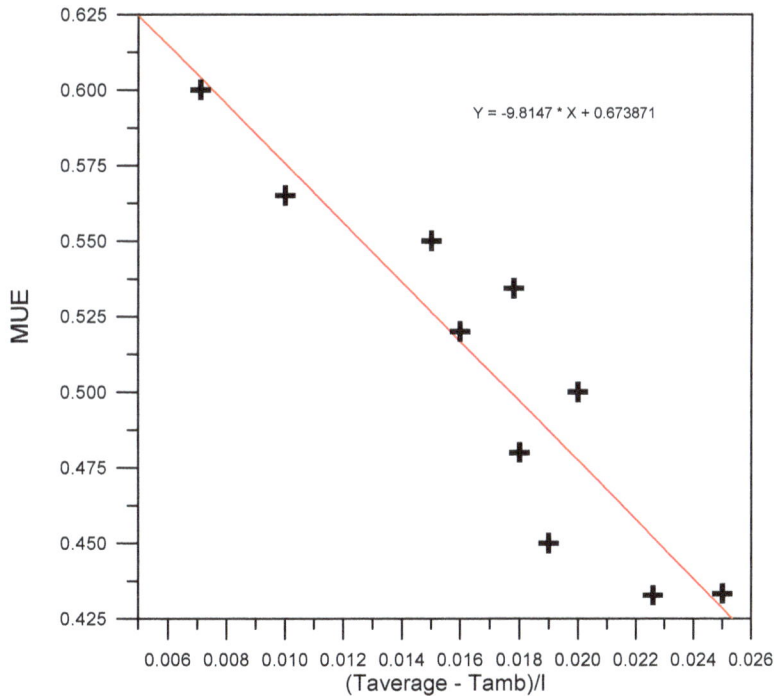

Fig (10). *MUE graph for the zigzag storage collector under study.*

8. Efficiency of Storage Collector

From the result of the previous section, the zigzag storage collector was the best to obtain a high temperature than the other two designs. The daily value of MUE of this type are, then plotted against $\dfrac{T_{avg} - T_{amb}}{I}$ as shown in fig.(10), the equation of this curve as below:

$$MUE = 0.673 - 9.8147\left(\frac{T_{av} - T_{amb}}{I_T}\right) \quad (10)$$

In comparison eq.(10) with eq.(6), we find that $-F_E U_L = -9.8147$, the value of F_E can be calculated from eq. (7) and equal to 0.96, that mean the heat transfer losses coefficient U_L equal to 10.22 W/m2.K, this value is high and to enhancement the performance the collector, reduction of this value is very necessary.

9. Conclusions and Recommendations

1. The performance of the zigzag flat plate collector was best than the wavy and flat plate collector.
2. The maximum value of temperature difference across the zigzag collector with 80 liters storage was 27 $^{\circ}$C in summer and 20 $^{\circ}$C in autumn during clear sky and no load conditions. With continuous loading test for zigzag flat plate collector the maximum temperature difference between the outlet and inlet temperature is 14 $^{\circ}$C at 12 p.m. and 10 $^{\circ}$C at the end of the day.
3. Studying water motion inside the collector experimentally by using the heliography techniques.
4. Experimental analysis of performance for parallel, series, and parallel-series connection of a bank of storage collectors.

Notation

Symbol	Description	Units
A_a	Irradiated face of collector	m^2
CW	Specific heat of water	J/kg.$^{\circ}$C
C_c	Specific heat of metal of tank	J/kg.$^{\circ}$C
F_d	Fouling factor	-
F_{sh}	Shading factor	-
h	Convection heat transfer coefficient	W/m^2.$^{\circ}$C
I_T	Total solar radiation on a tilted surface	W/m^2
m_c	Mass of the empty collector	kg
$\dot{m}_{fw}$	Mass flow rate of water	kg/s
m_w	Mass of water inside the collector	kg
T_a	Temperature of the ambient air	$^{\circ}$C
T_P	Temperature of the absorber plate	$^{\circ}$C
T_{wo}	Outlet temperature of load water from collector	$^{\circ}$C
T_{wi}	Inlet temperature of load water from collector	$^{\circ}$C
T_{cf}	Final temperature of the metal of tank	$^{\circ}$C
T_{ci}	Initial temperature of the metal of tank	$^{\circ}$C
UL	Collector over all coefficient	W/m2.K
T_{wsf}	Final temperature of the water inside tank	$^{\circ}$C
T_{wsi}	Initial temperature of the water inside tank	$^{\circ}$C
$T_{sunrise}$	Temperature of water inside tank at the sunrise time	$^{\circ}$C
UL	Collector over all coefficient	W/m^2.K
η_c	Collector efficiency	-

$\tau\alpha$	Transmittance – absorbtance product	-
$\Delta\tau$	Change in the period time	sec

References

[1] Duffie, J. and Beckman, W., 1991, Solar energy thermal processes, John Wiley and Sons, New York.

[2] Garge, H.P and Rani, U., 1982, Theoretical and experimental studies on collector/storage type solar water heater, Solar energy, Vol.29, pp.467-478.

[3] Vaxman, M. and Sokolov, M., 1985, Experiments with an integral compact solar water heater, Solar energy, Vol. 34, No. 6, pp. 447-454.

[4] Muneer, T., 1985, Effect of design parameters on performance of built-in storage solar water heater; Energy conservation and Management Vol. 25, No. 3, pp. 277-281.

[5] Khadiar ,M, 1987, Solar storage by using a bed of rock, M.Sc thesis, Mechanical engineering department, University of Basra.

[6] Mohamed, A.A.,1997, Integrated solar 'collector-storage tank system with thermal diode; Solar Energy, Vol. 61, pp. 221-218.

[7] Farhan, .A, 2002, Computational model for a prism shaped storage solar collector with a right triangular cross section, M.Sc thesis, Mechanical engineering department, University of Baghdad.

[8] Alawi, W.H, 2004, Numerical and experimental study of the solar collector storage pyramidical with right angle, M.Sc. Thesis, University of Technology, Baghdad.

[9] Junaidi HA, Henderson D, Muneer T, Grassie T, Currie J,2002, Study of stratification in a (ICSSWH) integrated collector storage solar water heater, Ninth AIAA/ASME Joint Thermophysics and Heat Transfer Conference, San Francisco, California.

[10] Jose´ M.S. Cruz, Geoffrey P. Hammond, Albino J.P.S. Reis,2002, Thermal performance of a trapezoidal-shaped solar collector/energy store, Applied Energy; 73 pp.195–212.

[11] Buchberg, H., Catton, I., and Edwards, D.K., 1976, Natural convection in enclosed spaces- A review of application of solar collection, Trans of the ASME, J. of Heat transfer, Vol. 98, pp. 182-188.

[12] Test, F.L., 1976, Parametric study of flat-plate solar collectors, J. of Energy conversion, Vol. 16, pp.23-33.

[13] Whillier, A, 1977, Prediction of performance of solar collector, ASHREA GRP 170, Application of solar energy for heating and cooling of building, Edited by Jordon, R.C. and Liu, B.Y.H., ASHREA.

[14] Faiman, D., 1984, Towards a standard method for determining the efficiency of integrated collector – storage solar water heaters, Solar energy 33, pp. 45-463.

14

Natural Gas Pricing in Eastern Europe

Valentyna Novosad

Scientific company "MAE", Horyva st, Kyiv, Ukraine

Email address:

mae2010@meta.ua

Abstract: Energy pricing, especially natural gas pricing is very important for all organizations and households. In many cases natural gas pricing has a significant effect on the formation of tariffs for electricity and the prices for heating and hot water. The formation of natural gas prices in Eastern Europe has its destructive features. This is due to shortage of domestic production and supplies of natural gas from Russia. In this article we examined these features and tried to give recommendations how to use these features to support own economics and prevent a fall of living standards of people in these countries.

Keywords: Economy, Marketing, Natural Gas Pricing, Natural Gas Market

1. Introduction

Energy pricing has a significant impact on activity of all organizations. Each household budget is greatly affected by the cost of energy. Therefore consideration of energy pricing is so important and relevant. [3].

Because the prices for natural gas have an independent value to the budgets of industry and households as well as a significant impact on the price formation for electricity and heating, the issue of the natural gas pricing is especially important for the every country.

At the same time in Europe natural gas prices are heavily dependent on external supplies of natural gas, particularly from Russia. Gazprom supplied Europe with 161, 5 billion cubic meters of natural gas in 2013. Today Gazprom remains the key gas supplier to the Europe. [6].

2. Main Part

In recent years the prices of natural gas have substantially decreased, but in comparison with the United States [tab.1] and Russian prices, [tab.2]European natural gas prices are still high [tab.3].

Table 1. Export-import prices of natural gas in US.

	Nov.2014		Apr.2015	
	$/fit3	$/m3	$/ft3	$/m3
Import price	3,97	140	2,59	92
Export price	4,44	157	2,88	102

*A source of information [7, 8, 10]

Table 2. Wholesale price in Russia.

	Rub/1000m3	$/1000m3
Moscow Region with tax,2015	3761	68

*A source of information [6, 8, 9, 11]
The dollar exchange rate as of June2015

Table 3. Prices for natural gas to Europe (average).

Years	$/MMbtu	$/1000m3
2011	10,3	390
2012	11,69	443
2013	12,29	465
2014	10,2	387
2015	7,27	276

*A source of information [5, 8, 10, 11]

High natural gas prices have a negative impact on the competitiveness of goods and welfare of European countries. If European countries with-developed economy can withstand this high level of the imported natural gas prices, in Eastern European Countries such prices may destroy their economy.

If we consider the prices of natural gas of the each country for end users, all European countries can be divided into 4 four groups.

1. Natural gas prices in the developed European countries.
2. Natural gas prices in the Eastern European Countries, which are part of the European Union (EU).
3. Natural gas prices in the Eastern European Countries, which are not part EU.
4. Natural gas prices in the Russia.

Table 4. Natural gas prices in well developed European countries[*].

Countries	Industry		Household		%
	Euro/ kWh	$/ 1000m3	Euro/ kWh	$/ 1000m3	Household to Industry
EU28	0,0373	392	0,0544	572	146
UK	0,0344	362	0,0560	589	162
Germany	0,438	460	0,519	546	119
France	0,0376	395	0,0606	657	162
Italy	0.0352	370	0,0621	658	178
Spain	0,0371	390	0,0714	751	193

[*]A source of information [1, 2, 8, 10, 11]
The dollar exchange rate as of June2015

Table 5. Natural gas prices in the Eastern European Countries, which are part of the European Union (EU)[*].

Countries	Industry		Household		%
	Euro/ kWh	$/ 1000m3	Euro/ kWh	$/ 1000m3	Household to Industry
Czech Republic	0,0320	307	0,0475	500	148
Estonia	0,0335	352	0,0374	393	112
Poland	0,0363	382	0,0414	436	114
Romania	0,0207	218	0,0161	169	78
Greece	0,0438	461	0,0728	766	165

[*]A source of information [1, 2, 8, 10, 11]
The dollar exchange rate as of June2015

Table 6. Natural gas prices in the Eastern European Countries, which are not part EU[*].

Countries	Industry		Household		%
	NV/ 1000m3	$/ 1000m3	NV/ 1000m3	$/ 1000m3	Household to Industry
Belarus, NV=BYR	-	276	2588000	168	61
Ukraine, NV= UAN	8975	427	7188	342	80
Moldova, NV=MDL	6301	333	6718	355	108

[*]A source of information [12, 14, 15, 16, 18, 20]
The dollar exchange rate as of June2015

Table 7. Natural gas prices in the Russia[*].

Country	Industry		Household		%
	RUB/ 1000m3	$/ 1000m3	RUB/ 1000m3	$/ 1000m3	Household to Industry
Russia	4043	73	4334	78	107

*A source of information [13, 9]
The dollar exchange rate as of June2015

The tables of natural gas prices for end users in Europe reflect the policy of different countries in the pricing and effects of this policy.

Simple copying by Greece of high European prices and high correlation between prices for industry and households has not led to the best result. More reasonable prices and the ratio between the prices for industry and households in other European countries allow to gradually raise the country's economy and improve the society welfare.

Countries that don't hurry to set high energy prices, especially for natural gas, taking into account own

participation in the Eastern markets, including Russian, make its industry as competitive and provide opportunities for the people to develop their households and to quickly reach the European standards of living.

The main characteristics of Eastern European Countries are Geographical location, historical development through the stage of socialism and the social consequences of this in the form of lower living standards for the population On the one hand they have a European markets with high prices and high quality products. On the other hand they have markets of Russia and other countries, which could have demand to their goods, but with a lower price than goods of highly developed countries.

Thus, economy of these countries depends not only on prices of imported energy resources, but also on the cost of energy resources in Russia and other countries. At the same time, these countries should have lower prices for imported natural gas from Russia considering lower transport costs.

These features can be an advantage for the development of their economy or an obstacles for it Blind adherence to the rules and prices of developed European countries can lead to a drop in the economy and impoverishment of their people faster than the approximation to the level of economy and life of the developed countries.

Before establishment of prices for the widely used in all spheres of countries life natural gas, its necessary to conduct the marketing research of all consumers to encourage them to economical use it, but don't harm them. The use of the principles of marketing in these cases is most effective means of helping to regulate the relationships between supplies and consumers.

In all countries there are two main categories of natural gas consumers. This is industry and households. We will consider these indicative categories of natural gas consumers in Eastern European countries.

2.1. Industry

Many years have passed since "era" of socialism, therefore industry of these countries has a very diverse structure. Some companies received investment and using the cheap labor, produce high-quality products and sell it with a good profit in Western countries. They are able to pay the cost of energy at a price corresponding to the European level of price.

Companies that sell their products to Russia and other countries with lower energy prices could lose its share on these commodity markets, if the prices of energy resources will be much higher than for products of other countries. When setting prices for natural gas for industrial enterprises, it is necessary to take into account these features in order to avoid the collapse of some businesses and increase unemployment in this regard. Therefore it is necessary to more carefully consider of all consumers in each category and to divide them into smaller groups.

2.2. Householders

Householders are another very important category of

natural gas consumers. This category of consumers has also within a variety of structure. Some consumer groups in this category are too sensitive to an increase in natural gas prices. High average European prices for natural gas and the cost of municipal services to the consumers don't stimulate to the savings of energy resources because households in any case will not be able to pay the bills. At the same time in all Eastern European Countries there are consumers who not only can afford to pay for the consumed natural gas, but also can increase their consumption even at the highest price. As well as with an increase of natural gas prices, the interests of those consumers, who use this energy resource just for cooking, will be affected less than the interests of consumers who use natural gas also for heating and for the preparing of hot water. At the same time the owners of the private houses where natural gas is used for heating and hot water are also very different in terms of consumption and in their capacity to pay bills.

There may be:

- Retirees, who consume a small amount of natural gas, but this amount exceeds their ability to pay;
- Large families in houses with area 300-500m2 with one breadwinner;
- Businessmen of the average level, who can regulate their consumption using the methods of energy conservation;
- Owners of huge property who are more concerned of better security of supply and technical services for gas supply system of their huge households than in lower prices.

All these supply conditions, technical requirements and the ability to pay bills, you can find using the methods of marketing research. Currently in Eastern European countries subsidies are only provided to the poorest members of society as well as in some countries there are different prices for different amounts of natural gas consumed by one household.

2.3. Regulation of Relationships

Marketing research of all consumers enables to have them countries not only debate whether"to raise or not raise" the prices of energy resources, but to regulate all relationships of customers and suppliers so that the interests of all participants of the process of the supply and consumption of natural gas are taken into account This means the authorities which engaged in the regulating the natural gas prices should be very attentive to the process of pricing natural gas and to balance the formula of equality interests (Ta) (1):

$$Ta = \frac{Cw.}{Gc.} = \frac{T1*G1+...+Ti*Ci}{GcTa=\frac{Cw}{Gc}\frac{T1*G1+\cdots+Ti*Gi}{Gc}} \qquad (1)$$

Where,
Cw- all money due to the suppliers;
Gc-all consumed natural gas;
Ti-a tariff for i-consumer;
Gi- consumed natural gas by i-consumer.

In this formula the optimal tariff of the supplier is equal to the weighted average tariff of all consumers. At the same time the tariff for a particular group of consumers would correspond to an opportunity to pay bills these groups of consumers.

Work of marketing specialists together with the specialists of the industry is necessary to regulate the relationships between suppliers and natural gas consumers so that the country's economics is successfully developed, the population is not impoverished and gas industry is reliable and secure.

3. Conclusions

High dependence on expensive imported natural gas of European countries has a negative impact on the competitiveness of goods and welfare of the people of these countries. At the same time Eastern European countries have some features that can help them if these features are used wisely.

One feature of Eastern European countries is their geographical location. These countries are located closer to the main supplier of natural gas in Europe -to Russia. Such location gives them lower prices for imported natural gas given lower transport costs.

However Eastern European countries have the second feature. It is a lower people's salary and as result a lower life level than in developed Western European countries.

Therefore the high levels of natural gas prices have very negative impact on welfare of Eastern European countries population.

At the same time the opportunities to pay the bills of industrial enterprises and home holders are very different.

Because Eastern European countries have such features the regulation of relationships between the natural gas suppliers and natural gas consumers should be especially attentive. Using of the marketing research in the process of natural gas pricing in these countries will allow identifying small groups of the natural gas consumers with the same conditions of supplies and opportunities to pay the bills. This also is the good way to find "gold points» of the pricing system and means to regulate in the best way all relationships in the sphere the natural gas supply and consumption.

References

[1]   Table14: Natural gas prices for industrial consumers. Euro stat (nrq_pc_203).

[2]   Table14: Natural gas prices for household consumers. Euro stat (nrq_pc_202).

[3]   Electricity and natural gas price statistics www.ec.europa.eu

[4]   Europe's Energy Portal www.energy.eu

[5]   European Union Natural Gas import price www.ycharts.com

[6]   Gazprom Www.gazprom.com

[7]   International Energy Agency http://www.iea.org/

[8]   Www.tetran.ru

[9]   Www.cbr.ru

[10] Www.dolgikg.com

[11] Swiss. Executive MBA www.freecurrencyrates.com

[12] Www.ch104.ua/for.clients/do-uvagi-klijentiv

[13] Www.newtariffs.ru/2015

[14] Www.bs-life.ru/nancy/nalogy/nds.html

[15] Www.nerc.gov.ua

[16] Www.mingas.bu/services

[17] Www.Eurobelarus/news/economy/2015

[18] Www.calk.ru/kurs BYR-USD.html

[19] Www.anre.nad.

Water-Pumping Using Powered Solar System - More than an Environmentally Alternative: The Case of Toshka, Egypt

Ahmed G. Abo-Khalil[1], Sameh S. Ahmed[2]

[1]Electrical Engineering Department, Assiut University, Assiut, Egypt
[2]Mining and Metallurgical Engineering Department, Assiut University, Assiut, Egypt

Email address:
a.abokhalil@mu.edu.sa (A. G. Abo-Khalil)

Abstract: In this paper, a solar powered water pumping system for supplying water to irrigate part of Toshka project is proposed. The design considers several factors including the crop, the size of the planting region, the number of peak sun hours, the efficiency of the solar array and its electronics, the pumping elevation and the pump efficiency. The pumping power to irrigate different crops that could be grown in Toshka is calculated. The proposed system consists of a vector control of induction motor coupled with a centrifugal hydraulic pump, the induction motor drive is fed by DC-DC converters with maximum power point tracking (MPPT) to extract the whole energy that the PV panels can generate. The effectiveness of the proposed system has been verified by simulation results and showed reasonable level for implementation that would assist the development at the region both economically and environmentally.

Keywords: Photovoltaic, Maximum Power Point, Toshka Project, Water Pumping

1. Introduction

Over the past 30 years, the population of Egypt has risen from 20 to nearly 90 million and it has been predicted that this trend will continue, reaching an anticipated 120 million in the next 10. The Nile valley in Egypt covers an area of about 11,000 km² with an average width of 12 km². The net irrigated area in the valley is 8600 km² [1]. Stretching the length of Egypt from north to south and occupying a valley around the Nile River, it covers 3 percent of Egypt's land but is home to 96 percent of its people. The majority of Egyptians live around the Nile. An estimated 60% live in urban areas - in cities which are growing faster than infrastructure to support them - while the mounting numbers in rural areas provide a ready supply of new migrants to the towns.

The Toshka New Valley project is meant to develop agricultural production and create new jobs away from the Nile Valley by creating a second Nile Valley. This includes redirecting water from Lake Nasser to irrigate part of the Western Desert of Egypt via canals. The project is meant to help Egypt deals with its growing urban population and was touted as the "New era of hope for Egypt". The Toshka Project is intended to house more than three million residents

and to increase Egypt's arable land area by 10%. The project is also designed to create a new delta south of the Western Desert parallel to the Nile, adding 520,231 acres to Egypt's cultivated area.

To irrigate the major part of this land, $1 billion was spent on creating the world's largest pumping station to extract five billion cubic meters of water from Lake Nasser annually, and digging the Sheikh Zayed Canal. The main canal (completed in 2002) is 50km long, 30m wide and 6m deep, with four branch canals totaling 159km in length, and a network of roads linking Toshka to another irrigation project at East Oweinat in the Western Desert [2].

A small area of Toshka New Valley relies on the underground water according to the developing plans of the Ministry of Irrigation and Water Resources. It is known that the presence of the Nubian Sandstone reservoir which economic studies have shown that rates of groundwater development in the South Valley region and the Western Sahara can be up to approximately 2.5 billion cubic meters annually, equivalent to almost 3,000 times the annual flow of the Nile,. It is shared with Sudan, Chad and Libya [3].

Nevertheless, most of the surface water from the Nile is nearly fully utilized; most of groundwater resource is almost undeveloped. In Toshka, beside the main canals driven from

Lake Nasser, there is a possibility of utilizing the groundwater at a western part of the project.

The number of targeted wells is 316 productive wells to grow about 36 thousand acres in scattered areas. The data collected shows that the depth to the water aquifer ranges from 80 to 100 m [4].

The Egyptian government plans to distribute these 36 thousand acres of Toshka lands to young people up to 100 acres. There is no significant irrigation infrastructure in place to support the irrigation needs of the proposed crop. Installation of irrigation systems that stores and delivers water to the crops is now necessary. As there is also no electrical grid infrastructure in these lands, installation of irrigation systems is a must.

2. Technical Aspects

Far from the electric grid, it is common to use fossil fuels to power generators in water pumping operations. While these systems can provide power where needed, there are some significant drawbacks, including [5]:

1) Fuel has to be transported to the generator's location, which may be quite a distance over some challenging roads and landscape.
2) Their noise and fumes can disturb communities.
3) Fuel costs add up, and spills can contaminate the land.
4) Generators require a significant amount of maintenance and, like all mechanical systems; they break down and some parts need to be replaced that are not always available.

Not to mention, points 1 and 2 have environmental impacts that need further efforts to overcome.

For many water pumping needs, the alternatives are: solar energy (photovoltaic system), wind power system, or nuclear station. This research is considering solar energy as an alternative. One more significant advantage for photovoltaic water pumps, they do not require the presence of an electric line in order to operate. This makes them extremely useful in rural locations such as ranches and farms, or in the developing world where electricity is often not available or it is far cheaper to purchase a solar panel than it is to run power lines. Therefore, the authors believe that solar water pump is a perfect solution to be used in the water pumping of Toshka project where power lines are not available and transport of fuel is expensive. Moreover its less environmental impacts.

Water pumping has regularly been a technical challenge, solving the problems of drinking water supply and regular irrigation was a prerequisite for the development of civilization in many of the ancient empires. The PV pumping systems are being installed worldwide, and there were approximately 60,000 unit such systems in 1998 which increased rapidly during the last few years. The ongoing efforts on performance improvement and modeling, system sizing and optimization, and performance of PV systems on the basis of experimental measurements have resulted in commercially acceptable, economically affordable, and easily maintainable with least possible expertise. These

developments have lead and are contributing to the improvement of the lives of remotely located dwellings [6].

Particularly in areas of average daily solar radiation intensity exceeding 5 kWh/m^3/day and of water requirements less than 100 m^3/day, PV systems had proven reliability and economic feasibility in comparison with diesel powered pumping systems. In addition, these systems need very limited maintenance, since they operate without storage batteries and they do not pollute the environment [7].

In this paper, an overall design and simulation of PV-powered pumping is developed. This work then investigates the feasibility of using solar power for the pumping power requirement for drip irrigation systems. The effectiveness of the proposed system has been verified by simulation results.

3. Toshka Project Characteristics and Energy Requirement

The Toshka project is located in a portion of Egypt's Western Desert known as the Toshka Depression. Situated to the west of Lake Nasser between the Upper (southern) Egyptian cities of Aswan and Abu Simbel as shown in Fig. 1. There are about 3200 km^2 of land lower in elevation than Lake Nasser and lower even than the River Nile prior to the creation of Lake Nasser. Before discussing the Toshka Project, it is important to first identify the geography of the project (see Fig. 2) [4]:

- Toshka Region;
- Toshka Depression;
- Toshka Bay;
- Toshka Spillway;
- Toshka Canal;
- Mubarek Pump Station (MPS) which is the largest pump station in the world.

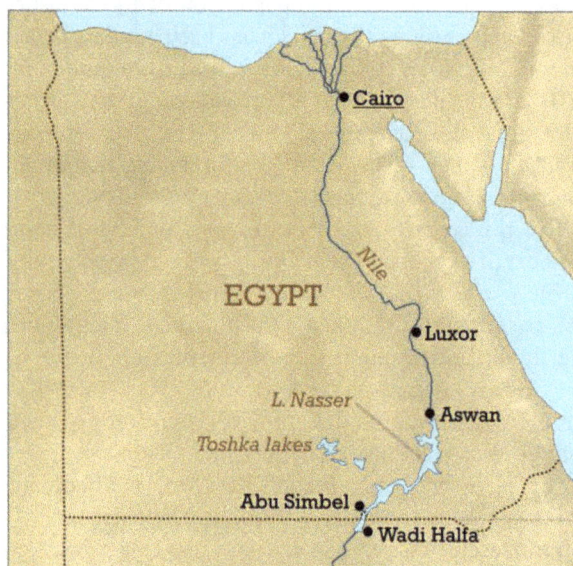

Fig. 1. *Location of Toshka region.*

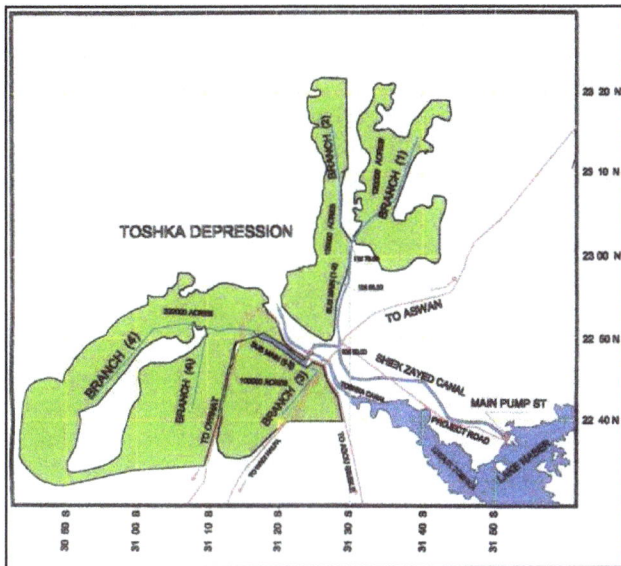

Fig. 2. *Outline of Toshka Project.*

Toshka Region is located southwest of Aswan, about 1000 km south of Cairo. Toshka Depression is a natural depression in that area with an average diameter of 22 km and a storage capacity of 120 billion m³. Toshka Bay is a shoot off Lake Nasser towards Toshka. Toshka Spillway is a free spillway discharging the water of Lake Nasser when it exceeds its highest storage level of 187m. It is a 22-km long, manmade canal connecting Toshka Bay with the Toshka Depression and works as a safety valve for Lake Nasser, upstream of the High Dam.

Toshka Canal is the heart and soul of the Toshka Project. It is a new canal conveying the excess water of Lake Nasser that is pumped into it through a giant pumping station that elevates the water to flow through the canal to reclaim and irrigate 520,231 acres.

The Toshka project begins with the main pumping station, also known as Mubarak pumping station (MPS), and is considered as the largest pump station in the world. The characteristics of MPS operation are as follow:

1- The required annual water for the total area of the project is $5.5*10^9$ m³ per year and the designed discharge capacity of a single pump is estimated to be about ($17*10^6$ m³/day).

2- Minimum and maximum suction levels (minimum and maximum operational water levels in Lake Nasser) are 147 m and 182 m, respectively.

3- Delivery level (Toshka Canal) is 201 m, consequently, minimum and maximum statically water raising heads are 19 m and 54 m respectively.

Many studies were performed to determine the required power and the energy for the project. Operation scenarios are different in each study; one scenario determines the peak power requirements for pumping stations by about 165 MW, of which 105 MW is allocated for main Mubarak Pumping Station. The associated annual energy requirements are about 950 GWh, of which 575 GWh is allocated for the main pumping station [8].

4. Photovoltaic Pumping System

Photovoltaic pumping system is a standard pump equipped with an electric motor, provided in electrical energy by photovoltaic panels installed on the. This pump is intended to pump water from the basement to make it accessible to users as shown in Fig. 3 [9]. Before selecting the system components, the system should be roughly sized to allow viewing of approximate component sizes. The sizing involved determining the area for implementing the solar powered irrigation system and the water needed for domestic use. It is also included calculation of the required pumping power the solar panel array area.

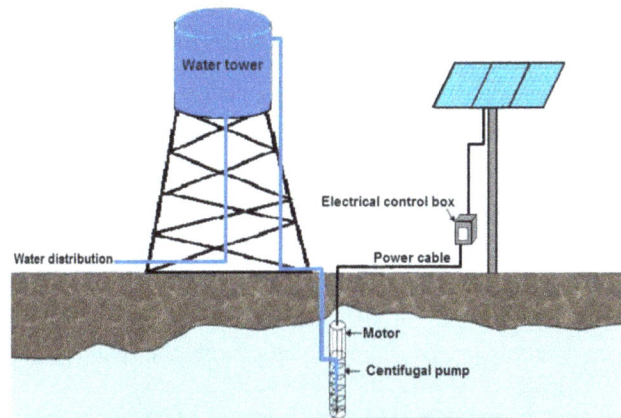

Fig. 3. *Photovoltaic water pumping without batteries with a tank to store water [8].*

4.1. Determining the Area to be Irrigated Using the PV System

The Egyptian government plans to distribute up to 100 acres for the Egyptian individual farmers.

Total land area (10 acres) = 40469 m²

Therefore, total crop area (taking into account area subtracted for PV arrays, roads, drainage and water tank) is 32000 m².

4.2. Crop Water Requirements

Toshka climate is characterized by warmth and drought which helps to speed the maturity of fruits and vegetables in the early times from their counterparts in neighboring countries. Different types of crops are grown in Toshka, such as: wheat, maize, citrus, peanuts, watermelon. These crops water requirements are shown in table 1 [10]. If we take of wheat as an example, to calculate the amount of water needed during the planting period, irrigation needs for period for wheat could be calculated as 32000m²x 0.6m = 19200m³. Therefore, Total Crop Water required (or Water deficit) for that period could be determined by Irrigation needs for period + 10% evaporation loss + 10% leakage loss = 23040 m³. Hence, total crop water requirements per day would be 23040m³/150 = 153.6m³.

Table 1. *Indicative Values of the Total Growing Period [10].*

Crop	Total growing period (days)	Crop water need (mm/total growing period)
Wheat	120-150	450-650
Bean Dry	95- 110	300-450
Maize	125-180	500-800
Tomato	135-180	400-800
Potato	105-145	500-700
Onion dry	150-210	350-550
Melon	120-160	400-700
Citrus	240-365	900-1200

4.3. Calculating the Required Pumping Power

Photovoltaic water pump sizing is the determination of the power of the solar generator that will provide the desired amount of water. The photovoltaic water pump sizing consists of [8]:

- Assessment of daily water needs of the population to know the rate flow required;
- calculation of hydropower helpful;
- determining of the available solar energy;
- determining of the inclination of the photovoltaic generator which can be placed;
- determination of the month sizing (the month in which the ratio between solar radiation
- and hydropower is minimum);
- sizing of the PV generator (determination of the required electrical energy);

The required PV size can be determined from:

$$P_{pv} = \frac{g * \rho_s * Q_s * T_H}{3600 * G * \eta_p * (1 - \eta_l)} \tag{1}$$

Where, P_{pv} is the electrical power (W);

g is acceleration of gravity (9.81 m.s^{-2});

ρ_s is water density (1000 kg/m^3);

Q_s is daily water needs (m^3/day);

T_H is the total head (m);

η_p is pump system efficiency;

η_l is the system loss due to the temperature and dust.

Assuming a 6 kW/m^2 irradiation, 153.6 m^3 daily water need, 100 m total head, 40% pump system efficiency, and 25% system loss. By using (1), the PV electrical power is 23kW.

5. System Description

Figure 4 shows a block-diagram of a stand-alone photovoltaic generator, boost converter for maximum power point tracking, DC/AC inverter, and induction motor coupled with a pump load. A MPPT is designed and implemented to extract the maximum power from a PV source using a boost converter. The MPP varies as a result of changes in its electrical characteristics which in turn are functions of irradiation level, temperature, ageing and other effects. The MPPT maximizes the power output from panels for a given set of conditions by detecting the best working point of the power characteristic and then controls the current through the panels or the voltage across them. The voltage and current, which is detected in real time, are used to extract the MPPT. The output of the boost converter is fed to the current controlled voltage source inverter (CC-VSI). The (CC- VSI) acts as a source of excitation with controlled frequency and it is responsible to feed power to the cage induction motor' drive. For effective utilization of the solar energy, optimum efficiency operation of the drive is desired throughout the operating range of the motor. This is achieved by maintaining a proper balance between the flux (i_{ds}) and torque (i_{qs}) components of the motor current, which in turn equalizes the fixed and the variable losses of the cage motor. For this purpose, the torque and flux component of the motor current are also logically related with the power at the output of PV array. Therefore, the available solar power is optimally converted into the useful mechanical power.

The induction motor d-axis current component is generally set to maintain the rated field flux in the whole range of speed, while the q-axis current component to control the motor torque based on the PV output power. The motor speed is estimated based on the motor voltages and currents. The detailed block diagram is shown in Fig. 5 [11].

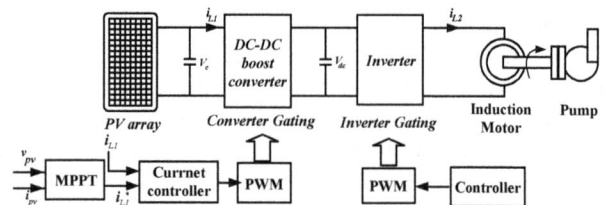

Fig. 4. *Control circuit for the PV system.*

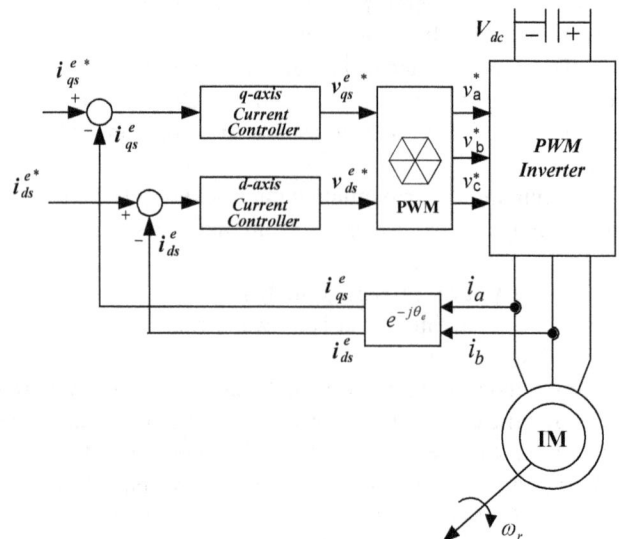

Fig. 5. *The block diagram of the controlled induction motor.*

6. Modeling of PV Module

Solar cells are devices that convert photons into electrical potential in a PN junction, of which equivalent circuit is shown in Fig. 6(a). PV module is composed of n of these cells in series, as shown in Fig 6(b), in order to reach a high voltage at the terminals. Due to the complex physical phenomena inside the solar cell, manufacturers usually present a family of operating curves (V-I) as shown in Fig. 7(a). These characteristics are obtained by measuring the array volt-ampere for a different illumination values. From these characteristics, the optimum voltage or current, corresponding to the maximum power point, can be determined. It is clearly seen in Fig. 7(b) that the current increases as the irradiance levels increase. The maximum power point increases with a steep positive slope proportional to the illumination.

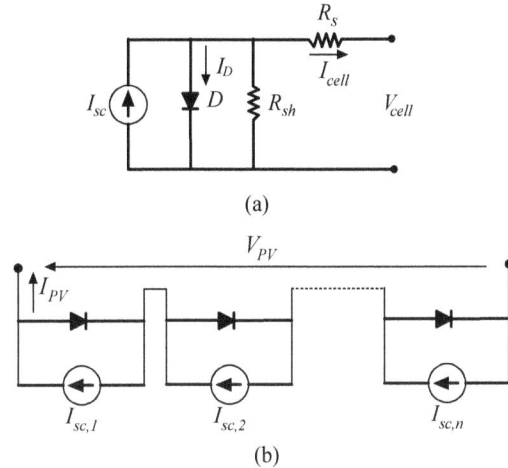

(a)

(b)

Fig. 6. *Equivalent circuit of PV array.*

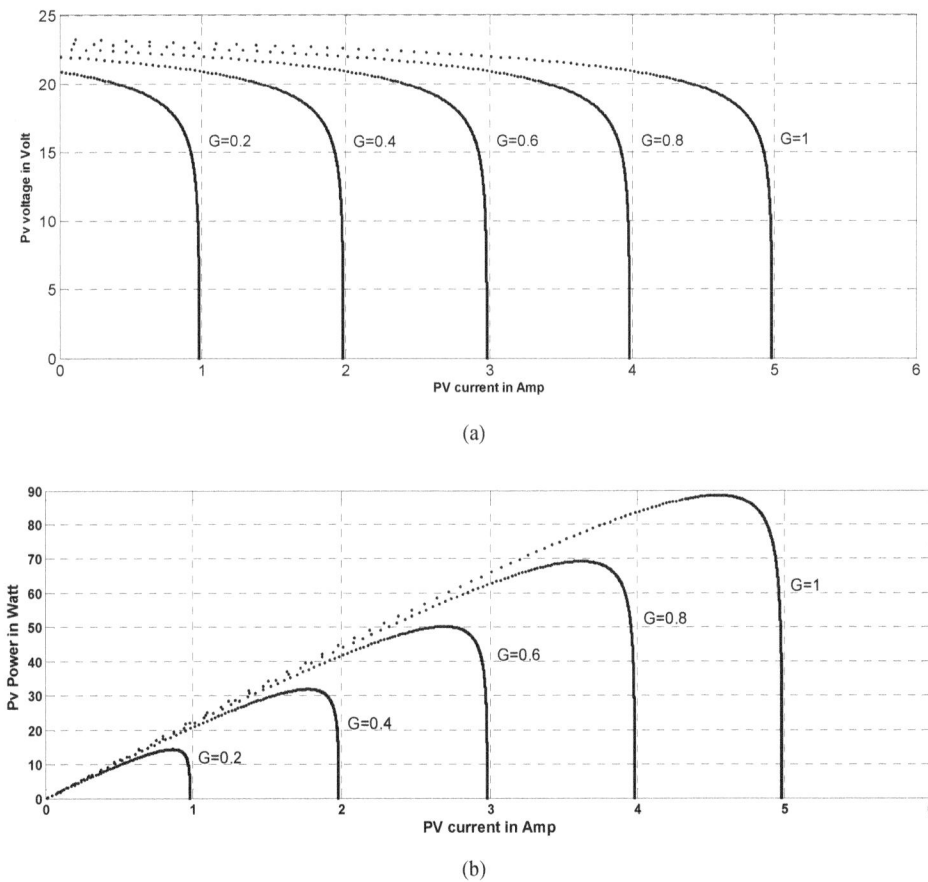

(a)

(b)

Fig. 7. *(a) V-I characteristics, and (b) P-I characteristics at constant temperature (25 °C).*

The main parameters which influence the illumination levels on a surface at a fixed tilt on earth are the daily and seasonal solar path, the presence of clouds, mist, smog and dust between the surface and the sunlight, and the shade of the object positioned such that the illumination level is reduced, etc.

The equation of the PV output current I is expressed as a function of the array voltage V

$$I = I_{sc} - I_o \left\{ e^{\frac{q(V+IR_s)}{KT_k}} - 1 \right\} - (V + IR_s)/R_{sh} \qquad (2)$$

where V and I represent the PV output voltage and current, respectively; R_s and R_{sh} are the series and shunt resistance of the cell (in Fig. 1); q is the electronic charge; Isc is the light-generated current; Io is the reverse Saturation current; K is the Boltzman constant, and Tk is the temperature in K.

Equation (2) can be written in another form as [11], [12]

$$I = I_{sc} \left\{ 1 - K_1 [e^{K_2 V^m} - 1] \right\} - (V + IR_s)/R_{sh} \qquad (3)$$

where the coefficient K_1, K_2 and m are defined as

$K_1 = 0.01175$,

$K_2 = K_4 / (V_{oc})^m$,

$K_4 = \ln((K_1 + 1)/K_1)$,

$K_3 = \ln[(I_{sc}(1 + K_1) - I_{mpp})/K_1 I_{sc}]$,

$m = \ln(K_3 / K_4)/\ln(V_{mpp}/V_{oc})$

and I_{mpp} is the current at maximum output power, V_{mpp} is the voltage at maximum power, I_{sc} is the short circuit current and V_{oc} is the open circuit voltage of the array.

Equation (3) is only applicable at one particular operating condition of illumination G and cell temperature T_c. The parameter variations can be calculated by measuring the variation of the short-circuit current and the open-circuit voltage in these conditions using the parameters at the normal illumination and cell temperature. Equation (3) is used for the I-V and P-V characteristics for various illumination and fixed temperature ($25 [^{o}C]$) in Fig. 7.

7. Simulation Results

In order to demonstrate the effectiveness of the proposed MPPT, some simulations have been carried out. The proposed design is implemented in PSIM software for a photovoltaic water-pumping system as depicted in Fig. 8. Fig. 9 shows that the output power changes according to the illumination variation from 800 [w/m^2] to 700 [w/m^2] to 600 [w/m^2]. The new incremental duty ratio varies according to the change in DC/AC inverter reference current. These relations between the input and output are determined in the base-rule form. It is noticed that the fast dynamic performance at step variation of illumination level is obtained.

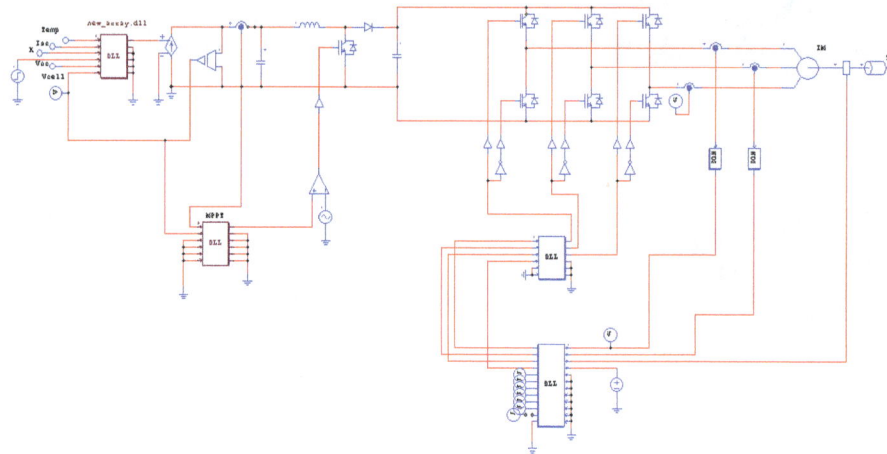

Fig. 8. *Detailed model and control scheme of the grid-connected PV solar system in the PSIM environment.*

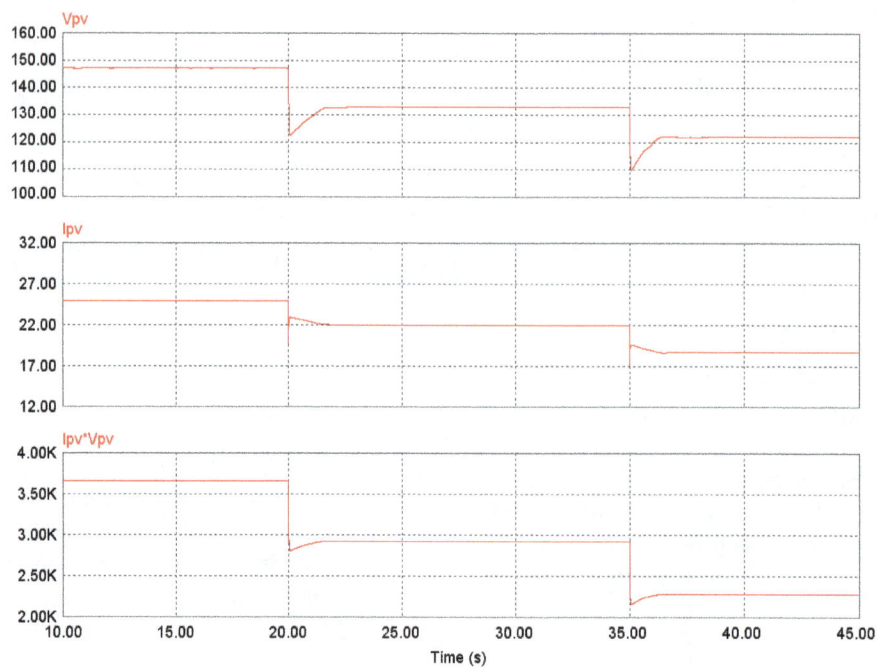

Fig. 9. *PV voltage, current and power at different illumination levels.*

8. Conclusions

In this paper, a demonstration for the application of solar pumping system to provide the required power for irrigation was presented. Crop chosen, the size of the planting region, the number of peak sun hours, the efficiency of the solar and pump system were taken into consideration to calculate the required solar power. To verify the validity and robustness of the proposed system, the simulation results have been obtained with a squirrel-cage induction generator pumping system.

References

[1] Fatma A. R. Attia, *et al.,* "A Hydrologic Budget Analysis for the Nile Valley in Egypt", *Groundwater*, Vol. 24, No. 4, pp 453-459, August 1986.

[2] A. F. Hassan, "The Toshka Project construction works, Toshka, Egypt," Toshka Research and Engineering Studies., 2001.

[3] K. Abu-Zeid, and A. Abdel Meguid, "Status of Integrated Water Resources Management Planning in Arab Mediterranean Countries", International Conference on Water Demand Management, World Bank Session, Dead Sea, Jordan, 2003.

[4] Ministry of Water Resources and Irrigation, Arab Republic of Egypt, "The Southern Valley Development Project (SVDP)," Cairo, Egypt, October, 2000.

[5] B. Eker, A. Akdogan, Protection methods of corrosion on solar systems, TMMOB Machinery Engineering Society, Mersin, Turkey, 2005.

[6] A. Z. Sahin and S. Rehman, "Economical Feasibility of Utilizing Photovoltaic for Water Pumping in Saudi Arabia," Hindawi Publishing Corporation, International Journal of Photo energy, Vol. 2012, Article ID 542416.

[7] N. M. Abdelsalam, M. M. Abdelaziz, A. F. Zobaa, and M. S. Aziz, "Toshka Project Electrical Power Demand (Mubarak Pump Station)" *12th International Water Technology Conference, IWTC12,Egypt,* pp. 503-517, 2008.

[8] M. Saïdou, K. Mohamadou, and S. Gregoire, "Photovoltaic Water Pumping System in Niger," Intec Open, pp. 183-193, 2013.

[9] C. Brouwer and M. Heibloem, "Irrigation Water Management," Food and Agriculture Organization (FAO) Training Manual No. 3.

[10] Byunggyu YU, Ahmed G. Abo-Khalil1, Mikihiko Matsui, Gwonjong Yu, "Sensorless Fuzzy Logic Controller for Maximum Power Point Tracking of Grid- Connected PV system" *IEEE International Conference on Electrical Machines and Systems* ICEMS, pp. 1-8, Nov. 2009.

[11] Ahmed G. Abo-Khalil, Y. S. Kim, and D. C. Lee, "Maximum Output Power Control of Wind Generation System Using Fuzzy Control," Korean Institute of Electrical Engineering Journal, vol. 54B, No. 10, Oct. 2005.

[12] Ahmed G. Abo-Khalil, "Gradient Approximation Based Maximum Power Point Tracking for PV Grid Connected System", *IEEE Power Electronics and Drive Systems Conference* PEDS, pp. 104-108, Dec. 2011.

16

Volume models for single trees in tropical rainforests in Tanzania

Abel Malyango Masota[1], Eliakimu Zahabu[2], Rogers Ernest Malimbwi[2], Ole Martin Bollandsås[3], Tron Haakon Eid[3]

[1]Tanzania Forest Services (TFS), P.O. Box 40832, Dar es Salaam, Tanzania
[2]Sokoine University of Agriculture, Department of Forest Mensuration and Management, P.O. Box 3013, Morogoro, Tanzania
[3]Norwegian University of Life Sciences, Department of Ecology and Natural Resource Management, P.O. Box 5003, 1432 Ås, Norway

Email address:
abelmasota@yahoo.com (A. M. Masota)

Abstract: The present study was the first to develop total tree, stem and branches volume models for rainforests in south-eastern Africa based on destructive sampling. The number of sample trees was 60 and diameter at breast height (*dbh*) and total tree height (*h*) ranged from 6 to 117 cm and from 6.4 m to 50 m, respectively. Large parts of the total volume and stem volume variations were explained by the models (Pseudo-R^2 ranged from 0.85 to 0.93) and they performed relatively well over different size classes. When considering the challenges in height measurements in rainforests, we in general recommend applying model 3 with *dbh* only as independent variable. For large trees we recommend model 2 (*dbh* and *h* as independent variables) because of the moderating effect *h* has on volume predictions. If accurate stem volumes are needed for forestry licensing or for calculating compensation of timber loss, we also recommend model 2. As long as the allometry of the trees obviously is not different from that of our study site, the developed models may also be applied for rainforests elsewhere in Tanzania, but further testing of the models is also recommended.

Keywords: Total, Branches and Stem Volume, Form Factor, Destructive Sampling

1. Introduction

In Tanzania rainforests are estimated to cover about 2 million ha, occurring mainly in mosaics (URT, 2009). These forests are rich in terms of flora and fauna diversity and have high catchment values (Frontier Tanzania, 2001; Munishi and Shear, 2004; URT, 2009), thus, most of them are protected for soil, water and biodiversity conservation. However, because of growing human needs for forest products and services, tropical rainforests are under threat due to high annual deforestation rates (Mayaux et al., 2013), and measures towards sustainable forest management are important (Castañeda, 2011).

Sustainable forest management requires among others knowledge on the total volume of the growing forest stock. Usually volume is estimated as total volume per unit area, whereby models predicting total tree volume of individual trees are used. According to the substantial review of volume models made by Henry et al. (2011) on rainforests in sub-

Saharan Africa, almost all efforts have been put into developing species-specific models predicting stem (merchantable) volume. Many of these models have been developed in west-African countries for commercial timber species (Akindele and LeMay, 2006). The only general (multiple species) model predicting total tree volume referred to in the review was developed by Alder (1982) in Ghana. In addition, we know that Munishi and Shear (2004) and Adekunle (2007) have developed general models predicting total tree volume based on inventory data from rainforests in Tanzania and Nigeria, respectively.

The only general total tree volume model for tropical rainforest that we know from east-Africa was developed by Munishi and Shear (2004) based on data from Eastern Arc Mountains in Tanzania. In this study they regressed individual total tree volume against diameter at breast height. However, the observed individual tree volume (*v*) used in this

model was not based on destructive sampling, but computed as a product of basal area (g) and total tree height (h) adjusted for taper by the cone formula ($v = g \times h/3$). It is quite obvious that uncertainty in volume estimation is larger for models based on such computations compared to if destructive sampling is applied, and volume is determined by summarizing accurately measured sections of trees.

In the lack of better alternative total volume models, the general volume equation ($v = g \times h \times f$) that relates the product of g and h to a form factor (f), has frequently been used to estimate v. A form factor is the ratio of measured volume for a tree to a cylinder volume based on diameter at breast height (dbh) and h of the tree. Although f vary between trees, $f = 0.5$ has routinely been applied in Tanzania across many different forest types (Kashaigili et al., 2013; Zahabu, 2008), including rainforests (Mpanda et al., 2011). This value appears to be correct for plantation species characterized by straight stems and small crowns (Malimbwi et al., 1998) but it has never been checked or verified for natural forest types in Tanzania. It is quite obvious that the uncertainty of volume estimation also related to such practices is large. Variations in the relationship between dbh and h, and consequently in f, are related to numerous environmental factors such as soil nutrients, climate, disturbance regime, successional status and topographic position, but also to tree species and several genetic factors (Feldpausch et al., 2011; Mugasha et al., 2013).

Tree volume models are generally scant in Tanzania, although a few exist for plantation forest (Malimbwi and Philip, 1989) and for miombo woodland (Malimbwi et al., 1994; Chamshama et al., 2004). Tanzania has recently carried out her first national forest inventory through a systematic sample plot design (National Forest Resources Monitoring and Assessment (NAFORMA) (URT, 2010; Tomppo et al., 2014). This will require reliable total tree volume models specific for different forest types, including rainforests. The country also needs total tree and stem volume models that can be applied for forest management planning in general, in the forestry licensing systems (Abbot et al., 1997; MNRT, 2007), for allocation of forest areas to harvest and for calculation of compensation of timber loss due to damages during for example, road construction. Models quantifying volume of branches that potentially can be used for firewood may also provide important information for the management of the forest resources (Dadzie, 2013).

The main objective of the present study was therefore to develop models for prediction of tree volume in tropical rainforests in Tanzania based on destructive sampling. Models for total volume, stem volume and branch volume were developed. Since the general volume equation using $f = 0.5$ is frequently applied in Tanzania, we also assessed the appropriateness of such practices. In addition, we compared the performance of the developed models with previously developed volume models for tropical rainforests in sub-Saharan Africa and elsewhere.

2. Materials and Methods

2.1. Study Area and Selection of Sample Trees

Data collection for this study was carried out between October 2011 and September 2012, in Amani Nature Reserve (ANR), which is situated at 5°05'-5°14'S and 38°40'-38°32'E in Usambara Mountains. These mountains are parts of the Eastern Arc Mountains. ANR covers 8,380 ha of tropical rainforests with elevation ranges between 190 and 1130 m above sea level (Frontier Tanzania, 2001). The mean annual rainfall ranges from 1800 to 2200 mm and the mean annual temperature is about 20°C, with mean daily minimum and maximum temperatures of about 16 and 24°C, respectively. The forest is dominated by trees of genera *Afrosersalisia*, *Allanblackia*, *Celtis*, *Drypetes*, *Ficus*, *Isoberlinia*, *Leptonychia*, *Macaranga*, *Myrianthus*, *Newtonia*, *Parinari*, *Sorindeia*, *Strombosia*, *Syzygium*, and *Tabernaemontana* (Newmark, 2001). Historically, ANR has been under commercial logging activities for more than 100 years and they continued until the mid-1980's (Frontier Tanzania, 2001). Also, ANR is a biodiversity hotspot, with high water catchment values and environmental conservation. Inspite of its importance, Mpanda et al. (2011) reported that ANR was subjected to illegal activities including excessive collection of medicinal plants, pit sawing, encroachment, mining and collection of building poles.

Selection of trees for destructive sampling was guided by the species distribution and dbh range observed on 142 systematically distributed permanent sample plots (PSPs) established in 1999 (Frontier Tanzania 2001) and remeasured in 2008 and 2009 (Mpanda et al., 2011; Mgumia, 2014). Almost 6,000 trees comprising 240 different species, with dbh range between 10 and 270 cm, were recorded on these plots. Based on this information we determined which tree species were most frequent so that we could focus our sampling on those. A total of 60 trees from 34 different tree species were selected to uniformly cover as much as possible of the observed dbh distribution of the 142 PSPs. Sample trees were measured both for dbh and h while standing. A calliper or diameter tape was used to measure dbh, while h was measured by using a Vertex hypsometer. For leaning trees on slope the breast height point was determined from the upper side of the tree (Dietz and Kuyah, 2011; URT, 2010). For trees with buttresses extending beyond breast height, dbh was measured at 30 cm above the buttresses. The sample trees were the same as those used by Masota et al. (2014) for development of biomass models. The details of sample trees (tree species, dbh and h) are shown in Appendix I. Summary statistics of the trees are shown in Table 1.

Table 1. Summary statistics of dbh and h of sample trees

Variable	n	Mean	Min.	Max.	St. dev.
dbh (cm)	60	50.8	6.0	117.0	25.6
h (m)	60	27.3	6.4	50.0	10.4

2.2. Destructive Sampling and Data Processing

The selected sample trees were cut at heights of 30 cm from the ground level. Felled trees were crosscut into two main components, namely stem (from the stump to the point where the first large branch protrudes the stem) and branches. Diameter cut-off between branches and twigs was 2.5 cm, and no volume from twigs was included. Stems and branches were crosscut into sections with lengths generally ranging between 1-2 m, but down to 0.5 m for the stems of the largest trees. Thereafter all stem and branch sections were measured separately for mid-diameter (cm) and length.

Data processing and analysis was carried out with SAS 9.2 software (SAS® Institute Inc., 2004). Volume (v) of individual stem and branch sections were calculated by Huber's formula (Philip, 1994; Abbot et al., 1997) and summed to obtain tree volume. Statistics of tree components volume are presented in Table 2. On average, the proportion of stem volume to total volume was 63% with a range between 11% and 92%. The average branch volume was 37% with a range from 7% to 89%. Plots of branches, stem and total tree volume (m^3) versus tree dbh (cm) are shown in Figure 1.

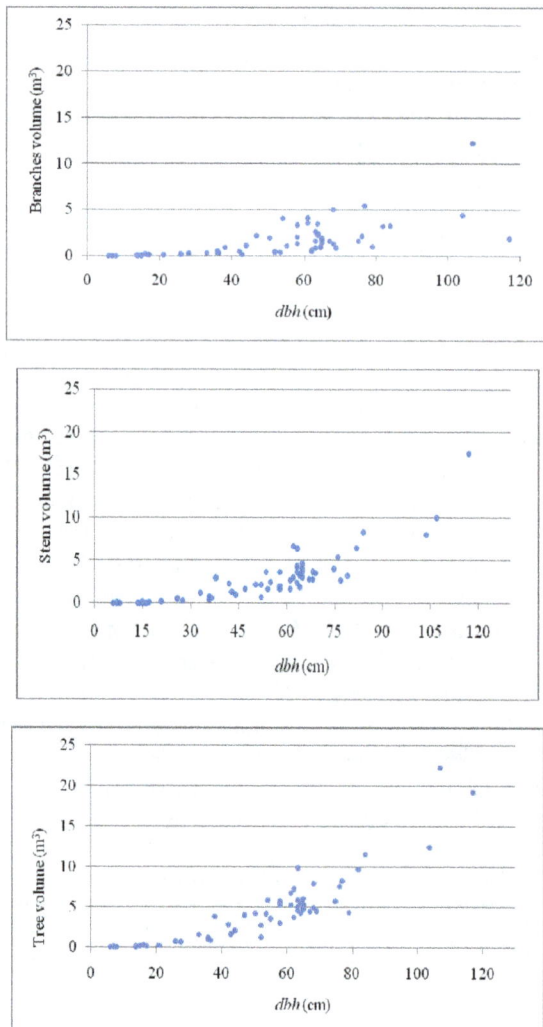

Figure 1. *Plots of branches, stem and total volume (m^3) versus tree dbh (cm)*

Table 2. *Statistics of tree components volume (m^3)*

Component	n	Mean	Min.	Max.	St. dev.
Stem	60	2.769	0.006	17.502	2.994
Branches	60	1.599	0.004	12.285	1.984
Total tree	60	4.367	0.017	22.372	4.358

2.3. Modelling of Volume

2.3.1. The Form Factor Approach

The shape of a tree can be approximated by a solid of revolution, or by a combination of several solids of revolution (neiloids, cones and paraboloids). Thus, if the basal area and the height are known, the volume of a tree can be estimated using a form factor ($g \times h \times f$). As mentioned earlier, it has been common to use a fixed f of 0.5, even though f varies with many factors, for example tree size. To assess the accuracy of this practice, we therefore used f=0.5 in prediction of total volume for the trees observed in our data, and then differences ($diff$) from the observed values were calculated (observed minus predicted). Furthermore, we also calculated the mean f for each of the volume components; total volume (f_{tm}), stem volume (f_{sm}), and branch volume (f_{brm}). These mean factors were also used in prediction of volume of their respective components, and $diff$-values were calculated. This approach is similar to using the fixed f, but it is calibrated for each component. However, since f is dependent on tree size, volume predictions using a mean f will be biased, at least for specific ranges of tree size. Thus, we also developed component specific models of f dependent on dbh (Model 1). As for the approach using a mean f, volumes were also in this approach predicted, but using predicted f dependent on dbh. Similar to the other approaches, $diff$-values were calculated.

$$f_x = a + b \times \ln(dbh) \qquad \text{(Model 1)}$$

where f= form factor, subscript x is either t(total volume), s (stem volume) or br (branch volume), a and b are parameters to be estimated. To represent v directly, model 1 can be incorporated in the general volume equation as in model 2.

$$v = g \times h \times (a + b \times \ln(dbh)) \qquad \text{(Model 2)}$$

2.3.2. Single Tree Volume Models

Although the general volume model relying on basal area, height and form factor, has some intuitive and nice properties, it is very common to fit empirical models dependent on h and/or dbh. Depending on allometry, different model forms may be appropriate. In the current study we initially considered a number of different model forms found in the literature. The alternative model forms were evaluated according to the root mean square error (RMSE). In addition, statistical significance and signs of the model parameter estimates were considered. We chose to present results for two (Models 3 and 4) nonlinear models adopted from Segura and Kanninen (2005) and Malimbwi et al. (1994). The models were fitted using the PROC NLIN procedure of SAS. A broad range of initial values for the model parameters were tested to ensure global convergence solutions.

$$v = \exp(a + b \times \ln(dbh)) \qquad \text{(Model 3)}$$

$$v = \exp(a + b \times \ln(dbh) + c \times \ln(h)) \qquad \text{(Model 4)}$$

where a, b and c are model parameters to be estimated. The model parameters are presented in Table 4.

2.3.3. Model Testing

The relative mean prediction error (MPE%)were computed from the calculated differences (*diff*) between the predicted and the observed volumes, both for the approaches based on the general volume equation and the empirical models. The MPE% was calculated both for all observations and for different size classes according to both *dbh* and *h*. Student *t*-tests were applied to test if the differences were significantly different from zero (Triola, 2012).

$$\text{MPE\%} = (\textstyle\sum (diff/n)/\text{MOV}) \times 100 \qquad (5)$$

where MOV is the mean observed volume (m^3).

The results from application of the general volume equation are displayed in Table 3, and the results from application of the empirical models can be found in Tables 5-7. For comparison, results from the use of the general volume equation with f dependent on *dbh* (Model 2) are shown in these tables as well.

2.3.4. Test of Previously Developed Volume Models

Similarly, we also applied a number of previously developed total tree volume models (Munishi and Shear, 2004; Adekunle, 2007; Adekunle et al., 2013) and then compared predicted volumes with observed volumes by means of MPE values that were tested for statistical significance by means of student *t*-tests. The results are shown in Table 8.

From Munishi and Shear (2004) we applied the following model:

$$v = 194.8803 \times dbh^{2.3982}.$$

The model was based on rainforest data from Usambara and Uluguru mountains in Tanzania. A total of 120 trees from 30 different tree species ranging from 13.5 cm to 195.0 cm in

dbh were included in their modelling data. The volume determination of the trees were not based on destructive sampling, but computed as a product of g and h adjusted for taper by the cone formula ($v = g \times h/3$). In their study, h was defined as the height from ground to 90% of the crown length.

From Adekunle (2007) we applied the following volume model:

$$v = 43.36(1 - e^{-0.067 \times g})^3,$$

where e is the base of the natural logarithm. The model was based on data from a rainforest in southwest Nigeria. A total of 421 trees from 61 different tree species ranging from 20 cm to 200 cm in *dbh* were included in their modelling data. The volume determination of the trees was not based on destructive sampling, but computed from Newton's formula (Husch et al., 2003) based on basal area of the trees at the base, at middle of the bole height and at the top of the bole height, and *h*. In their study *h* was defined as the height from ground to the top of the crown and bole height was defined as from ground to where the first large branch protrudes the stem.

From Adekunle et al. (2013) we applied the following model:

$$\ln(v) = 2.76 + 1.33 \times \ln(g),$$

The model was based on data from tropical moist forest in the state of Uttar Pradesh in northern India. A total of 535 trees from 25 different tree species, with *dbh* range from 10.2 to 63.5 cm were included in the modelling data. The volume determination and definitions of *h* and bole height was exactly the same as described above for Adekunle (2007).

3. Results

The f_{tm} was 0.59 and ranged between 0.15 and 1.15, f_{sm} was 0.36 and ranged between 0.06 and 1.0and f_{bm} was 0.23 and ranged between 0.03 and 0.70 (Figure 2). The model fits were relatively poor (low Pseudo-R^2 values), but all form factors were significantly decreasing degressively with tree *dbh*.

Table 3. Performance of the general volume equation over dbh and h classes by using different form factors

Class	n	Total volume (m^3)				Stem volume (m^3)			Branches volume (m^3)		
		Observed	$f=0.5$ MPE%	f_t MPE%	f_{tm} MPE%	Observed	f_s MPE%	f_{sm} MPE%	Observed	f_{br} MPE%	f_{brm} MPE%
dbh≤28	14	0.231	-31.2*	-3.9	-18.6*	0.136	-3.6	-16.5	0.094	-4.3	-21.3
dbh≤28	15	2.533	-14.2	-0.4	1.5	1.577	-2.7	-1.1	0.956	3.4	6.3
28<dbh≤55	15	5.435	-11.3	-4.2	5.0	3.375	-5.1	2.6	2.059	-2.7	9.5
55<dbh≤64.5	16	8.705	8.2	6.4	28.1*	5.620	2.9	20.3**	3.085	12.7	42.8
h≤20.4	14	0.282	-32.9*	-9.3	-20.6*	0.166	-7.9	-17.9	0.116	-11.2	-24.2
20.4<h≤27.6	16	3.215	-9.2	-0.8	7.5	1.943	0.7	7.8	1.271	-3.2	7.3
27.6<h≤34	15	5.239	-6.4	-0.6	10.8	3.176	1.2	10.9	2.064	-3.2	11.1
h>34	15	8.537	5.4	5.0	24.7*	5.672	-1.5	13.8	2.864	17.9	46.9
All	60	4.367	-1.6	-0.9	16.5	2.769	-1.5	3.7	1.598	6.1	25.7

* = $p<0.05$, ** = $p<0.01$, *** = $p<0.001$

Table 3 shows the MPE-values when applying the general volume equation ($v = g \times h \times f$) using $f=0.5$ and predicted f_t, f_s

and f_{br}, in addition to the observed f_{tm}, f_{sm} and f_{brm}. The results are distributed over different tree size classes (*dbh* and *h*) and for all trees, irrespective of size. For the f=0.5, total volume of small tree size classes was significantly underpredicted and magnitudes of underprediction decreased with increase of tree size classes. For the f_t, f_s and f_b, the overall MPE values increased from total to branches volume, but were not significantly different from zero. The table shows that application of the f_{tm} significantly underpredicted and

overpredicted total volume in small tree size and larger tree classes, respectively. For f_{sm} it significantly overpredicted stem volume in larger tree size classes. For the f_{bm}, there was no significant over- or underpredictions; however, the magnitudes of underprediction and overprediction were higher in larger tree size classes both in terms of dbh and h.

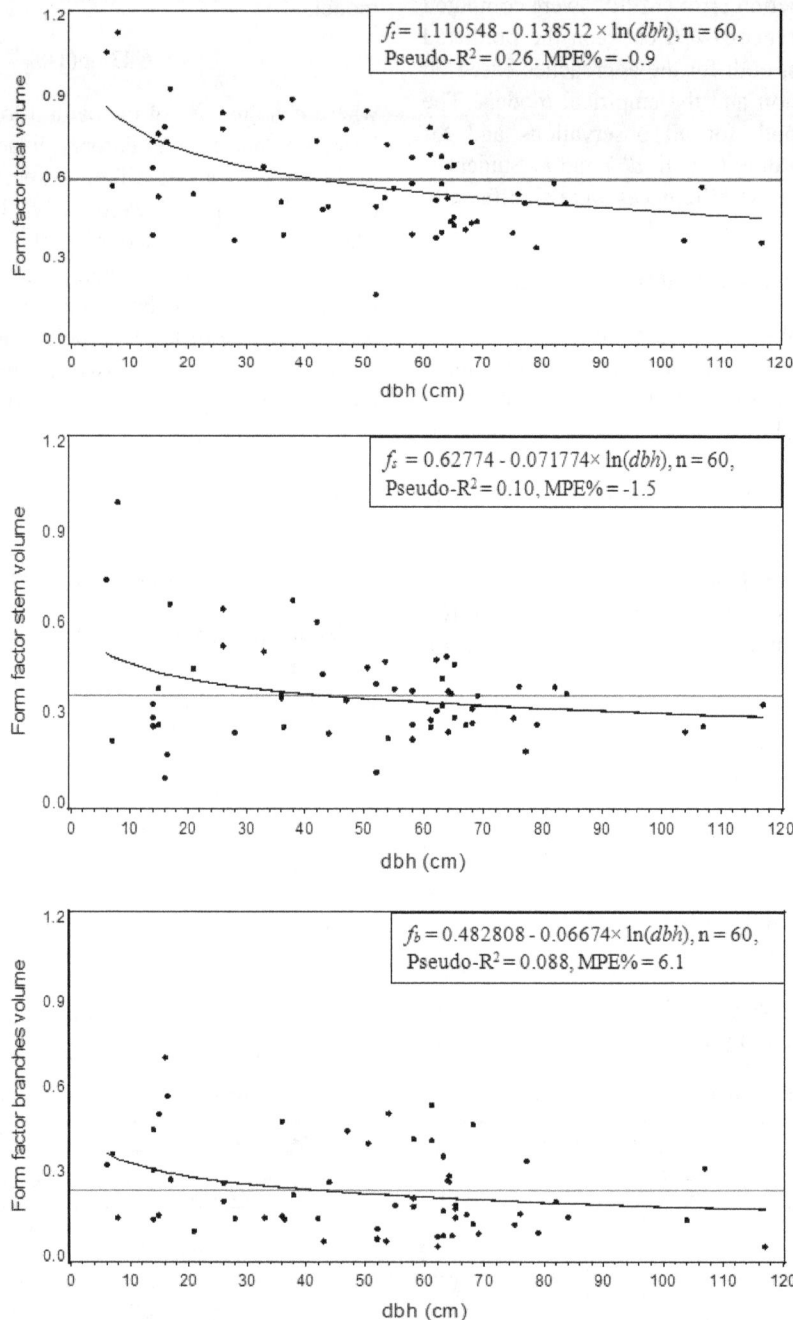

Figure 2. Plots of form factor for total, stem and branches volume over tree dbh (cm)

The fit of all alternative models are presented in Table 4. Based on RMSE, Pseudo-R^2, and the significance of parameter estimates, models 2, 3 and 4 were judged to be the best for total and stem volume, while models 2 and 3 were the best for branches volume. For branches volume, model 4 with both dbh and h as predictor variables, had insignificant

parameter estimate ($p<0.05$) for h. Generally when considering the tree components, the Pseudo-R^2 values decreased from total tree to branches.

The performances of best models for total, stem and branches volume are presented in Tables 5 to 7. The overall MPE values for the total volume models ranged between -

0.9% and 0.0%. In tree size classes, models 3 and 4 significantly overpredicted volume of small tree size classes (*dbh* ≤28 cm and *h*≤34) (Table 5). For the stem volume, the overall MPE values for the models ranged between -2.0% to -0.9%.No significant under- or overpredictions were observed over tree size classes (Table 6). For branches volume, the overall MPE values were 6.1 % and 1.8% for models 2 and 3, respectively. However, model 3 significantly overpredicted volume of branches in small tree size classes (Table 7).

Table 4. Parameter estimates for the different models for prediction of total (vt), stem (vs)and branches volume(vbr)

Dependentvariable	Model		RMSE (m3)	Pseudo-R2	MPE%
vt	2	g×h × (1.414741 -0.21174 ×ln(dbh))	1.343	0.91	-0.9
vt	3	exp(-7.41201 + 2.1901527×ln(dbh))	1.537	0.88	-0.3
vt	4	exp(-8.12477+ 1.653497×ln(dbh) + 0.852048×ln(h))	1.355	0.91	0.0
vs	2	g×h × (0.6251-0.07064×ln(dbh))	0.881	0.91	-1.5
vs	3	exp(-8.89962 + 2.428768×ln(dbh))	1.176	0.85	-2.0
vs	4	exp(-10.4281 + 1.434108×ln(dbh) + 1.63197×ln(h))	0.823	0.93	-0.9
vbr	2	g×h × (0.789641-0.14111×ln(dbh))	1.508	0.43	6.1
vbr	3	exp(-6.88089+ 1.831149*ln(dbh))	1.469	0.46	1.8
vbr	4	exp(-6.80703 + 1.894992×ln(dbh) -0.09859NS×ln(h))	1.481	0.46	1.7

NS: Insignificant parameter estimate

Table 5. Performance of best models (models 2, 3 and 4) for total volume over dbh and h classes

Class	n	Observedvolume (m³)	Models		
			2	3	4
			MPE%	MPE%	MPE%
dbh≤28	14	0.231	-3.9	24.1**	24.6**
28<*dbh*≤55	15	2.533	-0.4	-5.7	4.4
55<*dbh*≤ 64.5	15	5.435	-4.2	-8.3	-4.4
dbh>64.5	16	8.705	6.4	5.3	1.9
h≤20.4	14	0.282	-9.3	33.9*	16.3*
20.4<*h*≤27.6	16	3.215	-0.8	19.4	3.7
27.6<*h*≤34	15	5.239	-0.6	1.1	-0.4
h>34	15	8.537	5.0	-10.1	-0.5
All	60	4.367	-0.9	-0.3	0.0

* = *p*<0.05, ** = *p*<0.01, *** = *p*<0.001

Table 6. Performance of best models (models 2, 3 and 4) for stem volume over dbh and h classes

Class	n	Observed volume (m³)	Models		
			2	3	4
			MPE%	MPE%	MPE%
dbh≤28	14	0.136	-3.6	-4.1	-7.2
28<*dbh*≤55	15	1.577	-2.7	-14.8	2.0
55<*dbh*≤ 64.5	15	3.375	-5.1	-10.8	-3.5
dbh>64.5	16	5.620	2.9	6.5	-0.1
h≤20.4	14	0.166	-7.9	9.5	-19.5
20.4<*h*≤27.6	16	1.943	0.7	18.4	-12.0
27.6<*h*≤34	15	3.175	1.2	2.4	-2.9
h>34	15	5.672	-1.5	-12.2	4.8
All	60	2.769	-1.5	-2.0	-0.9

* = *p*<0.05, ** = *p*<0.01, *** = *p*<0.001

Table 7. Performance of best models (models 2 and 3) for branches volume over dbh and h classes

Class	n	Observed volume (m³)	Models	
			2	3
			MPE%	MPE%
dbh≤28	14	0.094	-4.3	80.6**
28<*dbh*≤55	15	0.956	3.4	8.2
55<*dbh*≤ 64.5	15	2.059	-2.7	-6.2
dbh>64.5	16	3.085	12.7	2.9
h≤20.4	14	0.117	-11.2	80.3**
20.4<*h*≤27.6	16	1.272	-3.2	18.9
27.6<*h*≤34	15	2.064	-3.2	-2.6
h>34	15	2.864	17.9	-6.0
All	60	1.599	6.1	1.8

* = *p*<0.05, ** = *p*<0.01, *** = *p*<0.001

The results of applying the previously developed volume models (Munishi and Shear, 2004; Adekunle, 2007; Adekunle et al., 2013) to our modelling dataset are presented in Table 8. Generally all models significantly underpredicted

total tree volume. The display of the models developed in the present study (models 3 and 4) and the model developed by Munishi and Shear (2004) over extrapolated *dbh* are shown in Figure 3.

Table 8. *Performance of total volume models developed by Munishi and Shear (2004), Adekunle (2007), and Adekunle et al. (2013) over dbh and h classes*

Class	n	Observed volume (m³)	Munishi& Shear 2004	Adekunle (2007)	Adekunle et al. (2013)
			MPE%	MPE%	MPE%
dbh≤28	14	0.231	-26.0*	-19.8	-45.7**
28<dbh≤55	15	2.533	-32.6*	-49.1**	-38.1**
55<dbh≤ 64.5	15	5.435	-30.3**	-53.9***	-30.9**
dbh>64.5	16	8.705	-14.3	-51.6***	-7.3
h≤20.4	14	0.282	-16.7	-16.6	-35.4**
20.4<h≤27.6	16	3.215	-9.8	-39.5***	-11.1***
27.6<h≤34	15	5.239	-22.0**	-15.7***	-21.2***
h>34	15	8.537	-27.2**	-58.5***	-21.5***
All	60	4.367	-22.1***	-51.6***	-19.5***

* = $p<0.05$, ** = $p<0.01$, *** = $p<0.001$

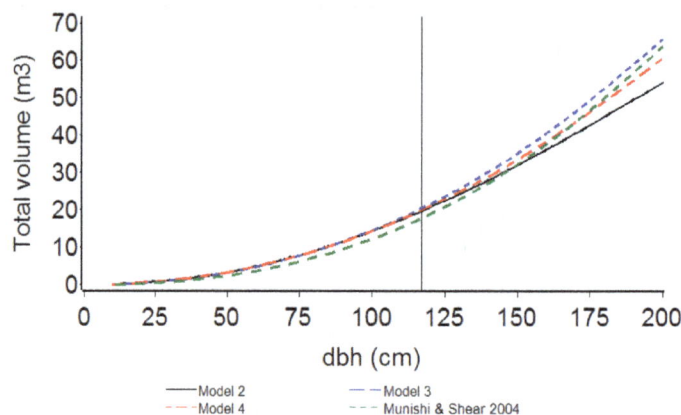

Figure 3. *Display of the total tree volume models developed in this study (Models 2, 3 and 4) and the model developed by Munishi and Shear (2004) over dbh. When applying models 2 and 4, h was predicted by means of diameter-height relationship models developed by Mugasha et al. (2013). Vertical line shows dbh range of the modelling data*

4. Discussion

The number of sample trees used for modelling in the present study (60) was relatively high compared to many studies on rainforests in sub-Saharan Africa (Henry et al., 2011). Some of the previously developed volume models were based on more sample trees than used in the present study, however the volume determination was in many cases not based on destructive sampling, but computed from different tree parameters (Munishi and Shear, 2004; Adekunle, 2007). The selection of sample trees in this study was based on a previous systematic sample plot inventory to secure species distribution to be representative and that the *dbh* range was covered. Although 34 different tree species is a relatively low number in relation to the over 200 species that are present in the study area (Frontier Tanzania, 2001), we were able to include the most frequently occurring species in addition to some of the rare species (see Appendix).

To our knowledge, no general model for total tree volume has previously been developed based on destructive sampling

for rainforests in Africa. When considering the high diversity of tree forms in rainforests (Feldpausch et al., 2011; Mugasha et al., 2013), it is quite obvious that the accuracy of observed volume will be more accurate with destructive sampling as compared to if volume is computed analytically from tree parameters measured on standing trees like Munishi and Shear (2004) and Adekunle (2007) did. In the data collection we also emphasized the accuracy in the measurements of the independent variables (*dbh* and *h*) by for example considering prevailing rules regarding leaning trees and buttresses (Dietz and Kuyah, 2011). Measuring tree height in closed-canopy forests is of course generally challenging (Larjavaara and Muller-Landau, 2013; Hunter et al., 2013), but at least we used a Vertex hypsometer in the measurements. This instrument is more flexible and probably provides more accurate height measurement than most other instruments.

As previously mentioned, the general volume equation with *f* = 0.5 routinely has been applied in Tanzania across many forest types (Zahabu, 2008; Mpanda et al., 2011; Kashaigili et al., 2013). Since the form factor depends on the diameter-height relationship of trees, it will not only vary between forest types but also between sites within the same forest type. The mean form factor based on total volume for our site was 0.59. Thus, by applying a form factor of 0.5 for the trees we would immediately produce a bias in total volume. However, also when applying the mean tree component specific form factor, biases were produced (Table 3). In addition to a general overprediction of total, stem and branches volume (MPE%=16.5%, 3.7 % and 25.7%), we also got a ~20% underprediction of volume for the smallest trees and a ~25%and 28% overprediction for the largest trees. Also for stem volume overprediction of 20.3% was found in *dbh* class of 55<*dbh*≤64.5 while for branches volume MPE values in tree size classes ranged between ~-24% to ~47%. Generally there will always be biases between the dependent (form factor) and the independent (*dbh*) variables if the relationship is non-linear (Gertner, 1991). By applying values predicted from the form factor model (Figure 2), the MPE values for total, stem and branches volume was reduced to -0.9%, 1.5% and 6.1%, respectively (Table 3). Continued practices of applying a form factor of 0.5 or a mean form factor derived from a particular sample of trees should therefore be avoided. However, if the form factor models developed in the present study are applied, the volume determination may be reasonably accurate as long as tree allometry of rainforests is not very different from that of our data.

Large parts of the variation in total volume were explained by models 2, 3 and 4 (Table 4). The overall MPE values for the models were less than 1% when tested on the modelling data, but significant differences between observed and predicted values (overprediction and underprediction) were seen for the smallest and largest trees for models 3 and 4 (Table 5). For model 2, no significant MPE% values were seen in any of the size classes. When comparing models 2 and 4, that both require *h* as input in addition to *dbh*, model 2 is therefore recommended. We initially tested several alternative model forms, in addition to model 3, to see if we could reduce MPE% for the smallest trees, but the model presented had the best performance. Since the overprediction of total volume for the smallest trees in absolute terms is small when applying model 3 (and also model 4), and since large trees usually account for a very large part of the total volume in rainforests, volume per unit area will probably in most cases be appropriate. In rainforests under early succession stages with many small trees there might be a slight overprediction of total volume when applying these models.

According to the review of volume models made for sub-Saharan Africa by Henry et al. (2011), both models with *dbh* only and models with *dbh* and *h* are commonly used. However, difficulties related to tree height measurements (e.g. Abbot et al., 1997; Segura and Kanninen, 2005; Larjavaara and Muller-Landau, 2013; Hunter et al., 2013) have made models with *dbh* only as independent variable the most commonly used. For the total volume models developed in the present study little was gained in terms of model fit when

including *h* as independent variable in addition to *dbh* (Table 4). Therefore, when considering the challenges in height measurements in rainforests, we in general recommend applying the model with *dbh* only as independent variable (model 3). However, for trees with very large *dbh*, it will probably be safer to apply the model 2 because *h* in such a model has a moderating effect of *dbh* on volume (see Figure 3). For such cases, recently developed general height-diameter models for different forest types in Tanzania including rainforests (Mugasha et al., 2013) could be applied if *h* is not available from the inventory.

For the stem volume models, the gain in model fit when including *h* (models 2 and 4) was larger than for the total volume models (Table 4). Since model 2 generally gave lower MPE% values over size classes than model 4, especially for the smallest size classes, we recommend this model. Although the use of the height-diameter models developed by Mugasha et al. (2013) will accumulate errors, inclusion of *h* when predicting stem volume will probably improve accuracy as compared to applying the model with *dbh* only (model 3). In the rare cases where all trees are measured for height, for example if accurate stem volumes are needed for forestry licensing systems or for calculating compensation of timber loss due to for example road construction, we clearly recommend to apply model 2.

The model fit of the branches volume models were generally poor. Pseudo-R^2 was much lower than compared to total and stem volume models (Table 4). In addition, for model 4 the parameter estimate of *h* was not significantly different from zero. The high variation in branches volume (see Table 2, Figure 1) is mainly due to a generally high diversity in branching patterns in tropical rainforest as a result of high species diversity, and large differences in light availability, succession stages and tree density (Sterck and Bongers, 2001). The high variation, however, is also affected by criterion used for separating stem and branches volume; stem comprised of volume from the stump to the point where the first large branch protruded the stem, while branches comprised the volume of remaining of the tree up to a branch diameter of 2.5 cm. The point where the first large branch protrudes the stem of course varies a lot between trees, and is thus affecting the variation in branches volume. Similar large variations in branches volume have also been observed by Dadzie (2013) in Ghana's tropical rainforests. Although the variation in branches volume was large in our modelling data, the overall MPE% of model 3 was near zero (Table 7), although it significantly overpredicted branches volumes for smaller trees. This model 3 may therefore still provide useful information on branches volume quantities in rainforests.

The application of the previously developed total tree volume models on our modelling data revealed significant underpredictions of volume (Table 8). We cannot rule out different site conditions and tree allometry as reasons for the underprediction when applying the model developed by Munishi and Shear (2004), but since their model was partly based on data from Usambara Mountains in Tanzania, where also our data were collected, this explanation is probably not the most important. The method applied to determine observed volume is a more likely explanation for the underprediction, especially since the cone formula, indicating a form factor of 0.33, which is much lower than the one observed for our data, was applied. It is therefore reason to believe that our models predict a more realistic volume than the previously developed model for Tanzania. The two other models (Adekunle, 2007; Adekunle et al., 2013) tested for their potential application in Tanzanian rainforests clearly also demonstrated their inappropriateness, irrespective of whether this was due to different site conditions or allometry, or to the analytical (non-destructive) method used to determine observed volume.

The developed volume models in this study are based on data from one rainforest site, and we know little about how well these data are representing rainforests elsewhere in Tanzania. However, most of the rainforests in the country, including Amani Nature Reserve where our data were collected, are parts of the Eastern Arc Mountains, and similarities regarding growth conditions and allometry are likely present. As long as the allometry of the trees obviously is not different from that of Amani Nature Reserve, we therefore believe that our volume models may be applied for rainforests outside this site. However, further testing of the developed models, if data from other rainforest sites in Tanzania becomes available, is recommended.

5. Conclusions

The present study was the first to develop total tree, stem

and branches volume models for rainforests in south-eastern Africa based on destructive sampling. The results showed that large parts of volume variation were explained by the models and that they performed relatively well when tested over different tree size classes. When considering the challenges in height measurements in rainforests, we in general recommend applying model 3 with *dbh* only as independent variable. For large trees we recommend model2because of the moderating effect *h* has on volume predictions. If accurate stem volumes are needed for forestry licensing or for calculating compensation of timber loss, we clearly recommend model 2. As long as the allometry of the trees obviously is not different from that of our study site, the developed models may also be applied for rainforests elsewhere in Tanzania, but further testing of the models is also recommended. Continued practices of applying the general volume equation with a form factor of 0.5 should be avoided.

Acknowledgements

Financial support for this work was provided by grants from the Climate Change Impacts, Adaptation and Mitigation (CCIAM) programme of Tanzania under the project "Development of biomass estimation models and carbon monitoring in selected vegetation types of Tanzania" at Sokoine University of Agriculture. We gratefully acknowledge the work of many people who were involved in data collection in difficult terrains. We also thank the Tanzania Forest Services (TFS) for permitting data collection in Amani Nature Reserve.

Appendix

List of trees used for development of volume models

Species	*dbh* (cm)	*h* (m)
Maesopsis eminii Engl.	63.6	41
Allanblackia stuhlmannii (Engl.) Engl.	21	15.6
Maesopsis eminii (Engl.)	33	30
Allanblackia stuhlmannii (Engl.) Engl.	36	24
Synsepalumcerasiferum(Welw.) T.D.Penn.	65	32
Cephalosphaerausambarensis(Warb.) Warb.	65	43.2
*Myrianthusholstii*Engl.	36	12.4
Synsepalumcerasiferum(Welw.) T.D.Penn.	28	27.5
*Myrianthusholstii*Engl.	26	18.5
Anisophylleaobtusifolia Engl. &Brehmer	36.5	24.8
*Zanhagolungensis*Hiern	55	26.4
Isoberlinia scheffleri (Harms)	67	31
Allanblackia stuhlmannii (Engl.) Engl.	42	27.9
Englerodendronusambarense(Harms)	26	16.3
*Funtumiaafricana (*Benth.) Stapf	14	11.1
Annickia kummeriae (Engl. & Diels) Setten & Maas	8	6.4
Maesopsis eminii Engl.	16	20.4
*Blighiaunijugata*Baker	14	10
Annickia kummeriae (Engl. & Diels) Setten & Maas	15	12.3
Isoberlinia scheffleri (Harms)	64.5	38.6
Quasia undulate (Guill. &Perr.) F.Dietr.	44	26.9
Alchorneahirtella(Benth.)	7	7.9

Species	dbh (cm)	h (m)
Macarangacapensis(Baill.) Benth. exSim.	6	8.4
Greenwayodendronsuaveolens (Engl. & Diels) Verdc.	14	15.8
Isoberlinia scheffleri (Harms)	15	17
*Parinariexcelsa*Sabine	53.5	35
Sorindeiamadagascariensis(Thou.) ex DC.	16.5	13.5
Cephalosphaerausambarensis (Warb.) Warb.	62	46.7
*Leptonychiausambarensis*K.Schum.	17	8
Lanneawelwitschii (Hiern) Engl.	65	24
*Croton sylvaticus*Hochst.	52	34.7
Newtoniabuchanani(Baker f.)	107	44
Cephalosphaerausambarensis(Warb.) Warb.	117	50
Sapiumellipticum(Hochst.) Pax	104	40
Anthocleista grandiflora (Gilg.)	63	36.8
Annickia kummeriae(Engl. & Diels) Setten & Maas	52	26.5
*Parinariexcelsa*Sabine	54	35.5
Newtoniabuchanani(Baker f.)	84	41.5
*Funtumiaafricana (*Benth.) Stapf	76	31
*Croton sylvaticus*Hochst.	64	25.6
Strombosia scheffleri Engl.	47	29
*Chrysophyllumperpulchrum*Mildbr.	75	33
Harungana madagascariensis Lam.ex.Poir.	61	23
*Polysciasfulva*Hiern	50.5	25
Xylopiaaethiopica (Dun.) A. Rich.	61	34
Quasia undulate (Guill. &Perr.) F.Dietr.	82	32
Annickia kummeriae (Engl. & Diels) Setten & Maas	62	33
Syzygiumguineense(Willd.) DC.	64	27
Isoberlinia scheffleri (Harms)	63	33
*Anisophylleaobtusifolia*Engl. &Brehmer	68	32
Allanblackia stuhlmannii (Engl.) Engl.	43	23
*Morindaasteroscepa*K.Schum.	79	26
Xylopiaaethiopica(Dun.) A. Rich.	58	38
Macarangacapensis(Baill.) Benth. exSim.	58	30
*Zanhagolungensis*Hiern	77	35
Antiaristoxicaria(Pers.) Lesch.	69	27
Parinari excels Sabine	58	30
Erythrophleumsuaveolens(Guill. &Perr.) Brenan	68	30
Maesopsis eminii Engl.	38	38
Synsepalummsolo(Engl.) Engl.	63	24

References

[1]   Abbot, P., Lowore, J, Werren, M. (1997). Models for the estimation of single tree volume in four Miombo woodland types. For. Ecol. Manage. 97: 25-37.

[2]   Adekunle, V.A.J. (2007). Nonlinear regression models for timber volume estimation in natural forest ecosystems, South Nigeria. Research Journal of Forestry 1:40-54.

[3]   Adekunle, V.A.J., Nair, K.N., Srivastava, A. K., Singh, N. K. (2013). Models and form factors for stand volume estimation in natural forest ecosystems: a case study of Katarniaghat Wildlife Sanctuary (KGWS), Bahraich District, India. Journal of Forestry Research 24: 217-226.

[4]   Akindele, S.O., LeMay, V. M. (2006). Development of tree volume equations for common timber species in the tropical rain forest area of Nigeria. For. Ecol. Manage. 226: 41-48.

[5]   Alder, D. (1982). Stock volume tables for indigenous trees on Subiri River Fr. UNDP/FAO/GHA/74/013, Takoradi, Ghana.

[6]   Castañeda, F. (2011). Current status and trends of forest management in tropical Africa. In: C.J. Geldenhuys, C. Ham, H. Ham (eds), Sustainable Forest Management in Africa: Some Solutions to Natural Forest Management Problems in Africa. Proceedings of the Sustainable Forest Management in Africa Symposium. Stellenbosch, 3-7 November 2008.

[7]   Chamshama, S. A.O., Mugasha, A.G., Zahabu, E. (2004). Biomass and volume estimation for miombo woodland at Kitulangalo, Morogoro, Tanzania. S. Afr. J. Sci. 200: 49-60.

[8]   Dadzie, P. K. (2013). Potential contribution of branchwood quantity, left after logging operations, towards reducing depletion rate and preserving Ghana's forest ecosystem. American Journal of Agriculture and Forestry 1: 32-39.

[9]   Dietz, J., Kuyah, S. (2011). Guidelines for establishing regional allometric equations for biomass estimation through destructive sampling. Carbon Benefits Project: Modelling, Measurement and Monitoring. World Agroforestry Centre ICRAF/GEF-UNEP, Nairobi, Kenya.

[10] Feldpausch, T.R., Banin, L., Phillips, O. L., Baker, T. R., Lewis, S.L., Quesada, C. A. et al. (2011). Height–diameter allometry of tropical forest trees. Biogeosciences 8: 1081-1106.

[11] Frontier Tanzania (2001). Doody, K.Z., Howell, K.M., Fanning, E. (eds), Amani Nature Reserve: A biodiversity survey. East Usambara Conservation Area Management Programme Technical Paper No. 52. Frontier Tanzania; Forestry and Beekeeping Division and Metsähallitus Consulting, Dar es Salaam, Tanzania and Vantaa, Finland. 117 pp.

[12] Gertner, G. (1991). Prediction bias and response surface curvature. For. Sci.37: 755-765.

[13] Henry, M., Picard, N., Trotta, C., Manlay, R. J., Valentini, R., Bernoux, M., Saint-André, L. (2011). Estimating tree biomass of sub-Saharan African forests: a review of available allometric equations. Silva Fennica 45: 477–569.

[14] Hunter, M.O., Keller, M., Vitoria, D., Morton, D.C. (2013). Tree height and tropical forest biomass estimation. Biogeosciences Discussions 10: 10491-10529.

[15] Husch, B., Beers, T.W., Kershaw, J.A. (2003). Forest Mensuration. 4th Ed. John Wiley and Sons, Inc. New Jersey, USA. 949 pp.

[16] Kashaigili , J.J., Mdemu, M.V., Nduganda, A.R., Mbilinyi, B.P. (2013). Integrated Assessment of Forest Cover Change and Above-Ground Carbon Stock in Pugu and Kazimzumbwi Forest Reserves, Tanzania. Advances in Remote Sensing 2: 1-9.

[17] Larjavaara, M., Muller-Landau, H.C. (2013). Measuring tree height: a quantitative comparison of two common field methods in a moist tropical forest. Methods in Ecology and Evolution 4: 793–801.

[18] Malimbwi, R.E., Philip, M.S. (1989). A compatible taper/volume estimation system for Pinus patula at Sao Hill Forest project, Southern Tanzania. For. Ecol. Manage.31: 109-115.

[19] Malimbwi, R.E., Solberg, B., Luoga, E. (1994). Estimation of biomass and volume in miombo woodland at Kitulanghalo Forest Reserve, Tanzania. Journal of Tropical Forest Science 7: 230-242.

[20] Malimbwi, R.E., Mugasha, A.G., Chamshama, S.A.O., Zahabu, E. (1998). Volume tables for Tectona grandis at Mtibwa and Longuza forest plantations, Tanzania. Forest Record 71: 1-23.

[21] Masota, A.M., Zahabu, E., Malimbwi, R., Bollandsås, O.M., Eid, T. (2014). Allometric models for estimating above- and belowground biomass of individual trees in Tanzanian tropical rainforests. Unpublished.

[22] Mayaux, P., Bartholome,´ E., Fritz, S,, Belward, A. (2004). A new land-cover map of Africa for the year 2000. Journal of Biogeography 31: 861–877.

[23] Mayaux, P., Pekel, J.F., Desclée, B., Donnay, F., Lupi, A., Achard, F., Clerici, M., Bodart, C., Brink, A., Nasi, R., Belward, A. (2013). State and evolution of African rainforests between 1990 and 2010. Phil Trans R Soc B., 368: 20120300, http://dx.doi.org/10.1098/rstb.2012.0300.

[24] Mgumia, F. (2014). Implications of forestland tenure reforms on forest conditions, forest governance and community livelihoods at Amani Nature Reserve, PhD Thesis, Sokoine University of Agriculture, Tanzania.280 pp. Unpublished.

[25] Ministry of Natural Resources and Tourism (MNRT) (2007). New royalty rates for forest products. Government Notice No. 231. Dar es Salaam, Tanzania. 12pp.

[26] Mpanda, M.M., Luoga, E. J., Kajembe, G. E., Eid, T. (2011). Impact of forestland tenure changes on forest cover, stocking and tree species diversity in Amani Nature Reserve, Tanzania. Forests, Trees and Livelihoods 20: 215–230.

[27] Mugasha, W.A., Bollandsås, O.M., Eid, T. (2013). Relationships between diameter and height of trees for natural tropical forest in Tanzania. Southern Forests 75: 221–237.

[28] Munishi, P.K.T., Shear, T.H. (2004). Carbon storage in afromontane rainforests of the Eastern Arc Mountains of Tanzania: their net contribution to atmospheric carbon. J. Trop. Forest Sci. 16: 78–93.

[29] Newmark, D.W. (2001). Tanzanian forest edge microclimatic gradients: Dynamic patterns Biotropica 33: 2-11.

[30] Philip, M.S. (1994). Measuring trees and forests. 2nd Edition. CAB International, Wallingford, UK. 310pp.

[31] SAS® Instute Inc. (2004). SAS Institute Inc., Cary, NC, USA.

[32] Segura, M., Kanninen, M. (2005). Allometric models for tree volume and total above ground biomass in a tropical humid forest in Costa Rica. Biotropica 37:2–8.

[33] Sterck, F.J, Bongers, F. (2001). Crown development in tropical rain forest trees: patterns with tree height and light availability. Journal of Ecology 89:1–13.

[34] Tomppo, E., Malimbwi, R., Katila, M., Mäkisara, K., Henttonen, H., Chamuya, N., Zahabu, E., Otieno, J.A. (2014). Sampling design for a large area forest inventory - case Tanzania. Canadian Journal of Forest Research, 10.1139/cjfr-2013-0490.

[35] Triola, M. F. (2012). Elementary Statistics. 12th Edition Pearson Publisher. 840 pp.

[36] United Republic of Tanzania (URT) (2009). Fourth national report on implementation of Convention on biological diversity (CBD). Division of Environment, Vice Presisent's Office, Dar es Salaam, Tanzania. 81 pp.

[37] United Republic of Tanzania (URT) (2010). National forest resources monitoring and assessment of Tanzania (NAFORMA). Field manual. Biophysical survey. NAFORMA document M01–2010, 108 pp.

[38] Zahabu, E. (2008). Sinks and Sources: A strategy to involve forest community in Tanzania in global climate policy. PhD-Thesis, University of Twente, Enschede, 235 pp.

17

Effect of water depth and temperature on the productivity of a double slope solar still

T. A. Babalola, A. O. Boyo, R. O. Kesinro

Department of Physics, Lagos State University, Ojo, Lagos State, Nigeria

Email address:

olakesinro02@gmail.com (R. O. Kesinro), nikeboyo@yahoo.com (A. O. Boyo), babalolataoheed@yahoo.com (T. A. Babalola)

Abstract: Drinkable water is a basic necessity for humanity, and the increase in human population growth has led to water pollution to the surface and underground water reservoirs. In order to meet the increasing demand for potable water, researchers have developed various technologies to meet this target. Solar distillation is a technology suitable for producing distilled water from brackish water. This is achieved by the use of a solar still. A solar still is a simple solar device used in converting salt/brackish water into potable water. In this research, the productivity of water by a double slope solar still was determined by varying the water depth and surrounding temperature for nine days in the premises of Lagos State University, Ojo, Nigeria at 6.5°N, 3.35°E. In this research embarked upon, it was observed that at a depth of 2.0cm the maximum output of the solar still was obtained and a maximum efficiency of 25.3%.

Keywords: Brackish Water, Productivity, Solar Still, Distillation

1. Introduction

The need for safe, clean drinking water is increasing rapidly due to rapid economic growth and increasing pressure on quality and quantity of water resources especially in developing countries. Water is readily available but it is brackish, salty and not safe for drinking. Emerging desalination technologies using renewable energy have made it possible to gain drinkable water as it can be cost effective.

In order to increase people's accessibility to potable water, some United Nations organizations such as the World Bank, World Health Organizations (WHO), United Nations Development Programme (UNDP) and United Nations Cultural and Educational Fund (UNICEF) have assisted various water projects for drinking. In the past, it was the arid regions that had brackish water that were known to be experiencing water shortage. However, the shortage has become phenomenal. To overcome this problem, there are various methods to produce fresh water from sea water, saline water or brackish water.

A solar still is a device that produces clean, drinkable water from dirty water using energy from the sun. Solar still is widely used in solar desalination. It has relatively low productivity but competitive to the other desalination methods for production of water due to its relatively low cost, simplicity in design and operation.

Several solar still designs have been proposed and many of them have found significant applications throughout the world. Solar desalination systems have low operating and maintenance costs and require large installation areas and high initial investments. There are two different types of solar still, those are; active solar still and passive solar still. Figure 1 indicates active type solar still; which contains the mechanical components like pump, valve etc. Figure 2 shows the passive type solar still; which don't require any mechanical components. Among active and passive solar stills passive solar still get more attractive comparing with active solar still. Because passive type solar still don't have moving elements, so no need of power consumption and no wear and tear problems.

By making necessary modifications to improve rate of heat transfer we can fetch maximum output from solar desalination. Reference [1] evaluated the distillate yield for a double slope laboratory still under controlled conditions for basin water and collector temperature within typical operating range. Reference [2] explored the current desalination technologies and their respective energy demands in Gulf Cooperation Council (GCC) countries with different alternatives to reduce energy consumption and analyze the present and the future prospective of water production rates and trends as well as the corresponding energy consumptions. Reference [3] developed an equation to predict the daily productivity of a single-sloped solar still. The developed equation relates the dependent and independent variables which control the daily productivity. Reference [4] was study of solar still using nanofluids and they found that using nanofluids in a solar still can increase the productivity of solar still. The effect of adding carbon nanotubes to the water inside a single basin solar still efficiency increases by 50%. Also, [5] studied the effects of orientation and depth of water in the basin of the still on the productivity of a double slope solar still and compare the same with that of a single slope solar still. Reference [6] conducted a study on solar distillation system by fuzzy sets. The study reveals that wind speed, ambient temperature, solar intensity, sprinkler, coupled collector, solar concentration, water depth etc affect on yield of solar still. Referene [7] developed single slope solar still with reflecting mirrors fixed on interior sides was coupled with a flat plate collector. He found that the daily productivity increased (5310 ml), 36% more than normal still operation (2240 ml) due to coupling with solar collector. He also observed that increased in basin water depth decreases the productivity and still productivity was proportional to the solar radiation intensity. Reference [8] compared the effect of desert climatic conditions on performance of a simple solar still with a similar one coupled to a flat plate collector. They tested whole day under clear sky conditions with different depth levels (2.5 to 3.5 cm.) of brackish water. The still productivity in summer varied from 4.01 to 4.34 l/m2/d for simple basin and 8.02 to 8.07 l/m2/d for the coupled one.

In this research work a double slope solar still was fabricated and distillate output was collected at different depths and temperature under normal climatic conditions in Lagos State University, Nigeria.

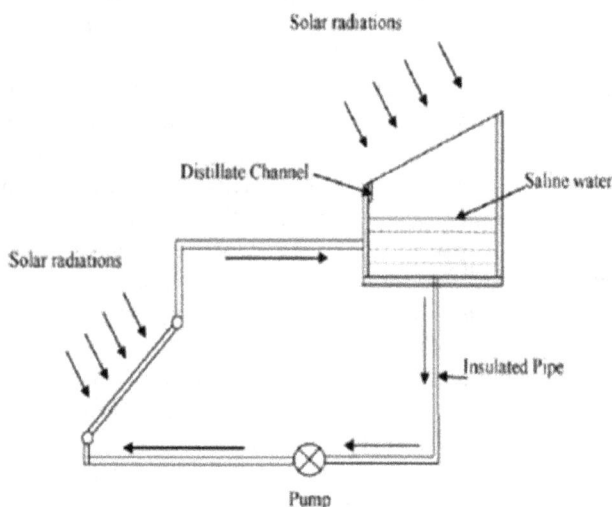

Figure 2. *Passive type solar still*

2. Methodology

The double slope basin-type solar still, constructed here, basically consists of an aluminum basin of dimensions 1.20m by 0.72m by 0.50m, placed in a wooden casing of dimensions 140cm by 74cm by 0.5cm; and roofed with tempered-glass cover plates of tilt angle 50° (to the horizontal).

The solar still rests on a robust plastic stand of approximate height of 45cm and top surface area roughly the same as that of the base of the casing. Fig.1 shows the diagram of the solar still. The level of the water in the basin was maintained by the use of a floater and the average mean ambient temperature for the day was obtained from the laboratory. The experimental setup for the research is as follows;

- The principal energy exchange mechanism in basin of solar still. A large part of solar radiation, direct and diffused, falling on still is absorbed in the blackened basin.
- The sensible heat absorbed by water is used to evaporate it and transferred to glass as vapour.
- The required output from the still is the condensed water on the bottom surfaces of the glass cover.
- The condensation is higher when condensing heat transfer from the basin of water is high.
- To maximize amount of water condensed.

A basin of solar still has a thin layer of water, a transparent glass cover that covers the basin and channel for collecting the distillate water from solar still. The glass transmits the sunrays through it and saline water in the basin or solar still is heated by solar radiation which passes through the glass cover and absorbed by the bottom of the solar still. In a solar still, the temperature difference between the water and glass cover is the driving force of the pure water yield. It influences the rate of evaporation from the surface of the water within the basin flowing towards condensing cover. Vapour flows upwards from the hot water and condense. This condensate water is collected through a channel.

Figure 1. *Active type solar still*

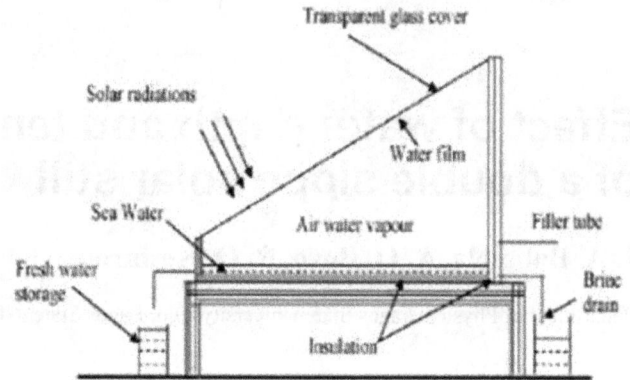

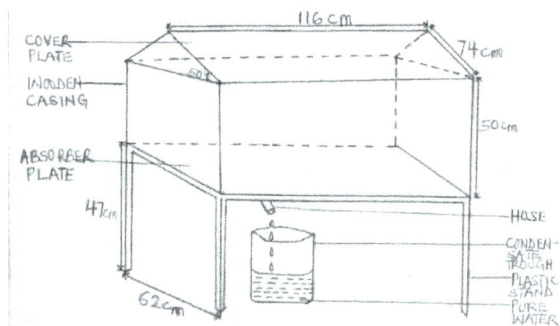

Fig. 3.2 – The constructed solar still

Figure 3. Schematics of the double slope still

Table 1. Technical specification of solar still

Specification	Dimensions
Aluminium basin	1.20m*0.72m*0.50m
Wooden casing	1.40m*0.74m*0.50m
Tilt angle of cover glass	50°
Width of cover glass	0.58m
Material of cover glass	Tempered glass

Figure 4. Double slope solar still

3. Result and Discussion

In this paper, the system is operated within 7 days with moderate and low sunshine of Ojo, Lagos state, Nigeria. The main target of this paper is to determine the effect of ambient temperature and depth of the brackish water on the output of the solar still. It was observed that maximum solar still output was recorded at the lowest water depth used (2.0cm). The graph of productivity against depth shown in graph 1 also shows that the output decreased non-linearly with increasing water depth, in concord with the findings of some earlier researchers like [9] and [10]. Consequently, the maximum instantaneous or daily efficiency η of the solar still is calculated from the data of distillate obtained at depth 2.0cm, using the equation

$$\eta = Q/H_T$$

Where Q = amount of solar energy utilized by the still and H_T = Solar insolation on horizontal surface.

Therefore, the maximum average daily efficiency of the

still is 25.3%. This is rather low and confirms the remark of [11] that solar still is not popular even in the remote arid areas because of its low productivity.

Also, the ambient temperature was also used to determine the productivity of the solar still and it was observed that the highest output was obtained at the maximum ambient temperature of 29.5°C as shown in graph 2. This fairly underscores the fact the solar still productivity also depends on temperature.

Table 2. ambient temperature, depth and productivity

Days	Average Temp. (°C)	Depth (cm)	Productivity (mL/m²day)	Max. Solar energy (J)
1	29.5	2	298	1002
2	28.0	3	268	844
3	28.0	4	255	872
4	27.0	5	242	733
5	26.0	6	236	884
6	27.5	7	228	953
7	27.5	8	225	888

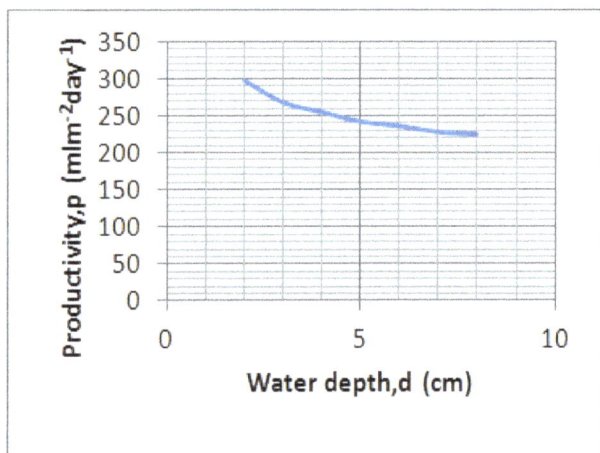

Graph 1. Graph of productivity against depth.

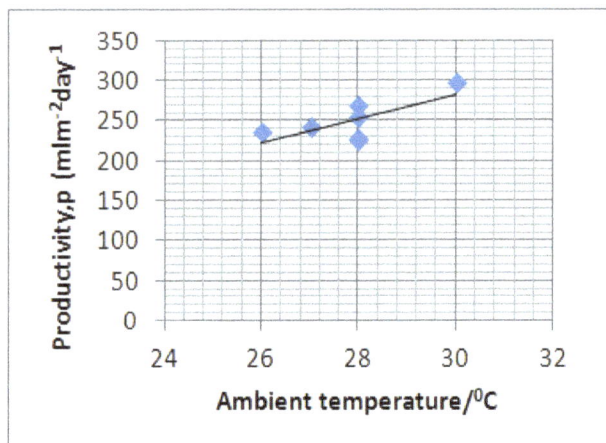

Graph 2. Graph of productivity against ambient temperature.

4. Conclusion

On the basis of this research the effect of water depth and ambient temperature on the productivity of the solar still was observed. It can be concluded from the paper that the increase in depth of the brackish water will result in low distillate output and increase in ambient temperature could result in increasing output. Also, more research has to be embarked upon to determine ways and materials to be used to increase the output of solar stills and make them more efficient.

References

[1] Rubio, E., M.A. Porta, J.L. Fernandez, Cavity geometry influence on mass flow rate for single and double slope solar stills, Applied Thermal Engineering 20 (2000) 1105–11, doi:10.1016/S1359-4311(99)00085-

[2] H. Fath, A. Sadik and T. Mezher, Present and Future Trend in the Production and Energy Consumption of Desalinated Water in GCC Countries, Int. J. of Thermal & Environmental Engineering Volume 5, No. 2 (2013) 155-165, doi: 10.5383/ijtee.05.02.00

[3] A.S. Nafey, M. Abdelkader, A. Abdelmotalip and A.A. Mabrouk, Solar still productivity enhancement, Energy conversion and management,42 (2001) 1401-08, doi:10.1016/S0196-8904(00)00107-2

[4] Gnanadason M K, Kumar P S, Rajakumar S and Yousuf M H S (2011), "Effect of nanofluids in a vacuum single basin solar still", IJAERS, 1, pp: 171-177, 2011.

[5] M.R. Rajamanickam, A. Ragupathy, Influence of Water Depth on Internal Heat and Mass Transfer in a double slope solar still, Energy procedia, 14 (2012) 1701-08, doi:10.1016/j.egypro.2011.12.887

[6] Mamlook R and Badran 0,(2007) Fuzzy sets implementation for the evaluation of factors affecting solar still production, Desalination, 203, pp. 394-402.

[7] Badran 0.0. and Al- Tahaineh H.A.,(2005) The effect of coupling a flat-plate collector on the solar still productivity, Desalination, 183, pp. 137-142.

[8] Boukar M. and Harmim A.,(2001) Effect of climatic conditions on the performance of a simple basin solar still: a comparative study, Desalination, 137,pp. 15-22.

[9] XAbdul Jabbar, N. Khalifa & Ahmad, M. Hamood (2009). On the Effect of Water Depth on the Performance of Basin-type Solar Stills. Solar Energy 83: 1312-1321.

[10] Garg, H.P., Mann, H.S., 1976. Effect of climatic, operational and design parameters on the year round performance of a single slop and double slop solar still under Indian arid zone conditions. Solar Energy 18, 159-164..

[11] Kalidasa, K., and Srithar, K. (2012). Performance Study on Basin-Type Double-Slope Solar Still With Different Wick Materials and Minimum Mass of Water. Renewable Energy. 36(2): 612-620

18

Enhancing Biomass Energy Efficiency in Rural Households of Ethiopia

Dagninet Amare, Asmamaw Endeblhatu, Awole Muhabaw

Bahir Dar Agricultural Mechanization and Food Science Research Centre, Bahir Dar, Ethiopia

Email address:

dagnnet@gmail.com (D. Amare), asmaende@yahoo.com (A. Endeblhatu), awolemuhaba@gmai.com (A. Muhabaw)

Abstract: The rural population of Ethiopia entirely depends on biomass for everyday energy needs except for light. The traditional system, particularly during cooking, incurs among others huge energy loss that could have been used otherwise. The system has been recognized as having significant effect on natural resource degradation, harmful health hazards and negative economic consequences. As a result, the government has been encouraging the use of energy saving technologies. Mirt and Gonze stoves are the two most dominantly promoted technologies. Promotion and efficiency evaluations were conducted. The result of the evaluation confirmed that households that use Mirt and Gonze stoves can save more than 33% and 20% of wood biomass that could have been used if traditional open stove was used, respectively. The time efficiency, length of time the stoves gave energy to bake additional Injera for Mirt and Gonze was increased by 63% and 50%, respectively. Thus, Mirt and stoves are efficient than the traditional open stoves. Due to durability, farmers preferred Mirt stove over Gonze. Utilization of Mirt stove can save 15% of wood biomass over Gonze. Promotion of Mirt stove in rural Ethiopia is vital to enhance biomass energy efficiency.

Keywords: Wood Biomass, Local Stove, Gonze, Mirt, Charcoal, Efficiency

1. Introduction

The use of wood as fuel source for heating and cooking is as old as civilization itself. Almost all African countries still rely on wood to meet basic energy need. Wood fuels account for 90-98% of residential energy consumption in most sub-Saharan Africa. Ethiopia consumed 0.566 million m³ of wood accounting for 9.1% of total African cooking and heating wood consumption[1]. Fuel wood accounts for around 78% of the total energy demand in Ethiopia [2].In general, average energy consumption of African households is significant. The average per capita firewood consumption in some African countries for families of 2-6 members was estimated at 1.14-1.36 tons. Families with seven and greater members consume on average 1.12 tons per capita with the annual to total consumption for an average family of 4.7 persons being 6.4 tons[3].

1.1. Local Energy Consumption and Sources

Biomass fuels (firewood, agricultural residues, animal wastes and charcoal) account for up to 90 percent of the energy supply of Ethiopia[1].The households of Zege, on average, had a total annual tree based wood harvest of 3.12 tons per annum where 1.31 tons per annum is extracted for fuel wood by families of 4.2 individuals. The average annual tree based wood consumed was 0.34 tons on adult equivalent and 0.32 tons on per capita bases [4]. In Dega and moist Woina Dega agro-ecological zones of Ethiopia, the annual per capita fuel wood consumption is estimated to be 609kg and 882kg respectively [5].

1.2. Source of Fuel Wood

In a study undertaken in Dera woreda, fuel wood was used as source energy for 87.3% of the households while in combination with animal dung for 12.2% of the households. Homestead eucalyptus plantation was the major (56.6%) source [6].Church forests also provide fuel wood for the population while trees in the farm land provide part of the demand. In other studies, approximately 48% of the households collect fuel wood from common areas [2].

1.3. Impact of Traditional Energy Utilization System

Energy utilization in the developing world is a major threat to the environment and health aspects occurring in the rural and poor urban households. Lack of clean and affordable energy is recognized as a significant barrier to development and major contributor to a host of environmental and human health problems [1].Reliance on traditional energy sources of biomass brought threat from overuse, creating additional environmental challenges ranging from local land use to global climate change and applications in smoky kitchens[1].If current fuel wood utilization trends continue, most developing countries are predicted to experience severe shortage of fuel wood by 2025. In sum, it leads to depletion of tree stocks or threat to biodiversity, desertification, reduced water quality, sedimentation, dust storms, air pollution and health problems such as respiratory illnesses and allergies [1] [2] [5][7].

1.4. Energy Utilization and Efficiency

In a scenario of climate change adaptation and biodiversity degradation, reducing the burden on forests is a truthfully critical intervention given the proportion of the population dependent on fuel wood extraction for household energy consumption. At the moment it is impossible to avoid the dependence on wood repay to the lack of alternative energy sources especially electricity. However, increasing the efficiency of the available resource or maximizing the energy available [8] will be a considerable contribution to the agenda of the UN's biodiversity conservation as most are produced from natural forests of indigenous species. It is also important for the country to catch up the rapid development through facilitation of the conservation and management of the natural resource base.

1.5. Defects of Traditional System

Given the available traditional energy utilization system, there is extravagancy in energy utilization [8]. High biomass energy consumption along with inefficient utilization has contributed for deforestation, biodiversity loss and land degradation[2]. In Ethiopia the common and dominant energy system is the open stove system. This system has been described as having several defects. For example much of the energy is lost without purposes owe to its openness and wind condition. Women are exposed for dual health problems. At first the smoke coming from the stove does not have a specific direction, it moves all ways resulting in open exposure of the women for the smoke heat. It is not uncommon to see significant population of the rural women with leaking eyes. Moreover the heat coming from the stove does heart the front leg of the women. It is also common to see darkened and dry front legs of women. The family is also in danger of the health effect as most are done inside or around the house where baby children are also victims of this technology. Thus, the traditional energy production system for baking and cooking is a basic economical and health issue problem at the household level.

1.6. Strategies Recommended

Household energy is one of the major problems to the deforestation and subsequent degradation rural areas. Hence, an appropriate energy supply and utilization should be part of the development strategy of a project intervention [1][5].Therefore, large-scale distribution of improved stoves will help to reduce pressure on the biomass resources, eases the conservation of forests, increase land productivity by reducing crop residue and dung usage for fuel wood, and helps to avoid rural health problems arising from smoke and heat during food preparation. The way out is provision of cheap and affordable fuel, afforestation and environmental consciousness through environmental education [1]. Further, policies that enhance integrated rural development and promote sustainable energy utilization in rural communities need to be put in place and implemented [9].Increasing efficiency of biomass energy utilization and reduction of wastage in Ethiopia is an important intervention [8][10].Hence, this research activity was designed to evaluate, select and promote the most viable energy saving technology to the rural population of the country.

2. Objectives

The objective of this research was to evaluate, select and promote the most efficient energy saving technology to the farmers

2.1. Materials and Methods

The demonstration and evaluation was conducted in Enqulal watershed of Dera woreda, Amhara region, North West Ethiopia. Overall 29 female farmers were selected randomly and trained. Out of which 3 female farmers were selected based on accessibility and willingness for evaluation of the stoves.

2.2. Technology Description

The Mirt and Gonze molds has totally eight different materials each, it was manufactured from 1.5mm sheet metal, round iron, flat iron and square pipe. It has two half circles, the external diameter of the circle has 325mm and the internal side of diameter 285mm which manufactures the sides of the stoves. The mold also has two materials the wood intern and the smock exit.

Mirt stove: A Mirt stove has groves that the components fit to each other. The average price of the Mirt stove in much of the markets in the region is 100 ETB. The components of Mirt stove energy saving stove mold are side mold, exit smock mold, wood intern mold and mold for dish.

Gonze stove: Gonze is made with mold but with no groves rather each closing another component. Thus the Gonze thou has a maximum diameter, to the size of the mold, does not have a minimum diameter. It can be reduced to suit the purpose or size of the stove.

Local stove: it is an open stove where three medium sized stones are used to put the "Mitad". It is open except the spaces occupied by the stones.

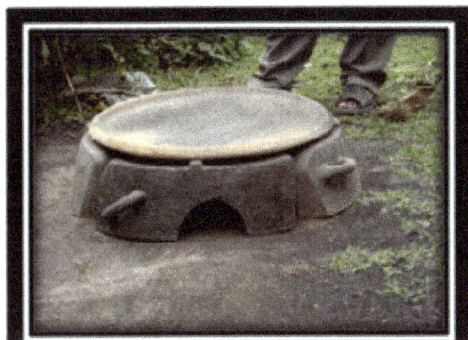

Fig.1. "Gonze" wood saving stove.

Fig. 2. Mirt" wood saving stove.

2.3. Methodology of Training and Evaluation

2.3.1. Initial Survey

Assessment on the status of kitchens and the availability of local materials and other improved fuel saving technologies will be carried out.

2.3.2. Selection of Farmers

As to the plan a site was selected in Dera woreda. The area selected is Enqulal watershed whereby 29 female farmers from both male and female headed households from 7 kebeles were selected. Women were selected from Enqulal watershed comprising 7 kebeles (29) with the help of the project technicians. The women selected were from both men (90%) and women headed households (10%).

2.3.3. The Experimental Women's Kitchens

The women selected for this evaluation were married and 40 years of age on average. They are experienced bakers of Injera spanning for more than half of their age. They were selected due to their houses accessibility (nearest to the road) and willingness to undertake the evaluation by sacrificing their time. The procedure provided to the evaluation women were installing in their kitchens, preparation of yeast according to the tradition and acquaintance with the new stoves for two weeks.

2.3.4. Test Condition

Given the requirements of the test the team decided to take major energy using stages like the first firing and the last heat that still helps to bake. Thus, it was reached that a sample for each of the three stoves on three women for Injera baking,

the most energy consuming activity, was considered vital[2][5].Besides, the least energy consuming activities like coffee and stew "wot" preparation were not considered separately. This is due to the priority that much of the energy consumption and health problems occur in Injera baking, which is the sole dominant bread for the Northern Ethiopian population.

2.4. Data Collection and Analysis

2.4.1. Data Collection on Measurements and Time

During data collection, three replications for each stove model were conducted. Each stove was given a chance of to bake 10 Injera but time was recorded when there is also additional heating value. Consequently, with the acquisition of Mirt and Gonze stove, evaluations were done one after the other in two cases and two parallels in one case. The weight of wood before baking at first for each stove, the size of wood left and amount of charcoal obtained after the time where there was not possible to bake any more Injera was measured using a spring balance of size 50kgs and the length of time of energy available for baking additional Injera. Farmers feedback was collected both during the training and evaluation sites.

2.4.2. Data Analysis

Data was analyzed using simple descriptive statistics on excel 2007 and Stata 11 software.

3. Result and Discussion

3.1. Household Energy Source Condition in the Area

The area belongs to the highlands of Ethiopia, and from observation it is highly degraded. Thou mixed farming takes place, due to the feed problem, the amount of dung extracted from animals is barely enough. The major energy source in the area is eucalyptus tree. As a result, the women participants expressed the presence of acute shortage of fuel wood for cooking. The energy source for lightening in most households, is gasoline thou there is extensive expansion rural electrification all over the country. Besides, it is optimistic [5] to assume that the rural community will use electricity for cooking activities given the supply of kitchen appliances along with a price level deemed and the knowledge of utilization that is also low in the urban areas.

3.2. Energy Shortage

The vast majority of rural people who are also dependent on traditional fuels use primitive and inefficient technologies [1].As a result, 25.4% of the respondents indicated that energy shortage is a problem in the area. However, only 4% of the respondents use energy saving technology (fuel stoves made from mud) as a means to compensate for energy shortage. Focus group discussion participants, in Dera woreda, demanded delivery of cement made energy saving stoves. Locally made energy saving stoves, mud, has been used by some households but they break up within days resulting in reluctance to use them.

3.3. Training of the Women Farmers

3.3.1. Theoretical

House hold and husbanded females were trained on the concepts of fuel wood crises and its effect on ecology, by displaying other region or country's experience supporting with pictures photos and movies using laptop and LCD using the training manual prepared.

3.3.2. Practical Production of the Stoves

Two models of Mirt and Gonze wood saving stoves were used for training. During the practical training, the team produced six Mirt and four Gonze stoves. Ultimately, the

stoves were given to farmers training center for display including one model made of local material. For the purpose of encouragement some 33 "Mitad"s were bought from Addis Zemen and given to them for further works.

3.3.3. Stove Energy Efficiency Evaluation

The type of energy used in this locality is biomass energy resources except for lightening at night which is either kerosene or electricity for some households. They are entirely dependent on wood biomass. The energy biomass from dung is insignificant due to herd and feed reduction. The following table shows the evaluation condition at the beginning of the experimentation.

Table 1. Fuel efficiency of the stoves.

Variable	Observation	Mean	Std. Dev.	Min	Max
Total fuel wood available at first (kg)	9	8.49	0.75	7	9.2
Amount of fuel wood burnt (kg)	9	3.76	1	2.4	5
Amount of time burnt (minutes)	9	59.11	15.01	39	80
Amount of fuel wood saved after burning (kg)	9	4.88	1.09	3	6.6
Size of charcoal obtained after the burn (kg)	9	0.10	0.09	0	0.24

3.3.4. Amount Wood Consumed

The following graph shows the performance of the stoves under three sample households (women). The weight of

wood burnt by three stoves to bake ten Injera (Ethiopian tradition flat breads) was evaluated at three different kitchen cells of by three women.

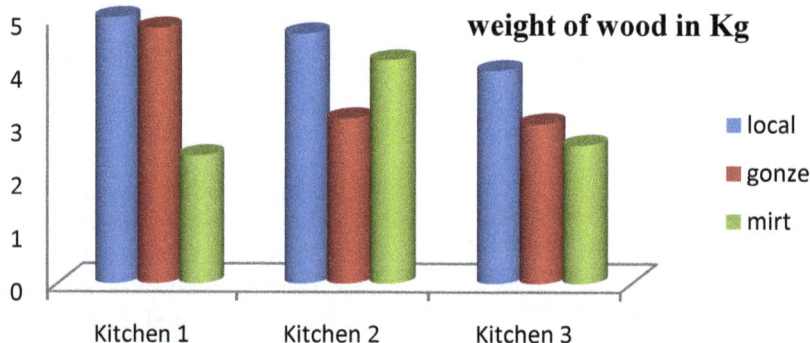

Fig. 3. Weight of wood burnt by the respective stoves in each woman's kitchens.

The weight of wood burned in the first woman's kitchen, the amount of wood burned by Gonze and local stoves looks more of equal while the weight was more than half by Mirt

stove. In the second woman's condition the amount of wood burnt using local stove and Mirt was comparably similar while Gonze was the smallest.

Table 2. LSD of wood burnt in kg.

(I) type of stove	(J) type of stove	Mean Difference (I-J)	Std. Error	Sig.	95% Confidence Interval	
					Lower Bound	Upper Bound
Traditional	Gonze	0.933	0.709	0.236	-0.801	2.667
	Mirt	1.50	0.709	0.079	-0.234	3.234
Gonze	Traditional	-0.933	0.709	0.236	-2.667	0.801
	Mirt	0.567	0.709	0.454	-1.167	2.301
Mirt	Traditional	-1.50000	.70868	.079	-3.2341	.2341
	Gonze	-.56667	.70868	.454	-2.3007	1.1674

In the third woman's kitchen the weight of wood burnt progressively increased from Mirt to Gonze and then local. In general, the amount of wood burnt to bake 10 Injera was averagely low for Mirt and the highest for local stove. One way ANOVA showed the absence of statistical difference (F=2.28, p= 0.1829) on the weight of wood consumed both by the three stoves and the women's kitchens. In general, Mirt and Gonze consumed 67.15% and 79.56% of the wood consumed by traditional stove. Mirt burned only 84.40% of the amount of wood consumed by Gonze stove.

3.3.5. Length of Time

The length of time refers to the number of minutes the three stoves under the three experimental women have performed. Gonze stove was providing heat for baking Injera for longer periods in the second woman's kitchen.

More importantly, the length of time Mirt stove provided heat to bake additional Injera after the 10th Injera was baked and the wood is avoided was almost constantly similar. Similarly, the length of time the local stove provided heat to

bake additional Injera was comparatively the lowest in all three conditions. This condition does not considered the amount of heat the stove provided parallel to stew making and water boiling in adjacency.

Table 3. ANOVA for burning time.

Source	SS	df	MS	F	Prob > F
Between groups	1216.22	2	608.11	6.22	0.0345
Within groups	586.67	6	97.78		
Total	1802.89	8	225.36		

Bartlett's test for equal variances: chi2(2) =2.0244 Prob>chi2 = 0.363.

ANOVA showed the presence of significant difference in length of time the wood burned among the three stoves. The length of time the stoves gave energy sufficient to bake additional 'Injera' was 1.63 and 1.5 times the length of traditional stove burned, for Mirt and Gonze respectively. Mirt gave a burning time of 0.08 times longer than Gonze.

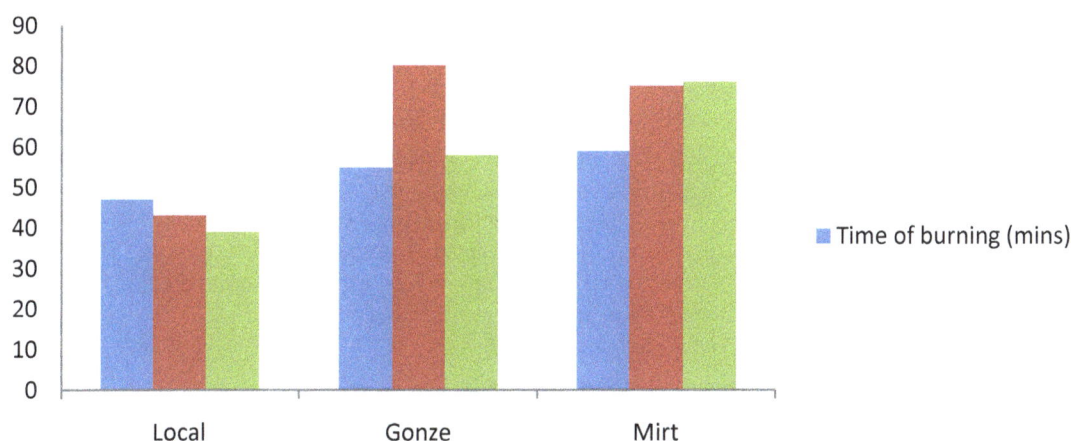

Fig. 4. Length of time of burning after baking of the 10th Injera.

3.3.6. Amount of Charcoal Extracted

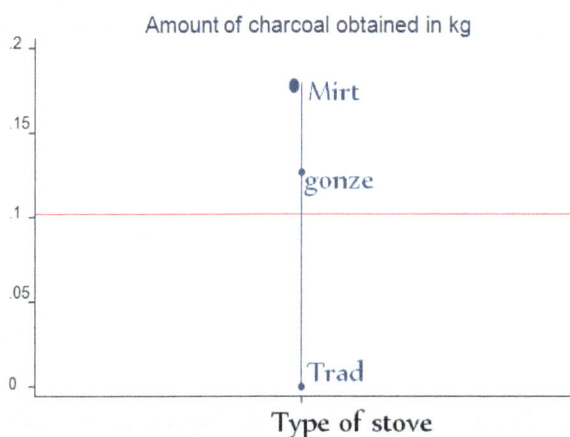

Fig. 5. Amount of charcoal extracted.

The amount of charcoal extracted after termination of the evaluation was measured. The local stove for Injera baking has no charcoal extract. The weight of charcoal extracted

from Mirt stove was averagely higher than that of Gonze. The amount of charcoal extracted from burned wood of traditional stove was nil due to the use of small size sticks that are highly blazing. This is due to the inability to continuously flap the heat to burn that the wood should be mostly sliced thinly. This incurs damage to what would have been saved as a by-product (charcoal) to be used for stew and coffee making. Thou, the amount is zero here, it is extracted in most cases however very grainy in observation from experience.

Table 4. ANOVA for charcoal extracted.

Source	SS	df	MS	F	Prob > F
Between groups	0.051	2	0.026	7.99	0.0204
Within groups	0.019	6	0.0032		
Total	0.071	8	0.009		

Bartlett's test for equal variances: chi2(1) =0.0102 Prob>chi2 = 0.920

One way ANOVA showed the presence of significant difference (F=7.99, p=0.0204) in amount of charcoal extracted from the three stoves among the three stoves.

3.3.7. Overall Evaluation

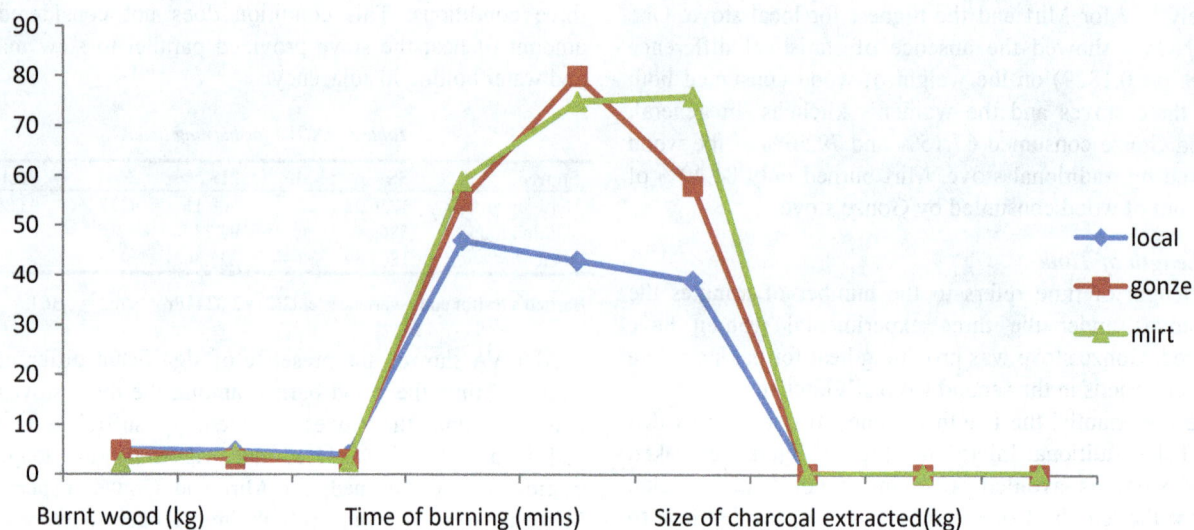

Fig. 6. Overall efficiency.

According to Dagninet *et al.* (2012)[1], the annual wood consumption for fuel will be 0.22 tons and 0.26 tons per capita bases. On adult equivalent bases the annual wood consumed will be 0.23 tons and 0.27 tons respectively for Mirt and Gonze stoves. On the other hand annual consumption efficiency can be calculated by considering the area as a moist Woina Dega livelihood zone [11]. Thus, annual per capita fuel wood consumption will roughly be 596kg and 703kg for Mirt and Gonze stoves respectively. It helps to save 290kg and 180kg of fuel wood annually on per capita bases. In general Mirt stove showed a comparatively better performance in terms of burning time and charcoal extracted relative to Gonze and traditional stoves. Gonze has a comparatively energy efficiency and wood reduction advantage compared to local stove. The Mirt stove has a better energy efficiency compared to both traditional and Gonze stoves.

Table 5. LSD of length of time wood burnt in minutes.

| (I) type of stove | (J) type of stove | Mean Difference (I-J) | Std. Error | Sig. | 95% Confidence Interval | |
					Lower Bound	Upper Bound
Traditional	Gonze	-21.33*	8.07	0.038	-41.09	-1.58
	Mirt	-27.00*	8.07	0.016	-46.76	-7.24
Gonze	Traditional	21.33*	8.07	0.038	1.58	41.09
	Mirt	-5.67	8.07	0.509	-25.42	14.09
Mirt	Traditional	27.00*	8.07	0.016	7.24	46.76
	Gonze	5.67	8.07	0.509	-14.09	25.42

*Mean difference is significant at the 0.05 level.

3.4. Women's Feedback

The test women reflected during the trial on the benefits and shortages of each stove. They preferred the Mirt stove due to longevity of age and ability to contain the heat for longer periods. Dissatisfaction on Gonze stove rose mainly due to loss of energy and breakability. Compared to the local stove, Mirt requires longer heating period and more solid wood while it is possible with leaves and thin wood slots that Injera is baked in local stove. All of the women preferred Mirt stove due to the benefits of energy saving, avoidance of smokes and front leg burning. Alongside women, male partners showed a heightened interest for the Mirt stove. Overall, they ranked from first to last Mirt, Gonze and traditional according decreasing trend of preference.

3.5. Opportunities for Promotion

The most valuable opportunities for promotion and adoption of the technology are the ever increasing shortage of biomass energy resources, the consciousness of the male partners about the health problems faced by their wives, especially their eyes and legs and the willingness to pay the market price of the stove (100-150birr). The policy inclination towards stall feeding is forcing people to own fewer sizes of cattle than used to be. This reduces the amount of dung available for energy. Thus, the importance of energy saving technologies will be crucial that makes promotion easier. However, for this specific area, there was mismatch in diameter of "Mitad" available in the local market and the diameter of the stove. Hence there is a need to modify to the required size.

4. Conclusion

The stoves increase efficiency of available energy utilization. Mirt and Gonze stoves help to save 33% and 20% of the average annual energy consumption per household, respectively. Efficiency could also be improved since there was a slight difference between the "Mitad" and the Mirt stove. Hence, reduction of the diameter of Mirt stove for that particular locality to encourage more adoption is vital. The farmers explained that while Gonze is useful as it does not permanently occupy a space, there is a problem of loss of heat and hence uses more fuel wood. In the case of Mirt stove, they explained that it takes longer time and relatively high amount of wood at the beginning for heating but after heating it helps to bake more than that can be done with local stove while the avoidance of smoke is considered vital.In general, the supply of a mold or Mirt stove design that fits to the diameter of the local "Mitad"s (averagely 56cm from a sample of 28 "Mitad"s compared to 62cm of the Mirt mold availed) helps to improve rural energy efficiency in Ethiopia.

References

[1] David. J Idiata, Mitchell Ebiogbe, Henry Oriakhi, and Osazuwa. L Iyalekhue, 2013.Wood Fuel Usage and the Challenges on the Environment. International Journal of Engineering Sciences, 2(4).Pps: 110-114.

[2] Abebe Damte, Steven F. Koch, and Alemu Mekonnen, 2012. Coping with Fuelwood Scarcity: Household Responses in Rural EthiopiaEnvironment for Development, discussion paper series (EfD DP 12-01).

[3] Bembridge, T.J. and J.E. Tarlton, 1990. Wood fuel in Ciskei: A Head load Study. SouthAfrican Forestry Journal, 154: 88-95.

[4] Dagninet Amare, Assefa Seyoum and Fekadu Beyene, 2012. The Remnants Forest Patches of Zege Peninsula, Ethiopia: Livelihood Strategies, Institutional Arrangements and Forest Products Extraction. Lambert Academic Publishing AG & Co KG.

[5] Susanne Geissler, Dietmar Hagauer, Alexander Horst, Michael Krause, Peter Sutcliffe,2013. Biomass Energy Strategy Ethiopia.23rd December, 2013. Energy for development Pdf.

[6] Beyene Belay and Dagninet Amare, 2014. Integrated Forest Management Plan in Five Selected Church Forests In Dera Woreda: Implication For Ecological Restoration and Biodiversity Conservation. Forest Management Plan Report Submitted to NABU Bahir Dar Office. Bahir Dar, Ethiopia. April 2104, unpublished.

[7] Audu, E.B., 2103. Fuel wood consumption and desertification in Nigeria. International Journal of Science and Technology, 3(1). ISSN 2224-3577

[8] Melis Teka, 2006. Energy Policy of Ethiopia, Ministry of Mines and Energy. Geothermal Energy Conference, Addis Ababa, Ethiopia.24-29 November 2006.

[9] Masekoameng K.E., Simalenga T.E. and Saidi T,2005. Household energy needs and utilization patterns in the Giyani rural communities of Limpopo Province, South Africa Journal of Energy in Southern Africa.16 (3).

[10] Dessie Tarekegn Bantelay, Nigus Gabbiye. Design, Manufacturing and Performance Evaluation of House Hold Gasifier Stove: A Case Study of Ethiopia. American Journal of Energy Engineering. Vol. 2, No. 4, 2014, pp. 96-102. doi: 10.11648/j.ajee.20140204.12.

19

Optical Modeling of Thin-Film Amorphous Silicon Solar Cells Deposited on Nano-Textured Glass Substrates

Mohammad Ismail Hossain, Wayesh Qarony

American International University-Bangladesh, Kemal Ataturk Avenue Banani, Dhaka, Bangladesh

Email address:

wayesh@aiub.edu (W. Qarony), m.hossain@aiub.edu (M. I. Hossain)

Abstract: We have introduced an approach to establish a methodology for 3D optical simulation that allows analyzing optical losses in the individual layers of a thin-film solar cell structure. Using commercial Finite-Difference Time-Domain (FDTD) tool, where Maxwell's Curl equations were rigorously solved for optimizing such cells, a computer modeling has been performed. We have reported the ways to investigate efficient light-trapping schemes by using periodically textured transparent conductive oxide (TCO) in thin-film amorphous silicon solar cells. The optical effects in small area thin film silicon p-i-n solar cells deposited on glass substrates coated with aluminum doped zinc oxide (ZnO:Al) have been addressed. In order to enhance the efficiency, TCO surface morphology has been analyzed, where pyramidal and parabolic textured surfaces have been used. For these cells, the quantum efficiency, short-circuit current, total reflectance, and all absorption losses have been successfully computed and analyzed. The investigation was carried out based on our proposed model that exhibits maximum current density of 17.32 mA/cm^2 for the absorbing layer thickness of 300 nm.

Keywords: FDTD, Light-trapping, Amorphous Silicon Solar Cell, p-i-n Thin Film, Quantum efficiency, Short-Circuit Current, Textured Surface, ZnO: Al, TCO

1. Introduction

The most promising thin-film technology has tremendous potentiality in reducing fabrication cost, robustness as well as excellent ecological balance sheet [1], [2]. Among thin-film solar cells, hydrogenated based amorphous silicon (a-Si:H) solar cells hold the best growing prospects due to its higher absorption coefficient, large open-circuit voltage (V_{oc}), and less material requirements. But, better performances can only be achieved for a few hundreds of nanometers due to the lower carrier diffusion length. With the synthesis of amorphous silicon materials and the implementations of new design and fabrication technologies over time, the amorphous PV efficiency has led to recent and significant improvements. So far the best conversion efficiency can be seen around 10.86%, along with a 16.52 mA/cm^2 of short-circuit current density have been reported for a-Si:H solar cell in 2012 [3]. But, it is necessary to overcome the existing conversion efficiency limitations for the better light trapping or photon management, which reduces the reflection losses, enhances scattering inside the cell, and increases optical path length in thin film solar cells [4]. Different surface texturing techniques, deposited on transparent conductive oxides (TCO) are used for better light trapping in order to maximize the absorption in the thin absorber layer of amorphous thin film solar cells, which increases the short-circuit current of the cells as well. Simulations based researches are always treated as the best approaches to optimize cell's structures as well as to get best performance fitting parameters, since the complexities, expenditures, and processing time are very high in conventional experimental or manufacturing processes [5].

In this manuscript, the optics in thin-film amorphous silicon (a-Si:H) solar cells with a hydrogenated Silicon Oxide (SiO:H) p-layer was investigated based on optical simulations.

The investigation was carried out both for pyramidal as well as parabolic surface textured. In order to study the influences of the introduction of pyramidal and parabolic texturing on external quantum efficiency, short-circuit current, and approximated conversion efficiency, results of smooth substrates were used as a reference. All of the simulations were performed for the nano-textured glass substrate.

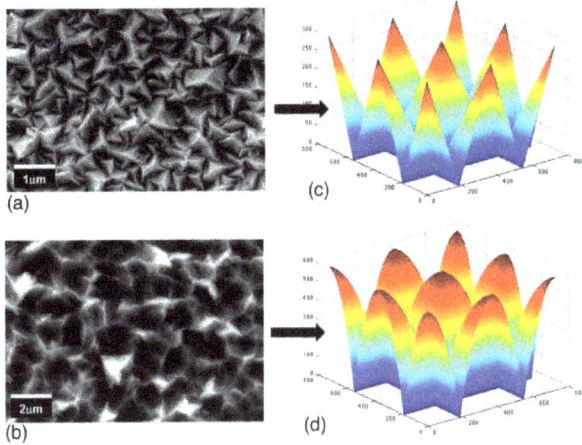

Fig. 1. *SEM images of two typical surface morphologies of (a) the standard front LPCVD ZnO optimized for amorphous cells, (b) the new optimized LPCVD ZnO after surface treatment, respectively [taken from 6]; and the corresponding (c) pyramidal and (d) parabolic shapes for the simulations.*

2. Transparent Conductive Oxide Surface

In order to have an efficient and transparent light trapping in thin-film silicon solar cells, a material with comparatively lower concentration of defects and band-gap above 3.0 eV is required. As the un-doped low-pressure chemical vapor deposition (LPCVD) zinc oxide (ZnO) material has a band-gap of 3.37 eV and the corresponding wavelength is approximately 370 nm, ZnO is mostly used as highly transparent conductive oxide (TCO) surface in the visible and near-infrared region [7]. As a consequence, the interface region of transparent conductive oxide absorbs photon energies above band-gap. It also helps to reduce free carrier absorption and increases light-trapping capability, especially for longer wavelengths [6]. Aluminum doped zinc oxide is successfully used in the optical simulations for the front and back contacts to get the efficient light trapping. Depending on the texture surface of TCO, thin-film solar cell efficiency is also realized. From the investigation of experimental surface roughness of TCO, it is possible to get an idea of surface profile dimension. Figure 1(a) and 1(b) show the SEM images of two typical surfaces of chemical vaporization decomposition (CVD) grown of zinc oxide [6]. In order to optimize the solar cell, surface analysis and optical simulations are performed simultaneously. In reality, the surface textures are randomly ordered which is basically the combination of pyramids and asperities can be seen from the Fig. 1(a). By doing the periodic texturing, investigation of optical properties in the solar cells can be achieved.

The correlation between surface roughness and the wave propagation of the optical simulations exhibits the pyramid like textures, where Fig. 1(a) and (c) show the real and simulated surface shapes of textures for this material. In addition, doing surface modification can enhance light trapping or solar cell optimization. In surface morphology treatment, the asperities and the tips of the pyramids are removed and the textured surfaces are turned into mostly parabolic shapes [6]. For the simulation purpose, parabolic

shape texturing environment has been created and implemented in our investigation. The experimental optimized pyramidal textured surface and the created parabolic shaped textures for the optical simulations are shown in Fig. 1(b) and 1(d), respectively.

3. Optical Simulation Model

A numeric method for computational electromagnetics, the Finite-Difference Time-Domain (FDTD) method has been applied for the optical modeling [8]. The complex refractive index is used to describe the optical properties of the different materials. The excitation source is a monochromatic, harmonic, planar wave with the wavelength that is irradiated perpendicular to the glass/TCO side. The input circular wave propagation or polarization is assumed for the simulation, which is the combination of the transverse electric (TE) and transverse magnetic (TM). At the lateral edges of the structure, periodical boundary conditions are assumed. The simulation domain is closed at top and bottom with a perfectly matched layer (PML). Throughout the simulation, p-i-n single junction thin-film amorphous silicon (a-Si:H) solar cell has been investigated. Schematic cross sections of a thin-film amorphous silicon solar cell with smooth substrate and textured interfaces are shown in Fig. 2(a), 3(a), and 4(a).

The amorphous silicon solar cell structure consists of a 420 nm of ZnO:Al front contact deposited on a thick (1000 nm) glass substrate. The zinc oxide layer is followed by a (p-i-n) silicon diode with a specific thickness of cell and a back contact consisting of a 100 nm of ZnO:Al layer and a perfect metal back reflector thickness of 1200 nm. The thicknesses of the p-layer and n-layer of the cell have been assumed as 10 nm. Depending on the structure, absorbing layer has been considered for the thickness of 300 nm and 500 nm. The dimensions of the textured structures are described by the period and height, and the base is assumed to be square shaped.

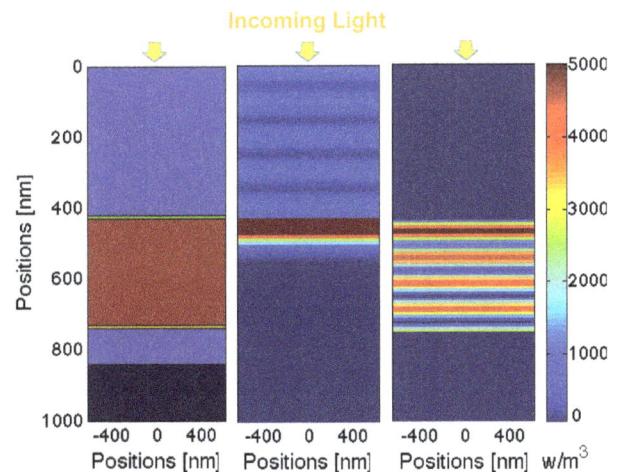

Fig. 2. *(a) Schematic of device structure; Simulated power loss profile under monochromatic illumination of wavelength (b) 400 nm and (c) 600 nm deposited on smooth glass substrate.*

The time average power loss within an amorphous silicon

solar cell on a smooth substrate is calculated by using equation (1) and shown in Fig. 2(b) and 2(c) for 400 nm and 600 nm wavelengths, respectively. For the shorter wavelengths (< 400 nm), photon energy is absorbed within the first few hundreds of nanometers of the solar cell. Due to the high absorption coefficient of amorphous silicon for the shorter wavelength energy, lights cannot reach to the back reflector. The longer wavelengths (> 600 nm) energies propagate through the entire silicon layer and very small amount of lights get reflected from the back. Contrastively, solar irradiance absorption in the solar cell decreases due to further increment of wavelength. Therefore, it is necessary to demonstrate a technique or idea that will enhance the light trapping for the longer wavelength energies as well as effective thickness. In order to increase the effective thickness, there are several techniques demonstrated in several literatures [5]-[6], [8]-[12].

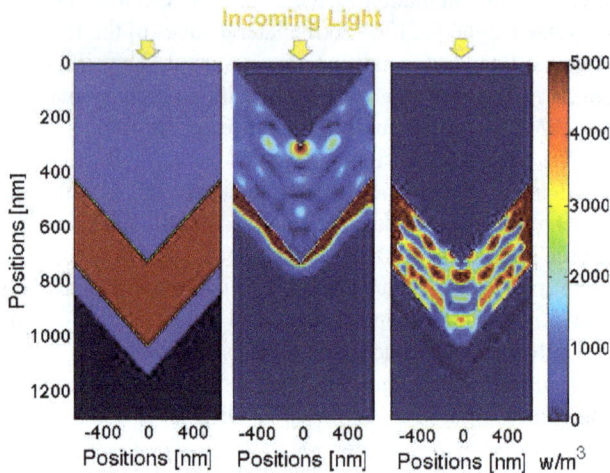

Fig. 3. *(a) Schematic of device structure; Simulated power loss profile under monochromatic illumination of wavelength (b) 400 nm and (c) 600 nm which deposited on pyramidal textured glass substrate.*

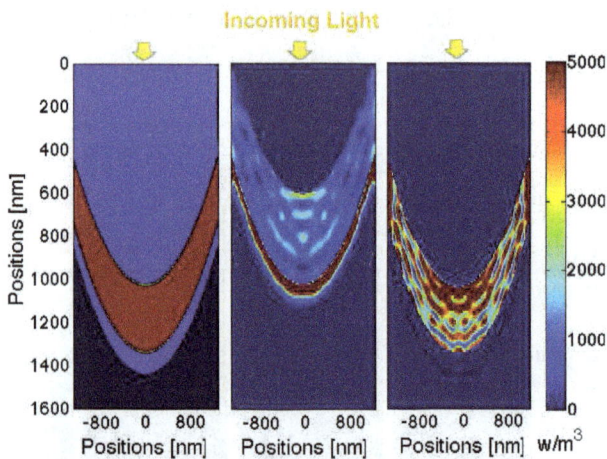

Fig. 4. *(a) Schematic of device structure; Simulated power loss profile under monochromatic illumination of wavelength (b) 400 nm and (c) 600 nm which deposited on parabolic textured glass substrate.*

By introducing the periodic nano-textured interfaces, scattering and diffraction of photons, and the light trapping can be improved. A higher short-circuit current and the efficiency indicators are achieved by textured surface than

smooth substrate, since the optical path-length of the cell is increased. The simulated power loss profiles for incident wavelength of 400 nm and 600 nm for the textured case are also shown in Figs. 3(b)–3(c) and 4(b)-4(c).

The power loss profile was determined for pyramidal texture with period of 1000 nm and height of 300 nm, which can be found from the surface analysis of zinc oxide in Fig. 1. In the case of parabolic texture, after surface treatment the period and height are 2000 nm and 600 nm, respectively. From the electric field distribution, power loss for the individual regions are calculated by the equation no. (1).

$$Q(x, y, z) = \frac{1}{2} c \varepsilon_0 n \alpha \left| E(x, y, z) \right|^2 \quad (1)$$

Where this term leads to find the quantum efficiency, which is expressed by equation no. (2).

$$QE = \frac{1}{P_{Opt}} \int Q(x, y, z) dx dy dz \quad (2)$$

Where P_{Opt} is optical power, c is speed of light in free space, ε_0 is permittivity of free space, α is the energy absorption coefficient which is related to extinction coefficient k ($\alpha = 4\pi k/\lambda$), with n being the real part of the complex refractive index, and E(x,y,z) is the distributed electric field.

After achieving quantum efficiency and power loss short-circuit current can be calculated by equation no. (3).

$$I_{SC} = \frac{q}{hc} \int\limits_{\lambda_{min}}^{\lambda_{max}} \lambda . EQ(\lambda) . S(\lambda) d\lambda \quad (3)$$

Here, h is Planck's constant, λ is the wavelength, and S(λ) is the spectral irradiance of sunlight (AM 1.5). In the above power loss profiles, total wavelength range has been considered from 300 nm to 800 nm. It was calculated for an incident wave with amplitude of 1 V/m. For the shorter wavelengths, most of the lights are absorbed within the vicinity of ZnO:Al/Si layer for the 400 nm wavelength. The power loss in the silicon layer is dominated by the constructive and destructive interferences of the forward and backward propagating waves for the case of 600 nm wavelength. Due to the low absorption coefficient of silicon in longer wavelengths, a large fraction of the light is reflected from the back contact leading to the formation of a standing wave in front of the back contact. If we compare the field distributions of two textured cases, most of the absorptions are realized in the border of the unit cell in pyramid textured, whereas for the parabolic textured it is done through the bulk of the cell in the longer wavelengths.

4. Results and Discussions

In this section, results are shown based on optical wave simulation within thin-film amorphous silicon solar cells for wavelengths ranging from 300 nm to 800 nm. Different structures of the solar cell such as smooth substrate, pyramidal textured, and parabolic textured substrates have been used for the investigation and the schematic cross sections for these cells are shown in Fig. 2(a), 3(a), and 4(a), respectively. At first, the investigations were performed on a smooth substrate used as a reference to compare the performance with textured cells. Of these three structures, solar cell with flat structure was achieved a short-circuit current density of 12.72 mA/cm^2, whereas both pyramidal and parabolic textured cells give approximately same short circuit-current density of 17.32 mA/cm^2. The calculated external quantum efficiency of solar cell with 300 nm of absorbing layer thickness are shown in Fig. 5, considering 100% internal quantum efficiency for an incident light spectrum from 300 nm to 800 nm. All of the three structures exhibit almost similar characteristics for wavelength spectrum shorter than 400 nm, since the incoming lights for shorter wavelength ranges are absorbed in the front of the amorphous silicon solar cell and only a fraction of light can reach in the back contact for diffraction. From the external quantum efficiency (EQE) spectrum of flat solar cell, it is seen that it peaks at almost 82.6% in the absorption band of around 500 nm, whereas it is about 7.4% and 6.1% higher at 600 nm of optical spectrum for pyramidal and parabolic structures, respectively.

The absorption profiles at these peaks are mainly determined by the interference and superposition of the diffraction modes of the front and back textures. For longer wavelengths, the solar cells with textured surfaces exhibit distinctly enhanced quantum efficiency compared to a solar cell on smooth substrates as the optical path length and effective thickness of the cell are increased by the introduction of the textures.

parasitic absorption losses as well as from the power loss profile plotting. Although the external quantum efficiency and absorbance for pyramidal and parabolic textured surfaces are almost same in 300 nm of absorbing layer cells, distinguishable changes are observed for 500 nm of photo-active layer depicted in Fig. 6. For blue and infrared part of the spectral wavelength, the absorption for parabolic textured cell is almost always slightly higher than pyramidal textured cell.

As a consequence, the short-circuit current density which is calculated based on external quantum efficiency is 18.48 mA/cm^2 for parabolic textured cell, which is 0.55 mA/cm^2 higher than pyramidal (17.93 mA/cm^2) structure.

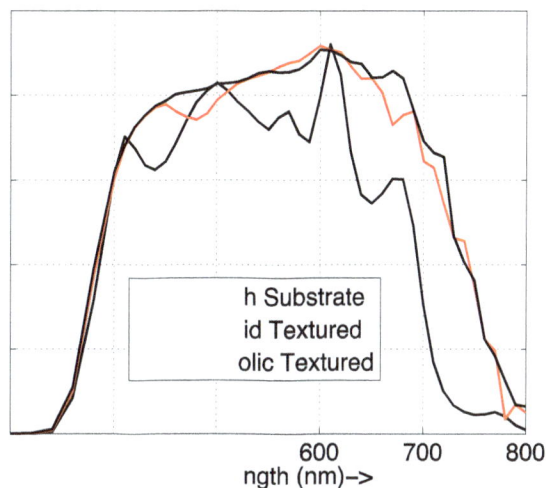

Fig. 6. *Comparison of Quantum Efficiencies of 500 nm intrinsic layer amorphous silicon solar cell for the flat, pyramidal, and parabolic textured of ZnO:Al deposited on glass substrate.*

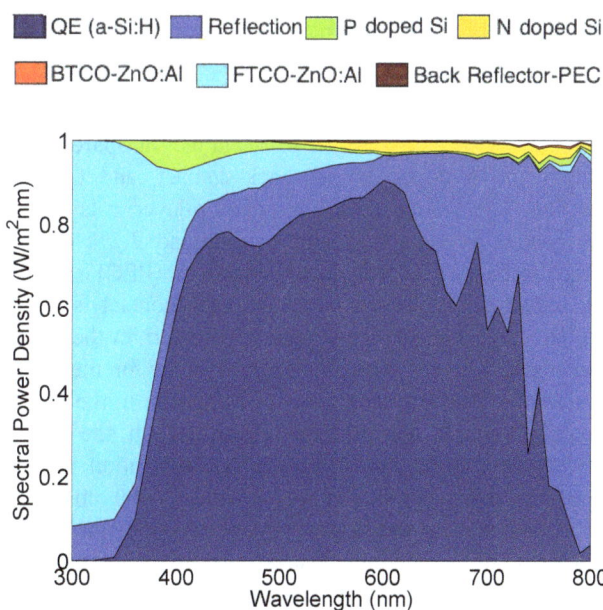

Fig. 7. *Layer specific AM1.5G spectral power distribution of a-Si:H pin solar cell with pyramidal textured deposited on glass substrate.*

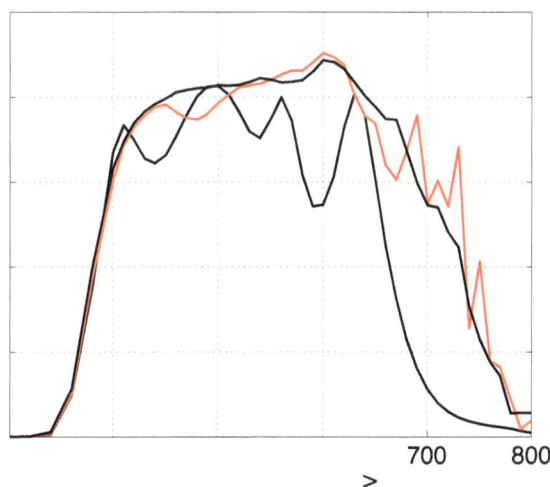

Fig. 5. *Comparison of Quantum Efficiencies of 300 nm intrinsic layer amorphous silicon solar cell for the flat, pyramidal, and parabolic textured of ZnO:Al deposited on glass substrate.*

However, a relative decrease in the EQE of all kind of cells is strongly pronounced in the very shorter wavelength (300 nm - 400 nm) due to the parasitic absorption loss in the ZnO:Al layer, which is clearly observed in Fig. 7 and 8, the

Although the thicker amorphous silicon cell becomes unstable for the realization of open-circuit voltage, an investigation was carried out from our own interest to analyze the improvement of the short-circuit current density that is around 1 mA/cm^2.

In order to define the optical losses for all layers in thin-film solar cell, better way to describe with area plot. They are depicted for pyramidal and parabolic in Fig. 7 and 8, respectively, where the influence of two different texturing on the optical properties are visualized.

As the parasitic absorptions in all other layers such as FTCO, p-layer, n-layer, and BTCO do not contribute to the overall short-circuit current as well as the conversion efficiency, these losses have to be minimized in order to boost-up solar cell.

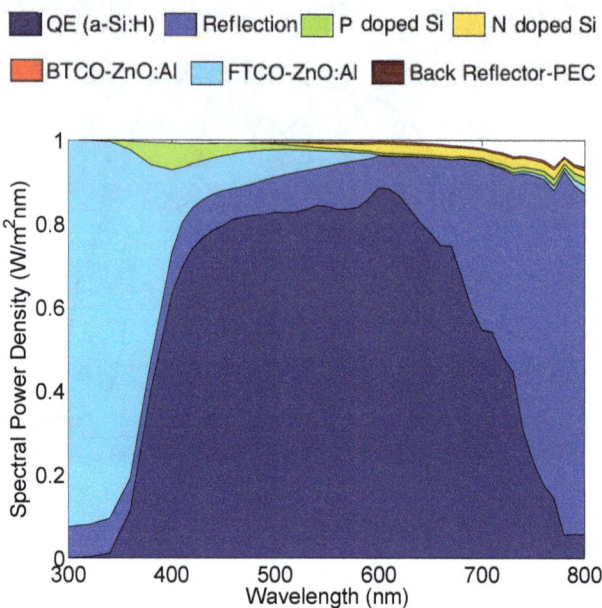

Fig. 8. *Layer specific AM1.5G spectral power distribution of a-Si:H pin solar cell with parabolic textured deposited on glass substrate.*

In the Fig. 7 and 8, the absorption in the i-layer for the diode is depicted along with parasitic losses in the transparent front TCO, p and n doped layers, back contact, and the total reflection. The absorption in the doped p-layer, n-layer, and back contact is very few, which is around 2.7% of total absorption. Since a perfect electric conductor (PEC) is used as back reflector, the absorption in the back contact is almost zero. The largest loss in the p-layer is observed for the shorter wavelength, whereas it is almost transparent for the longer wavelengths. But, the scenario is just opposite in the case of n-layer. Although the parasitic absorption in the shorter wavelengths is more than 50% of the total absorption, it is also almost transparent for shorter wavelengths (< 600 nm). It can be observed that for the longer wavelengths, a major amount of incident photon energy is lost as reflection. Our studies on different textured structures have shown that the amount of reflection is decreased in the case of parabolic structure, which enhances the external quantum efficiency. It is also clearly observed that all the grooves at shorter and longer wavelengths are removed and it gets smoother for parabolic surface texturing.

5. Conclusion

We have presented optical modeling for the p-i-n single junction thin-film amorphous silicon solar cell in superstrate configuration that is deposited on flat and textured glass substrates coated with transparent conductive oxide (TCO). Fundamental limitations for the single junction thin film amorphous silicon solar cell were discussed in terms of open-circuit voltage and short-circuit current. In addition, for the optimization of cell, we investigated the real TCO surface morphologies and that is characterized the consequence of using different types of periodic texture. Our simulated results provide good agreement with updated amorphous silicon solar cell. The improvement of short-circuit current is enhanced by 36.16% than smooth substrate in textured cases. The short-circuit current density was obtained as 17.32 mA/cm^2 for the intrinsic layer thickness of 300 nm. Contrastively, if we want to have an idea of conversion efficiency with reasonable open-circuit voltage of 900 mV and 73% of fill-factor, [3] it is 11.38% as analyzed in our work.

Acknowledgment

The authors would like to thank Prof. Dietmar Knipp, Dr. Rahul Dewan, and Mr. Vladislav Jovanov for their all kinds of support to learn the subject matter during our master study. The lecture notes from the graduate course "Computational Electromagnetics" offered at Jacobs University Bremen, Germany by Prof. Jon Wallace was of great help for the formulation of the FDTD algorithm and understand the subject matter.

References

[1] B. Rech and H. Wagner, "Potential of Amorphous Silicon for Solar Cells," Applied Physics A: Material Science and Processing, vol. 69, Issue 2, pp. 155-167, 1999.

[2] J. Meier, J. Spitznagel, U. Kroll, C. Bucher, S. Faÿ, T. Moriarty, and A. Shah, "Potential of Amorphous and Microcrystalline Silicon Solar Cells," Thin Solid Films, vol. 451-452, pp. 518-524, 2004.

[3] Ping-Kuan Chang, Ting-Wei Kuo, Mau-Phon Houng, Chun-Hsiung Lu, Chih-Hung Yeh, "Effects of Temperature and Electrode Distance on Short-circuit Current in Amorphous Silicon Solar Cells," pp. 175-178, 18-25 May 2012.

[4] A. V. Shah, H. Schade, M. Vanecek, J. Meier, E. Vallat-Sauvain, N. Wyrsch, U. Kroll, C. Droz, and J. Bailat, "Thin-Film Silicon Solar Cell Technology," Progress in Photovoltaics: Research and Applications, vol. 12, Issue 2-3, pp. 113–142, March–May 2004.

[5] S. Faÿ, L. Feitknecht, R. Schlüchter, U. Kroll, E. Vallat-Sauvain, A. Shah, "Rough ZnO Layers by LP-CVD Process and Their Effect in Improving Performances of Amorphous and Microcrystalline Silicon Solar Cells," Institut de Microtechnique (IMT), Rue A. - L. Breguet 2, vol. 90, no. 18-19, pp. 2960-2967, 2000 Neuchâtel, Switzerland, Nov 2006.

[6] J. Bailat, D. Dominé, R. Schlüchter, J. Steinhauser, S. Faÿ, F. Freitas, C. Bücher, L. Feitknecht, X. Niquille, T. Tscharner, A. Shah, C. Ballif, "High-Efficiency P-I-N Microcrystalline and Micromorph Thin Film Silicon Solar Cells deposited on LPCVD ZnO Coated Glass Substrates," Institut de Microtechnique, University of Neuchâtel, Breguet 2, CH-2000 Neuchâtel, Switzerland, 2006.

[7] N. Bouchenak Khelladi, N. E. Chabane Sari, "Simulation Study of Optical Transmission Properties of ZnO Thin Film Deposited on Different Substrates", American Journal of Optics and Photonics. Vol. 1, no. 1, pp. 1-5, 2013.

[8] S. Lo, C. Chen, F. Garwe, and T. Pertch, "Broad-Band Anti-Reflection Coupler for a : Si Thin-Film Solar Sell," Journal of Physics D: Applied Physics, vol. 40, no. 3, pp. 754, 2007.

[9] C. Haase and H. Stiebig, "Thin-Film Silicon Solar Cells with Efficient Periodic Light Trapping Texture," Applied Physics Letter, vol. 91, pp. 061116, no. 6, 2007.

[10] J. Yang, A. Banerjee, and S. Guha, "Triple-junction amorphous silicon alloy solar cell with 14.6% initial and 13.0% stable conversion efficiencies," Applied Physics Letters, vol. 70, no. 22, pp. 2975–2977, 1997.

[11] K. Söderström, F.-J. Haug, J. Escarré, C. Pahud, R. Biron, and C. Ballif, "Highly reflective nanotextured sputtered silver back reflector for flexible high-efficiency n-i-p thin-film silicon solar cells," Solar Energy Materials and Solar Cells, vol. 95, no. 12, pp. 3585–3591, 2011.

[12] H. Sai, T. Koida, T. Matsui, I. Yoshida, K. Saito, and M. Kondo, "Microcrystalline silicon solar cells with 10.5% efficiency realized by improved photon absorption via periodic textures and highly transparent conductive oxide," Applied Physics Express, vol. 6, no. 10, Article ID 104101, 2013.

20

A comparison between statistical analysis and grey model analysis on assessed parameters of petroleum potential from organic materials

Hsien-Tsung Lee

Department of Electrical and Information Technology, NanKaiUniversity of Technology, Nan Tou County, Taiwan

Email address:

t114@nkut.edu.tw

Abstract: This study aims to investigate the impact of changes in the T_{max} and Ro% on the assessed parameters (S1, S2, S1+S2, HI, QI, BI, PI, TOC) of petroleum potential of organic materials. The samples studied include coals and coaly shales of Mushan Formation, Shihti Formation and Nanchuang Formation in NW Taiwan, coals and an oil shale from Mainland China, the well-drilled chip samples from NW Australia, in addition to the data of samples were included from literatures. This work will get on the detecting data of 10 parameters (S1, S2, S1+S2, TOC, HI, QI, BI, PI, Ro%, Tmax) and progressing statistical analysis, and focus the study on comparison between grey forecast of grey relational grade and regression model forecast. The results from statistical analysis (include temperature-treated samples were individually subjected to Rock-Eval analysis) of the all parameters data for all samples in this research project, not only be executed a linear regression, curve regression between any two parameters, and multivariate regression, but also be carried on the forecast of grey correlation grade of grey theory (include grey relational generating (Nominal-the-better-:Ro%; Larger-the-better-: Tmax, HI, QI, BI, S2, S1+ S2, S1; smaller-the-better-: TOC, PI) and globalization grey relational grade). So far, obtain roughly the consistency of results from two type predictive analysis. The constructed HI, QI and BI bands were broad at low maturities and gradually narrowed with increasing thermal maturity. The petroleum generation potential is completely exhausted at a vitrinite reflectance of 2.0-2.2% or a T_{max} of 510-520°C. An increase in HI and QI suggests extra petroleum potential related to changes in the structure of the organic material. A decline in BI signifies the start of the oil expulsion window and occurs within the vitrinite reflectance range 0.75-1.05 % or a T_{max} of 440-455 °C. Furthermore, petroleum potential can be divided into four different parts based on the cross-plot of HI vs. %Ro. The area with the highest petroleum potential is located in section II with %Ro=0.6-1.0%, and HI>100. Oil generation potential is rapidly exhausted at section III with %Ro >1.0%. This result is in accordance with the regression curve of HI and QI with %Ro based on 97 samples with %Ro=1.0~5.6%.

Keywords: Vitrinite Reflectance (Ro %), Grey Relational Analysis, Grey Model, Rock-Eval Pyrolysis, Petroleum Potential, Statistical Analysis

1. Introduction

Petroleum potential of source rocks is described in terms of quantity, quality and level of thermal maturity of organic material (Bordenave et al., 1993; Lee, 2011). Rock-Eval pyrolysis (Espitalie' et al., 1977, 1985; Lee, 2011; Lee and Sun, 2013) is an effective method for the assessment of petroleum potential in source rocks as well as for the geochemical characterization of organic material.Rock-Eval pyrolysis is rapid, inexpensive, requires only small amounts of material, and can generate reliable data. Teichmüller (1973) and Vassoevich et al. (1974) proposed a vitrinite reflectance

range from 0.5–0.55%Ro to 1.3%Ro for the oil window, although it is expanded from 0.5–0.6%Ro to 1.3–1.35%Ro by authors such as Hunt (1996), Petersen (2002), Peters and Cassa (1994), Taylor et al. (1998), and Tissot et al.(1987).Furthermore, Vassoevich *et al.* (1974) determined that oil generation occurs at a vitrinite reflectance of 0.5%Ro, whereas oil cracking starts to occur at a vitrinite reflectance of approximately 1.3%Ro. The temperature at the peak of S2 is referred to as Tmax. The total genetic potential of the sample is defined as S1+S2. The total genetic potential of organic matter in the synthetic assessment of the petroleum potential is described using the hydrogen index (HI, S2/TOC), the bitumen index (BI, S1/TOC) (Killops et al., 1998) and the quality index (QI, [S1+S2]/TOC) (Pepper and Corvi, 1995). The commencement of the effective oil window corresponds to the maturity at which BI begins to decrease during maturation, leading to efficient oil expulsion.

The maturity of organic matter is one of the most important parametersin the evaluation of oil-gas (Tissot and Welte, 1984; Lee, 2011). In assessing maturity of organic matter, vitrinite reflectance (%Ro) is the most commonly examined parameter. Three major characters need to be studied in order to determine the petroleum potential of source rock: properties of organic material, process of thermal maturation, and abundance of hydrocarbon. However, high pressures during hydrocarbon formation increase the activation energies of organic matter in chemical reactions, which may lead to the suppression of vitrinite reflectance. In addition, vitrinite reflectance is strongly influenced by oxidation, so care must be taken to avoid over or under estimating the maturity. Therefore, the maturity parameters of hydrocarbon must be explored in order to construct effective indices for the assessment of petroleum potential. Vitrinite reflectance is a widely used parameter as a geothermometer for the estimation of the thermal maturity. But problems such as human mistakes in measurement, technical problems, and problems associated with the structural and compositional heterogeneity of organic matter. In most cases, the first two types of uncertainties can be handled by standardization (ASTM, ISO). The third problem needs to be solved by statistical analysis.

The anomalously high HI values of coal samples can lead to an overestimation of the hydrocarbon potential. Peters(1986) suggested that the hydrocarbon potential of coal is best determined by using elemental and petrographic analysis. Several succeeding studies evaluated HI and S2 as predictors of the hydrocarbon potential by correlating them with the atomic H/C. However, the results were not satisfactory. Powell et al (1991) found a relationship between HI and atomic H/C for Australian coals with atomic H/C in the range of 0.8-1.0. Petersen and Rosenberg (2000) studied the relationship between HI and %Ro but only found a weak and negative correlation. Other researches exhibited that the HI value displays a strong correlations to volatile matter, which can also be considered as a substitute for petroleum generation potential (Suggate and Boudou, 1993; Newman et al., 1999; Lee, 2011; Lee and Sun, 2013).

After performing the vitrinite reflectance measurement and pyrolysis analysis of organic material, ten parameters %Ro, Tmax, HI(S2/TOC), QI([S1+ S2]/TOC), BI(S1/TOC), PI(S1/[S1+ S2]), S2, S1+ S2, TOC, and S1 can be obtained. They will show the characteristics of the organic material in evaluating their petroleum potential. In avoiding overestimate or underestimate of the generation potential of organic material, we will investigate the correlation among above respective parameter with the datasets of 608 and 506 samples in this research. On the other hand, maturity and types of organic material are related to %Ro and all parameters, so we can estimate petroleum potential of organic material by the statistical analysis to evaluate

correlativity among all aforementioned parameters. For the majority of organic material, present a more complex correlation between Rock-Eval parameters and maturity (rank). So we can also assess the evolution and petroleum potential of organic material by using cross-plots of maturity (%Ro) and Rock-Eval parameters.

Meanwhile, the purpose of this study is to establish reliable indices for the synthetic assessment of organic matter in the evaluation of petroleum potential. The scope of this study will be focused on Rock-Eval pyrolysis, vitrinite reflectance measurement, TOC measurement, maceral composition analysis, regression model forecast, gray relational grade, multivariate statistical analysis and cross-plots of %Ro and Tmax vs. above parameters according to data obtained from vitrinite reflectance measurements, TOC, and Rock-Eval pyrolysis. This study presents new guidelines for improved assessment of generative potential and thermal maturity.

2. Methods of Experiment and Analysis

Maceral analysis and vitrinite reflectance measurements were performed on polished pellets under a Leitz MPV Compact Microscope (light source 12V, 100W; wavelength 546 nm; refractive index of soak oil, Ne=1.5180). Opticalmicroscope was used to identify three main maceral groups (exinite, vitrinite and inertinite), as well as inorganic minerals (pyrite and clay minerals) through point counting. At least 100 measurements were made for each sample. Elemental Analysis and Rock-Eval Pyrolysis were performed in the Precise Instrument Center of National Science Council and EDRI of Chinese Petroleum Corp. respectively. All analytical works were carried out according to ISO or ASTM standards procedures. This study was performed by using the following procedures. The dataset of study starts from the collection of sample data including from literature and our own historical archives. The total number of samples is 1140 (1020 from literature and 120were analyzed for this study; database cf. Table1). Out of the1140 samples, only 608(Table1) have all 10 parameters (%Ro, Tmax, HI, QI, BI, PI, S2, S1+ S2, TOC, and S1). The 120 samples include TW1-48 (from NW Taiwan), ML1-59 (from China), and AU1-13(from Australia).

The scope of this study will be focused on Rock-Eval pyrolysis, vitrinite reflectance measurement, TOC measurement, maceral composition analysis, elements analysis, regression model forecast, gray relational grade, and relative mathematical model. In addition, process new guidelines for improved assessment of the kerogen type, generative potential and thermal maturity using Rock-Eval parameters. Grey relational Analysis include grey generating, grey relational generating operation (Deng, 1988; Wen, 2004).

(1) Grey Generating $x_i^*(K) = \dfrac{x_i^{(0)}(K)}{\alpha}$

(2) GM (1, 1) Model $\dfrac{dx^{(1)}}{dt} + ax^{(1)} = b$

(3) Verhulst Model $\dfrac{dx^{(1)}}{dt} + ax^{(1)} = b(x^{(1)})^2$

Vitrinite reflectance measurement and pyrolysis analysis were performed to evaluate the relationship between the petroleum potential and the maturity for the aforementioned 120 samples. And then go on statistical analysis and drawing cross-plot of the parameters (%Ro, Tmax, HI, QI, BI, PI, S2, S1+ S2, TOC, and S1) in the datasets of 1140, 608, and 506 samples. In order to get the correlation between the aforementioned parameters and to evaluate synthetically the petroleum potential of organic material, can perform statistical analysis methods. Furthermore, the statistical analysis are performed with descriptive, correlation, independent samples T-test, linear and curve regression, factor, principal component, nonparametric tests by Excel and SPSS 16.0 (Keller, 2001; Zhang et al., 2007). All of the analyses were carried out on 10 parameters for the datasets of 608 and 506 samples, and 3 parameters (%Ro, Tmax, HI) for the dataset of 1140 samples. The total genetic potential of organic matter in the synthetic assessment of petroleum potential is evaluated by hydrogen index (HI, S2/TOC), bitumen index (BI, S1/TOC), and quality index (QI, (S1+S2)/TOC).

Statistical methods used on this study include: descriptive, multiple discriminant analyses, Pearson's correlation, K-Independence and 2-independent samples T-tests, linear and curve regression, Q and R mode hierarchical cluster analysis, and principal component analysis (PCA). The first step in the multivariate analyses was the construction of a Pearson correlation coefficient matrix using all of the geochemical parameters contained in the datasets. The most statistically significant variables in the datasets were thus identified.

Table 1. List of sample database (the dataset of 1140 samples).

Period	Region collected	Number	Notes
Cenozoic	Poland, Australia, Indonesia, USA, New Zealand, Taiwan, China	251	b, c, d, f, h, j, o, p, q, r, t, u, s
Cretaceous	Poland, Australia, Nigeria, China	99	b, c, h, i, m, p, t, u
Jurassic	China, Poland, Australia, NW China, New Zealand	106	a, b, c, e, h, l, m, p, t, u
Triassic	China, NW China	97	a, c, e, n
Permian	China, Poland, Australia, New Zealand, USA, Indonesia	90	a, b, c, h, l, m, p, q, t, u
Carboniferous	China, Ukraine, NW China, Netherland	497	a, e, g, k, t

Notes	Sample numbers		Parameters detected	References
a	CJ	1~38	%Ro, Tmax(°C), HI, QI, BI, PI, S2, TOC, S1+S2, S1	Chen, et al., 2003
b	KM	1~17	%Ro, Tmax(°C), HI, QI, BI, PI, S2, TOC, S1+S2, S1	Kotarba and Lewan, 2004
c	PT	1~82	%Ro, Tmax(°C), HI, TOC, QI, BI, S1+S2, S1, S2, QI	Powell et al., 1991
d	AH	1~19	%Ro, Tmax(°C), HI, QI, BI, PI, S2, TOC, S1+S2, S1	Amijaya and Littke, 2006
e	XM	1~48	%Ro, Tmax(°C), HI, QI, BI, PI, S2, TOC, S1+S2, S1	Xiao et al., 2005
f	CU	1~47	%Ro, Tmax(°C), HI, QI, BI, PI, S2, TOC, S1+S2, S1	Canonico et al., 2004
g	PA	1~71	%Ro, Tmax(°C), HI, QI, BI, PI, S2, TOC, S1+S2, S1	Sachsenhofer et al., 2003
h	KJ	1~80	%Ro, Tmax(°C), HI, QI, BI, PI, S2, TOC, S1+S2, S1	Kotarba et al., 2002
i	AS	1~35	%Ro, Tmax(°C), HI, QI, BI, PI, S2, TOC, S1+S2, S1	Akande et al., 1998
j	NC	1~22	%Ro, Tmax(°C), HI, QI, BI, S1+S2, S1, S2, QI, TOC	Norgate et al., 1999
k	VH	1~403	%Ro, Tmax(°C), HI, QI, BI, S1+S2, S1, S2, QI, TOC	Veld et al., 1993
l	MG	1~27	%Ro, Tmax(°C), HI, QI, BI, PI, S2, TOC, S1+S2, S1	Wang , 1998
m	MS	1~24	%Ro, Tmax(°C), HI, QI, BI, PI, S2, TOC, S1+S2, S1	Xiao et al., 1996
n	MT	1~16	%Ro, Tmax(°C), HI, QI, BI, PI, S2, TOC, S1+S2, S1	Xiao, 1997
o	TU	1~12	%Ro, Tmax(°C), HI, S2, TOC, S1+S2, S1, QI, BI, PI	Wu, et al.2003
p	MU	1~13	%Ro, Tmax(°C), HI, S1, S1+S2, S1, QI, BI, PI, TOC	Liu, et al., 2000
q	TK	1~25	%Ro, Tmax(°C), HI, QI, BI, PI, S2, TOC, S1+S2, S1	Chiu et al., 1993
r	TC	1~41	%Ro, Tmax(°C), HI, QI, BI, PI, S2, TOC, S1+S2, S1	Chiu et al., 1996
s	TW	1~48	%Ro, Tmax(°C), HI, QI, BI, PI, S2, TOC, S1+S2, S1	
t	ML	1~59	%Ro, Tmax(°C), HI, QI, BI, PI, S2, TOC, S1+S2, S1	
u	AU	1~13	%Ro, Tmax(°C), HI, QI, BI, PI, S2, TOC, S1+S2, S1	

※1. Sample data of a ~ r from the literature(Notes)
2. Sample data of s ~ u were detected in this study
3. HI(S1/TOC), QI([S1+S2]/TOC), BI(S1/TOC), PI(S1/[S1+S2])
4. The dataset of 1140 samples contain 3 parameters (%Ro, Tmax, HI)
5. The dataset of 608 samples contain 10 parameters (%Ro, Tmax, HI, QI, BI, PI, S2, TOC, S1+S2, S1)

2.1. Kruskal-Waillis Test

$$K - W = \frac{12}{N(N+1)} \sum_{i}^{k} n_i (\bar{R}_i - \bar{R})^2$$

where $\bar{R}$ represents mean rank, n_i represents the observation number of i group, k represents the number of groups, and N represent the total observation number of all groups.

If p (probability)$>\alpha$(0.05, significance level), accept null hypothesis (Ho).

2.2. 2-Independent T-Test

The 2-independentT-tests used here evaluates the probability that the mean value of a particular parameter exhibited by two data sets ($\bar{x}_1$ and $\bar{x}_2$). -The t-statistic is obtained using the following equation- :

$$t = \frac{\bar{x}_1 - \bar{x}_2}{\sqrt{\frac{S_p^2}{n_1} - \frac{S_p^2}{n_2}}} ,$$

where $\bar{x}_1$、 $\bar{x}_2$ represent the mean value,

$$S_p^2 = \frac{(n_1 - 1)S_1^2 + (n_2 - 1)S_2^2}{n_1 + n_2 - 1}$$

where n_1、 n_2 represent the number of measurements and S_1、 S_2 represent the standard deviation.

If p (probability)$>\alpha$(0.05, significance level), accept null hypothesis (Ho).

Chi-squared (x^2) statistics were used to assess the normality of the distribution of values in the data setfor a particular parameter. $x^2 = \sum_{i=1}^{r} \frac{(n_i - e_i)^2}{e_i}$

Where r is the number of categories, n_i are the observed frequencies and e_i are the theoretical frequencies.

2.3. K-Independence Samples Test

$$x^2 = \sum_{i=1}^{2} \sum_{j=1}^{k} \frac{(O_{ij} - E_{ij})^2}{E_{ij}}$$

where O_{ij} represents observation number, i is number of row , j is number of column, and E_{ij} represents expectation number, i is number of row , j is number of column.

If p(probability) > α(0.05, significance level), accept null hypothesis (Ho).

3. Results and Discussion

3.1. Grey Relational Analysis

The results from statistical analysis of the all parameters data for all samples in this research project, not only be executed a linear regression, curve regression between any two parameters, and multivariate regression, but also be carried on the forecast of grey correlation grade of grey theory (include grey relational generating of Chang's, Effect, Hsia's, and Lin's (Nominal-the-better-:Ro%; Larger-the-better-: Tmax, HI, QI, BI, S2, S1+ S2, S1; smaller-the-better-: TOC, PI; Tables 2-5) and globalization grey relational grade; Table 6). So far, achieve approximately consistent results from two modes predictive analysis. Petroleum potential can be divided into four different parts based on the cross-plot of HI vs. %Ro. The highest petroleum potential is located in the second part with %Ro=0.6-1.0%, T_{max}=430-450°C, HI>100, and QI>120. Oil generation potential is rapidly exhausted in the third part with %Ro >1.0%. This result is in accordance with the regression curve of HI and QI with %Ro based on 97 samples with %Ro=1.0~5.6%. The exponential equation of regression can thus be achieved: $HI = 994.8e^{-1.7Ro}$ and $QI = 1646.2e^{-2.0Ro}$ (R²=0.72).

Table 2. *The results of Chang's grey relational generating for the data of all parameters (focus the data from Ro%=1.0 to Ro%=5.6)*

Ro%	Tmax	HI	QI	BI	S2	S1+S2	S1	TOC
0.933	0.926	0.193	0.207	0.126	0.009	0.010	0.016	-28.583
0.933	0.930	0.177	0.187	0.082	0.002	0.002	0.002	-4.917
0.933	0.915	0.125	0.135	0.078	0.003	0.003	0.004	-11.167
0.933	0.926	0.154	0.168	0.128	0.001	0.001	0.001	-0.667
0.933	0.911	0.381	0.392	0.087	0.044	0.045	0.026	-72.417
0.933	0.922	0.041	0.041	0.000	0.000	0.000	0.000	-2.500
0.933	0.937	0.305	0.322	0.155	0.028	0.029	0.037	-56.333
0.933	0.930	0.061	0.125	0.664	0.000	0.001	0.010	-1.583
0.933	0.932	0.085	0.110	0.260	0.001	0.001	0.006	-3.667

Table 3. The results of Effect grey relational generating for the data of all parameters (focus the data from Ro%=0.55 to Ro%=1.02)

Ro%	Tmax	HI	QI	BI	S2	S1+S2	S1	TOC
0.933	0.945	0.428	0.467	0.368	0.010	0.011	0.023	0.065
0.933	0.926	0.193	0.207	0.126	0.009	0.010	0.016	0.033
0.933	0.930	0.177	0.187	0.082	0.002	0.002	0.002	0.145
0.933	0.930	0.291	0.305	0.128	0.088	0.092	0.101	0.005
0.933	0.915	0.125	0.135	0.078	0.003	0.003	0.004	0.076
0.933	0.926	0.154	0.168	0.128	0.001	0.001	0.001	0.375
0.938	0.922	0.041	0.041	0.000	0.000	0.000	0.000	0.222
0.938	0.937	0.305	0.322	0.155	0.028	0.029	0.037	0.017
0.938	0.930	0.061	0.125	0.664	0.000	0.001	0.010	0.279

Table 4. The results of Hsia' grey relational generating for the data of all parameters (focus the data from Ro%=0.17 to Ro%=5.6)

Ro%	Tmax	HI	QI	BI	S2	S1+S2	S1	TOC	PI
0.920	0.507	0.353	0.370	0.097	0.295	0.302	0.207	0.258	0.952
0.920	0.452	0.299	0.322	0.164	0.312	0.327	0.435	0.078	0.921
0.920	0.438	0.347	0.360	0.057	0.328	0.332	0.138	0.163	0.968
0.920	0.438	0.353	0.375	0.147	0.243	0.252	0.258	0.390	0.937
0.920	0.480	0.314	0.325	0.032	0.334	0.336	0.088	0.060	0.984
0.920	0.521	0.216	0.235	0.098	0.020	0.021	0.023	0.921	0.937
0.920	0.562	0.202	0.213	0.029	0.175	0.177	0.062	0.246	0.984
0.920	0.575	0.086	0.141	0.479	0.001	0.001	0.012	0.993	0.492
0.920	0.616	0.343	0.371	0.207	0.036	0.038	0.056	0.908	0.921

Table 5. The results of Lin's grey relational generating for the data of all parameters (focus the data from Ro%=0.17 to Ro%=0.6)

Ro%	Tmax	HI	QI	BI	S2	S1+S2	S1	TOC	PI
0.922	0.300	0.149	0.187	0.156	0.140	0.167	0.960	0.020	0.735
0.922	0.415	0.061	0.074	0.032	0.069	0.078	0.229	0.005	0.819
0.922	0.397	0.084	0.099	0.027	0.009	0.009	0.018	0.520	0.848
0.929	0.331	0.003	0.004	0.005	0.000	0.000	0.000	0.990	0.681
0.929	0.271	0.046	0.072	0.223	0.001	0.002	0.043	0.722	0.533
0.929	0.192	0.278	0.319	0.100	0.014	0.015	0.032	0.645	0.848
0.929	0.204	0.311	0.363	0.156	0.299	0.339	1.000	0.015	0.819
0.929	0.204	0.138	0.150	0.007	0.092	0.096	0.032	0.075	0.938
0.929	0.331	0.098	0.111	0.019	0.092	0.100	0.116	0.020	0.877

Table 6. The results of Globalization grey relational grade for the data of all parameters (focus the data " from Ro%=0.5 to Ro%=2.2")

	Hsia's	Nagai's	Wen's	Wu's
	0.0984	0.1148	0.0983	0.0916
	0.0909	0.1134	0.0922	0.0855
	0.0948	0.1148	0.0953	0.0861
	0.0958	0.1172	0.0962	0.0962
	0.0954	0.1062	0.0959	0.0959
	0.0998	0.1040	0.0993	0.1006
	0.0982	0.1114	0.0981	0.1000
	0.0998	0.1169	0.0993	0.1006
	0.0933	0.1104	0.0943	0.0931
Eigenvector	0.0999	0.1084	0.0995	0.1008
	0.0989	0.1017	0.0987	0.0996
	0.0985	0.1045	0.0983	0.0994
	0.0999	0.0963	0.0994	0.1001
	0.0997	0.1115	0.0993	0.1006
	0.0958	0.1115	0.0962	0.0959
	0.0977	0.1176	0.0977	0.0996
	0.0988	0.1152	0.0985	0.0994
	0.0969	0.1097	0.097	0.0982
	0.0967	0.1081	0.0968	0.0986
	0.0937	0.1127	0.0944	0.0863
Maximum Eigenvalue	96.7423	59.8616	97.6133	91.8172

3.2. Hierarchical Cluster Analysis of HI, QI

A vitrinite reflectance (%Ro) ranged from 0.5-0.55 to 1.3, 0.5-0.7 to 1.3 or 0.5-0.6 to 1.3-1.35 for the conventional "oil window" was proposed by Petersen (2002), Teichmuller (1993), Vassoevich et al. (1974), Tissot and Welte (1984), and Hunt (1996). Q mode hierarchical cluster analysis for the dataset of 608 samples confirmed that 61 data samples with all ten parameters belong to the same cluster, numbering from 2 ~ 14 with $0.5 \leq \%Ro \leq 1.0$, S2<2, and HI<80. In addition, the others 41 data samples with ten parameters are in a different group with HI>380. They show higher dispersion than those with lower HI values. Peters and Cassa (1994) stated that the source rock has poor petroleum potential if the organic material has $0.5 \leq \%Ro \leq 1.0$, S2<2, and HI<80. Therefore, the 102 data were removed from the cross-plots (HI vs. %Ro, QI vs. %Ro, BI vs. %Ro, HI vs. Tmax, QI vs. Tmax, BI vs. Tmax) for dataset of 608 samples, after deleting 102 data from dataset of 608 samples, another dataset of 506 samples could be obtained. Nonparametric tests (2 and k independent samples) between the datasets of 608 and 506 samples confirm that the distribution of values from respective parameters except TOC exhibit no significant difference ($p>0.05$, Table 7a, 7b). Nonparametric tests (k independent samples) among the datasets of 1140, 608, and 506 samples, showed that there was no significant difference ($p>0.05$, Table 7c) in the distribution of HI values. The distribution of the HI value for all of the datasets remains constant with increasing thermal maturity.

After conducting a Pearson's correlation analysis for datasets of 608 and 506 samples, exhibited same results and they had high correlation about the couple of parameters (%Ro vs. Tmax, HI vs. QI, TOC vs. [S1+S2], S2vs. [S1+S2], S2 vs. TOC; cf. Table 8a). So they were obtained again the evolution in the HI with increasing thermal maturity for a large worldwide datasets of 608, and 506 samples (Figures 1 and 2), and the evolution in the total generation (QI) with increasing maturity for the worldwide datasets of 608 and 506 samples (Figures3 and 4). From the results of Pearson's correlation analysis (Table 8b), during Ro=0.17~0.6% rise up to %Ro=0.61~1.0, declining the correlativity for HI and QI with BI from medium to low while BI increase with thermal maturity until the %Ro up to ~0.75%Ro (Figures 5-7). On the other hand, during Ro=0.17~1.0% rise up to 1.0~5.6%, the correlation for HI and QI with %Ro increased, with correlative coefficient up to r = -0.72 (Table 8c). They show clearly the apex of the upper and lower limit of the HI and QI band from the cross-plots (HI vs. %Ro, QI vs. %Ro, HI vs. T max, QI vs. T max) for the dataset of 506 samples. From these cross-plots, we can define more definitely to the lines of maximum HI and QI from ~ 0.6 to ~1.0%Ro or from ~ 430 to ~450°C, which in accordance with the results of descriptive analysis for the mean value of respective parameter in the interval of %Ro span. The parameters (HI, QI) have the highest mean value with increasing thermal maturity at the ~0.6 %Ro (within 0.5-0.7%Ro) and ~1.0%Ro (within 0.9-1.1%Ro), ~430°C Tmax (within 425-435°C Tmax) and ~450 °C Tmax (within 442-460°C Tmax) (Table 9).

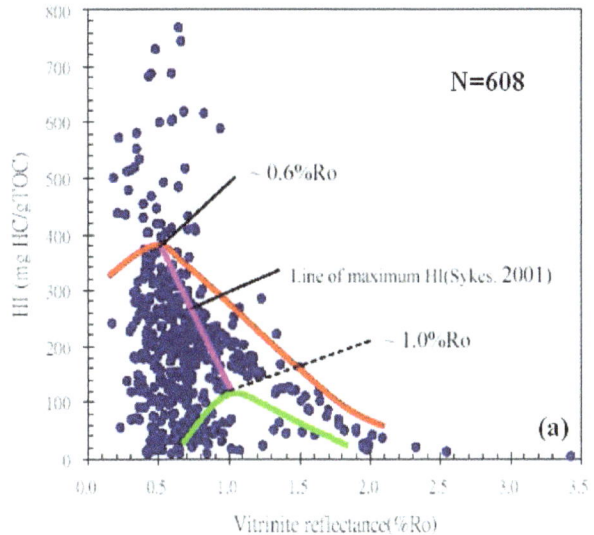

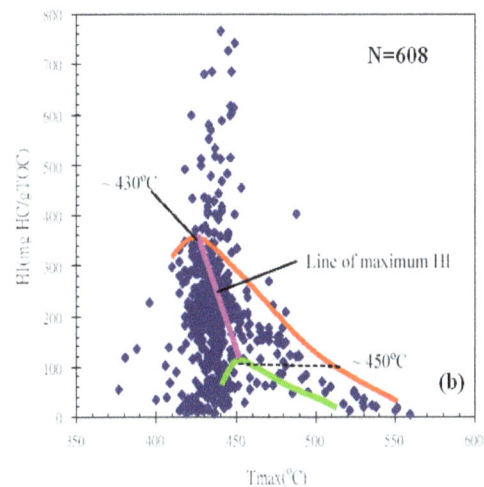

Fig. 1. The evolution in the HI with increasing thermal maturity for a large worldwide dataset of 608 samples: (a) HI vs. %Ro; (b)HI vs. Tmax. The sample data define an HI band that narrows with increasing maturity. A line of maximum HI can be confirmed from ~ 0.6 to 1.0%Ro or from ~ 430 to 450°C.

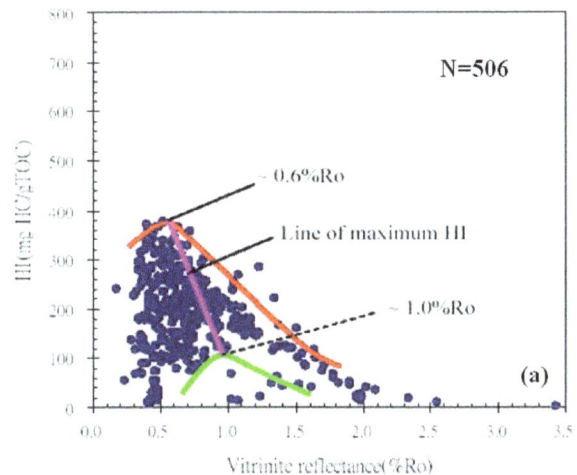

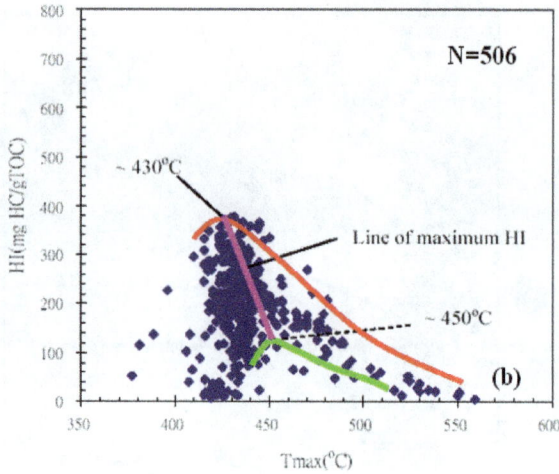

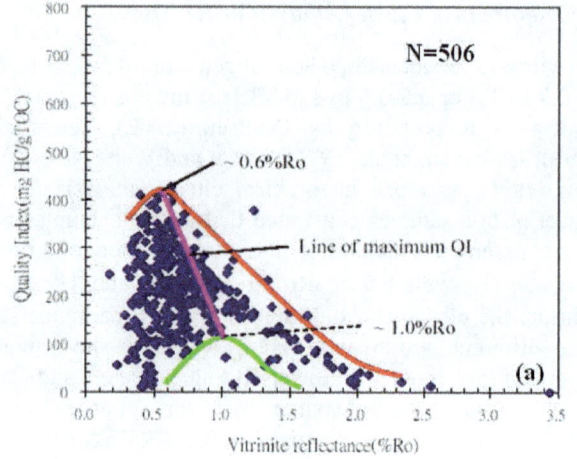

Fig. 2. *The evolution in the HI with increasing thermal maturity for a large worldwide dataset of 506 samples: (a)HI vs. %Ro; (b)HI vs. Tmax. After deleting 102 data from dataset of 608 samples, can present very clearly the upper and lower limit of the HI band. From this fig., We can define more definitely to a line of maximum HI from ~ 0.6 to 1.0%Ro or from ~ 430 to 450°C.*

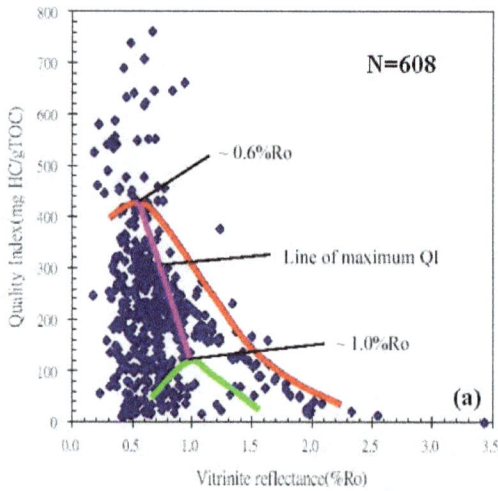

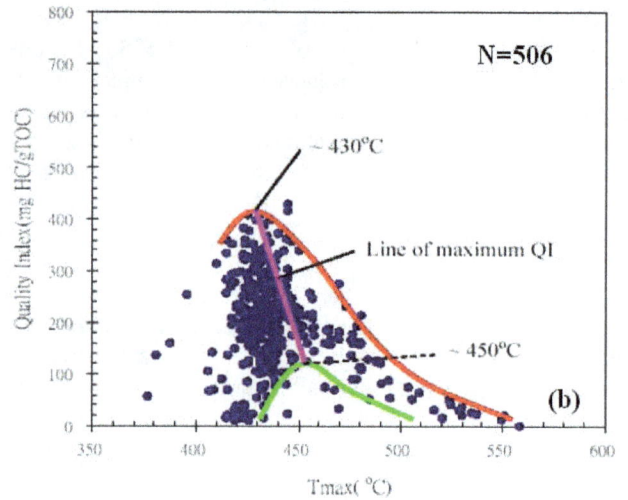

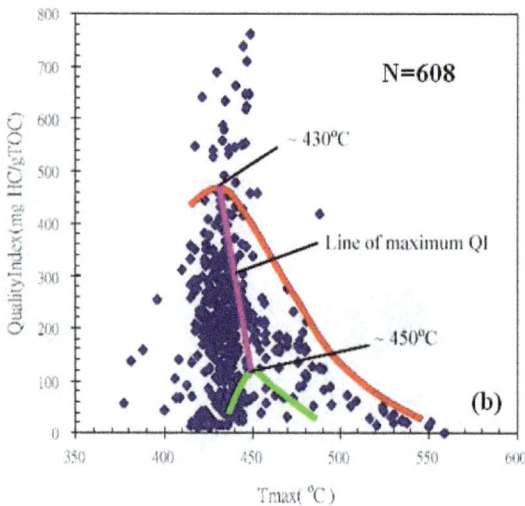

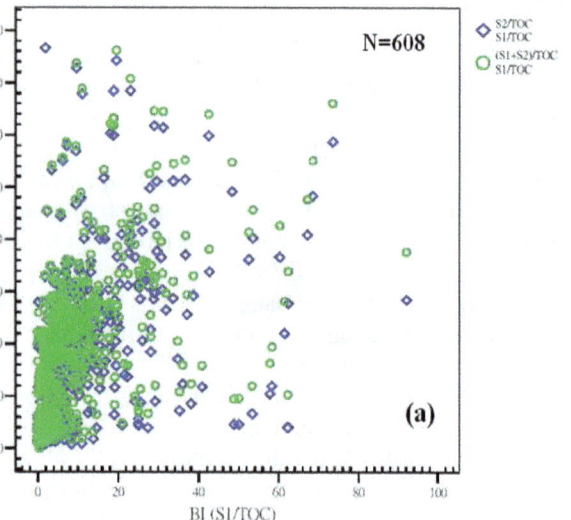

Fig. 4. *The evolution in the QI with increasing thermal maturity for a large worldwide dataset of 506 samples: (a)QI vs. %Ro; (b)QI vs. Tmax. After deleting 102 datas from dataset of 608 samples, can present very clearly the upper and lower limit of the QI band. From this fig., We can define more definitely to a line of maximum HI from ~ 0.6 to 1.0%Ro or from ~ 430 to 450°C.*

Fig. 3. *The evolution in the total generation (QI) with increasing maturity for the worldwide dataset of 608 samples: (a) QI vs. %Ro; (b) QI vs. Tmax. The decline in QI indicates the onset of initial oil expulsion.*

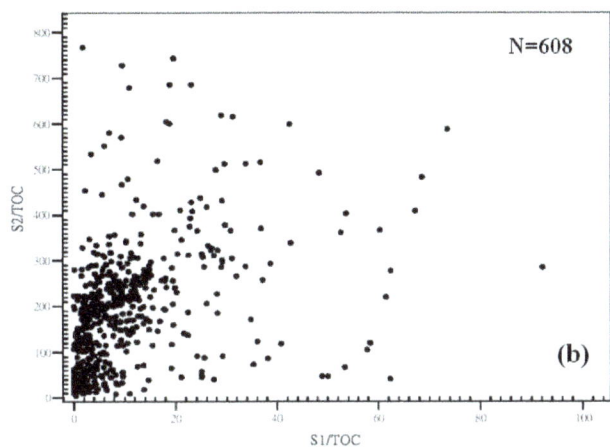

Fig. 5. (a) Plot of HI and QI (Y-axial) vs. BI; (b) Plot of HI vs. BI for dataset of 608 samples.

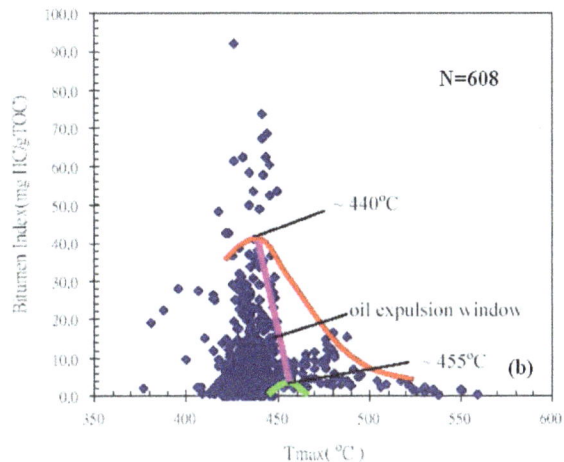

Fig. 6. (a) BI vs. %Ro and (b) BI vs. Tmax for the worldwide dataset of 608 samples. The majority of the samples confirm a BI band that illustrates the "oil expulsion window". The line between ~ 0.75 and ~ 1.05%Ro or 440 and 455°C defines the start of the effective oil window.

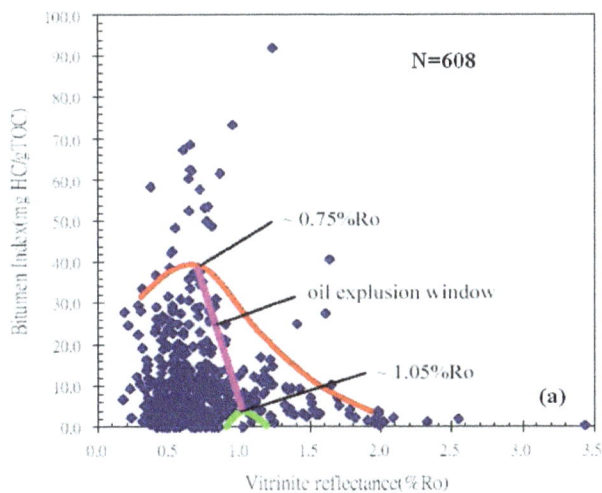

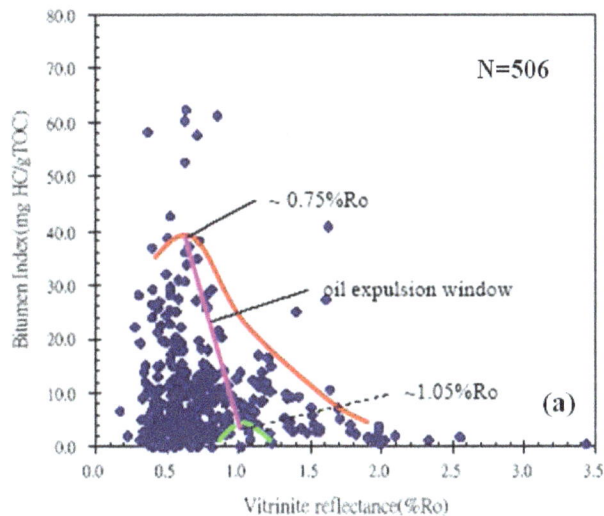

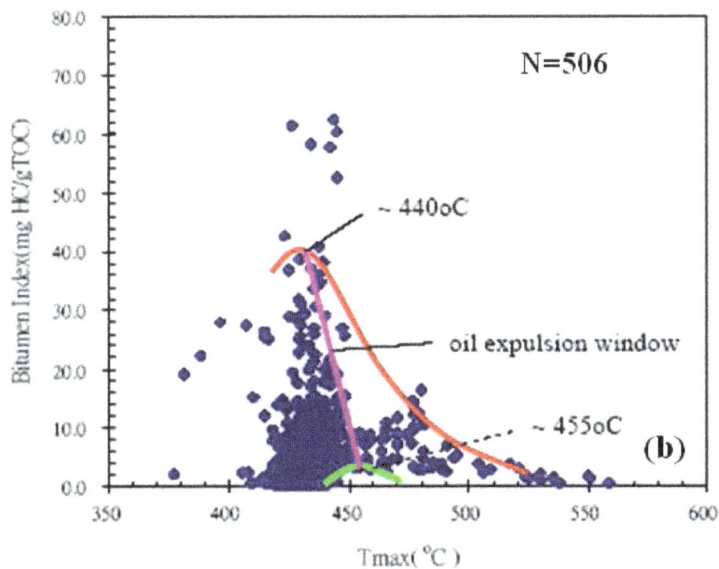

Fig. 7. (a) BI vs. %Ro and (b) BI vs. Tmax for the worldwide dataset of 506 samples. The majority of the samples confirm a BI band that illustrates the "oil expulsion window". The line between ~ 0.75 and ~ 1.05%Ro or 440 and 455°C defines the start of the effective oil window.

Table 7. The large amount of data allows statistically significant differences between the parameters from different datasets to be established.

(a)Test		parameters									
		%Ro	Tmax	HI	QI	BI	PI	S2	TOC	S1+S2	S1
A	Sig	0.58	0.18	1.00	1.00	0.96	0.00	0.97	0.63	0.98	0.49
B	Sig	0.95	0.99	0.05	0.08	0.87	0.36	0.05	0.04	0.05	0.16
C	Sig	0.65	0.64	0.61	0.59	0.33	0.09	0.03	0.01	0.03	0.05

※
1. A: MosesTest; B: Two-Sample Kolmogorov-Smirnov Test; C: Manm-Whitney Test.
2. Sig: p-value (accompanying probability; significance level $\alpha=0.05$).
3. If $p>\alpha$, accept null hypothesis (Ho). It represents no significant difference.

(b)Test		parameters						
		%Ro	Tmax	HI	QI	BI	PI	S1
D	(1) χ^2	0.05	0.47	0.39	0.47	0.49	2.13	1.93
	Sig.	0.82	0.49	0.53	0.49	0.48	0.14	0.16
	(2) χ^2	0.03	0.39	0.32	0.38	0.40	1.94	1.76
	Sig.	0.87	0.53	0.57	0.53	0.53	0.16	0.18
E	(1) χ^2	0.19	0.22	0.26	0.28	0.92	2.81	3.57
	Sig.	0.65	0.64	0.61	0.59	0.33	0.09	0.06
F	Sig.	0.660	0.64	0.61	0.59	0.34	0.09	0.06

※
1. D: Median Test; E: Kruskal-Wallis Test; F: Jonckheere-Terpstra Test.
2. (1) χ^2: chi-square statistics; (2) χ^2: Yales' continuity chi-square correction.

(c)Tests		parameters		
		%Ro	Tmax	HI
D	(1)x²	66.30	9.01	4.05
	Sig.	0.00	0.01	0.13
E	(1)x²	56.79	18.53	0.61
	Sig.	0.00	0.00	0.73
F	Sig.	0.00	0.00	0.60

※
1. D: Median Test; E: Kruskal-Wallis Test; F: Jonckheere-Terpstra Test.
2. χ^2: chi-square statistics.

(d)Tests		parameters									
		%Ro	Tmax	HI	QI	BI	PI	S2	TOC	S1+S2	S1
G	t	3.01	4.79	3.64	4.07	6.02	6.80	-5.28	-9.16	-5.10	0.31
	Sig.	0.01	0.00	0.00	0.00	0.00	0.00	0.00	0.00	0.00	0.76

※   1. G: 2 Independent Samples T-test   2. t: t-test for equality of Means

(a) The non-parametric tests (2 independent samples) for datasets of 608 and 506 samples. The distribution of data from respective parameters except TOC are not significantly different $(P>\alpha)$.

(b) The non-parametric tests (K independent samples) for datasets of 608 and 506 samples. The distribution of data from respective parameters are not significantly different $(P>\alpha)$.

(c) The non-parametric tests (K independent samples) for datasets of 1140, 608 and 506 samples show that the distribution of parameter (HI) values are not significantly difference $(P>\alpha)$ between the different dataset, i.e. the distribution of HI in different datasets does not change with increasing thermal maturity.

(d) 2 Independent Samples T-test   for TW1-48 samples and CJ1-38 samples, they exhibit the significantly different $(P<\alpha)$ for type of the distribution of values from respective parameters except S1.

Table 8. Distribution of Pearson's correlation coefficient of dataset of 608 and 506 samples

Datasets		Pearson' correlation coefficient (r) r (couple of parameters)			
		G	M	L	EL
a	0.17≦%Ro≦5.60 608 data	0.86(R vs T) 0.96(H vs Q) 0.99(S vs A) 0.82(C vs A) 0.81(S vs C)	0.58(H vs A) 0.51(Q vs A) 0.70(A vs D) 0.57(C vs D)	0.37(H vs B) -0.33(R vs H) 0.46(Q vs B) 0.36(B vs D)	(R vs H) (R vs B) (R vs S) (T vs B) (T vs S) (H vs P) (Q vs P) (B vs C) (B vs A) (R vs A) (R vs C) (T vs C)
	0.17≦%Ro≦5.60 506 data	0.89(R vs T) 0.98(H vs Q) 0.99(S vs A) 0.84(C vs A) 0.85(S vs C)	0.65(H vs A) 0.62(Q vs A) 0.72(A vs D) 0.54(C vs D)	0.36(H vs B) -0.38(R vs H) 0.46(Q vs B) 0.46(B vs D)	(R vs H) (R vs B) (R vs S) (T vs B) (T vs S) (H vs P) (Q vs P) (B vs C) (B vs A) (R vs A) (R vs C) (T vs C)
b	0.17≦%Ro≦0.60 245 samples	0.99(H vs Q)	0.61(H vs S) 0.60(H vs A) 0.53(Q vs B) 0.60(Q vs S) 0.60(Q vs A) 0.61(B vs D)	0.48(H vs B) 0.39(B vs P)	(R vs H) (R vs Q) (R vs B) (R vs P) (R vs S) (R vs C) (T vs H) (T vs Q) (T vs B) -0.21(P vs S)
	0.60≦%Ro≦1.00 266 samples	0.93(H vs Q)	0.53(H vs S) 0.53(H vs A) 0.54(B vs P)	0.43(Q vs B) 0.38(Q vs S) 0.39(Q vs A) -0.36(P vs S) -0.40(P vs C)	(R vs H) (R vs Q) (R vs B) (R vs P) (R vs S) (R vs C) (T vs H) (T vs Q) (T vs B) 0.200(B vs D)
c	0.17≦%Ro≦1.00 511 samples	0.88(S vs C)	0.56(R vs T) 0.57(H vs S) 0.57(H vs A) 0.50(Q vs S) 0.51(Q vs A)	0.32(H vs C) 0.40(H vs D) -0.34(P vs C)	(R vs H) (R vs Q) (R vs P) (R vs S) (T vs H) (T vs Q) (T vs B) (T vs C)
	1.00≦%Ro≦5.60 97 samples	0.83(R vs T)	-0.72(R vs H) -0.72(R vs Q) -0.65(Tvs Q) (T vs H) (H vs S) (H vs A) (H vs D) (Q vs S) (Q vs A) (S vs C) (R vs P)	-0.48 (R vs S) -0.47 (R vs A) -0.36 (R vs D) -0.45 (T vs B)	0.16(H vs C) -0.22(P vs C)

※
1. The symbol of respective parameter : %Ro(R), Tmax(T), HI(H), QI(Q), BI(B), PI(P), S2(S), TOC(C),[S1+S2](A),S1(D).
2. G represents high correlation (｜ r ｜ ≧0.8).
3. M represents medium correlation(0.5≦ ｜ r ｜ < 0.8).
4. L represents low correlation(0.3≦ ｜ r ｜ < 0.5).
5. EL represents extremely low correlation (｜ r ｜ <0.3).

Table 9. Mean values of various parameters in different intervals of %Ro and Tmax(℃) for datasets, with 95% confidence interval mean

(a) The mean value of respective parameter in the interval of %Ro span for dataset of 608 samples									
Interval of %Ro span	Tmax(℃)	HI	QI	BI	PI	S1	S2	S1+S2	TOC
0.3<%Ro≦0.4	426	195	206	7.8	0.04	1.7	68	71	29
0.4<%Ro≦0.5	429	202	207	7.9	0.05	3.2	76	79	27
0.5<%Ro≦0.6	433	*228*	*239*	11.2	0.06	*4.8*	*101*	*106*	39
0.6<%Ro≦0.7	435	221	225	11.6	0.06	2.9	89	93	36
0.7<%Ro≦0.8	438	178	191	*12.2*	*0.09*	2.7	78	81	34
0.8<%Ro≦0.9	440	176	186	10.3	0.08	3.1	89	91	42
0.9<%Ro≦1.0	448	171	181	9.4	0.06	3.2	80	83	39
1.0<%Ro≦1.3	460	166	177	9.9	0.05	*4.6*	*94*	*98*	56
1.3<%Ro≦1.7	472	104	113	8.9	0.08	2.9	62	65	51
1.7<%Ro<3.5	527	31	33	1.8	0.09	1.0	20	21	59
(b) The mean value of respective parameter in the interval of Tmax(℃) span for dataset of 608 samples									
Interval of Tmax(℃)span	%Ro	HI	QI	BI	PI	SI	S2	S1+S2	TOC
377≦Tmax≦420	0.46	164	174	8.7	0.07	3.2	68	71	34
420<Tmax≦425	0.47	180	188	8.1	0.06	2.3	67	69	27
425<Tmax≦430	0.59	*219*	*231*	10.8	0.05	4.9	117	121	48
430<Tmax≦435	0.62	213	225	10.8	0.06	3.4	93	97	35
435<Tmax≦438	0.65	194	202	10.6	0.07	3.1	76	79	31
438<Tmax≦442	0.67	200	206	12.9	*0.09*	2.9	80	83	29
442<Tmax≦450	0.77	*252*	*268*	*15.2*	0.06	3.7	88	92	38

450<Tmax≦460	0.93	176	182	6.9	0.05	2.7	_93_	_97_	44
460<Tmax≦500	1..24	142	147	7.2	0.05	_4.7_	89	93	59
500<Tmax≦559	2.06	31	33	1.8	0.09	0.9	20	21	58

(c)Dataset of 1140 samples. The mean value of respective parameter in the interval of %Ro and Tmax(℃) span

Interval of %Ro span	Tmax(℃)	HI	Interval of Tmax(℃)span	%Ro	HI
0.3<%Ro≦0.4	427	188	377≦Tmax≦420	0.45	189
0.4<%Ro≦0.5	430	216	420<Tmax≦425	0.47	193
0.5<%Ro≦0.6	434	_244_	425<Tmax≦430	0.56	_212_
0.6<%Ro≦0.7	435	230	430<Tmax≦435	0.73	199
0.7<%Ro≦0.8	437	181	435<Tmax≦438	0.74	201
0.8<%Ro≦0.9	440	183	438<Tmax≦442	0.77	205
0.9<%Ro≦1.0	443	171	442<Tmax≦450	0.82	_234_
1.0<%Ro≦1.1	454	_182_	450<Tmax≦460	1.02	182
1.1<%Ro≦1.2	462	173	460<Tmax≦480	1.24	151
1.2<%Ro≦1.3	466	152	480<Tmax≦500	1.40	118
1.3<%Ro≦1.7	473	117	500<Tmax≦559	2.05	31
1.7<%Ro<3.5	523	33			

3.3. Hydrocarbon Potential from Evolution of HI vs. %Ro

The evolution of HI with increasing %Ro for 608 and 506 samples are shown in Figures 1a and 2a and the evolution in HI with increasing Tmax for 608 and 506 samples in Figures 1b and 2b. In four figures, an area can be drawn around the vast majority of the samples. The HI band is widest (up to 305mg HC/g TOC broad) below Ro of approximately 0.6%, as greater than which the band width of HI will be gradually reduced down to 20mg HC/g TOC or less at Ro of about 2.1% (Figures 1a and 2a). Above this %Ro value, HI does not show any change with increasing maturity. The HI band is widest (up to 315mg HC/g TOC) below a Tmax of about ~430°C, which, according to the relationship between Tmax and %Ro, corresponds to ~0.54%Ro (Figure 8). With increasing maturity the band width of HI will be gradually reduced down to 40mg HC/g TOC or less at a Tmax above 510°C (corresponds to Ro ~2.05%, cf. Figure 8). The upper limit of the HI band reaches a maximum HI of about 375mg HC/g TOC at a %Ro of ~0.6 or a Tmax of ~430°C. Even though the lower limit (the organic materials with the lowest initial petroleum potential), reaches a maximum HI of about ~110mg HC/g TOC at a %Ro of ~1.0 or a Tmax of ~ 450°C. The broad bands up to a Ro of about 0.6% and Tmax about 430°C reflect a distinguished variation in petroleum generation potential of the organic material with similar maturities. This can be attributed to the heterogeneous chemical component of the organic material and the different depositional environment.

Petersen (2006) suggested that the HImax line was defined between the upper limit (0.6%Ro or 430°C Tmax) of the HI-band outlined by the majority of the humic coals and type Ⅲ kerogen and the lower limit (1.0%Ro or 450°C Tmax). Although the dissociation activation energy is higher and broader in distributed range for type Ⅱ kerogen than type Ⅲ kerogen, the onset of petroleum generation was similar at the early stage (Suggate et al., 1993; Hu, 2001; Sun et al., 2001). Therefore in this study, for samples of type Ⅱ/Ⅲ kerogen, we can obtain the same results in HImax (Sykes, 2001; Sykes

and Snowdon, 2002; Pedersen et al., 2006; Petersen, 2006). The line defined by the %Ro values [0.6 , 1.0] and the Tmax (°C) values [430 , 450] in the Figure 1 represents the line of HImax, and the slope of the HImax line suggests that the organic materials with the highest petroleum potential can reach their HImax at the lowest maturity. As illustrated by Sykes and Snowdon (2002), HI values of immature organic materials or of mature organic materials that have passed beyond the line of HImax, can be extrapolated along their maturation pathways into their HImax values so as to estimate their the true petroleum potential. The effect of the gradual exhaustion of the petroleum potential is evident from the narrowing of the HI band with increasing maturity. However, the organic materials may still possess petroleum potential at a Ro up to 1.3% (HI up to ~175 mgHC/g TOC) which is the end of the traditional oil window (Figures 1a, 2a). The HI shows that the generative potential for liquid petroleum of the organic materials is exhausted at a Ro of approximately 2.0–2.2% or Tmax of 510–520°C (Figures 1 and 2), corresponds to 2.05-2.23%Ro (Figure 8). For humic coals and kerogen type Ⅲ , liquid petroleum generation is negligible at a %Ro of approximately 1.8, and at 2.0%Ro the petroleum generative potential is exhausted (Petersen, 2002).

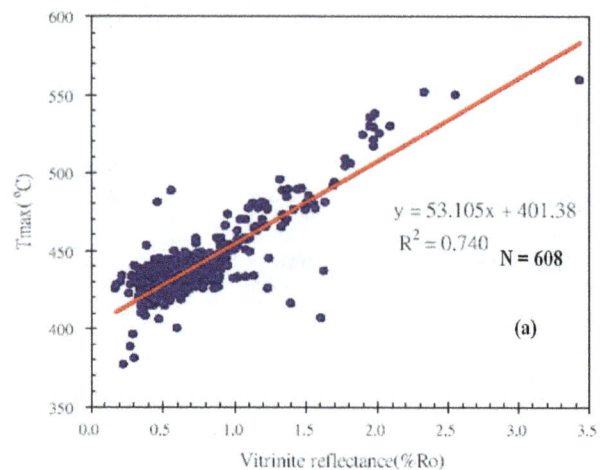

$y = 53.105x + 401.38$
$R^2 = 0.740$
$N = 608$

(a)

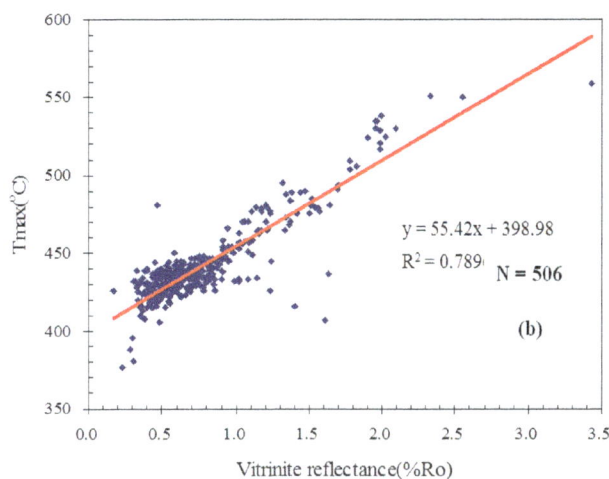

Fig. 8. Correlation between %Ro and Tmax based on the datasets of 608 (a) and 506 (b) samples.

3.4. QI-Band Defined by the Majority

The Quality Index (QI, Rock-Eval derived S1+S2 yields normalized to total organic carbon, TOC) can be used to represent the total generation potential (Wang, 1998; Sykes, 2001; Sykes and Snowdon, 2002; Petersen, 2006; Rabbani and Kamali, 2005; Kotarba et al., 2007; Pepper and Corvi, 1995). From the results of descriptive analysis for the mean value of respective parameter in the interval of %Ro or Tmax($^\circ$C) span, the QI has the highest value of mean at the ~0.6 %Ro (within 0.5-0.7%Ro) and ~1.0%Ro (within 0.9-1.1%Ro), ~430°C Tmax (within 425-435°C Tmax) and ~450 $^\circ$C Tmax (within 442-460°C Tmax) (Table 9). As noted previously, we can obtain the same result from the Pearson correlation analysis (Table 9b-c; Figures 5-7). The evolution in QI with increasing %Ro for 608 and 506 samples is shown in Figures 3a and 4a. and the evolution in QI with increasing Tmax for 608 and 506 samples in Figures 3b and 4b. The data are distributed in an area (QI band) which can be drawn around the vast majority of the samples. The QI band is broadest (up to 360mg HC/g TOC broad) below a Ro of approximately 0.6%, beyond which it gradually narrows to a band width of 35mg HC/g TOC or less at a Ro of about 2.2% (Figures 3a, 4a). Above this %Ro value, the QI does not show any change with increasing maturity.

A similar evolution in QI is shown by Figures 3b and 4b. The QI band is widest (up to 380mg HC/g TOC) below a Tmax of about ~430°C, which, according to the relationship between Tmax and %Ro, corresponds to ~0.54%Ro (Figure 8). With increasing maturity the band width of QI will be gradually reduced down to 40mg HC/g TOC or less at a Tmax above 510°C (~2.05%Ro, cf. Figure 5). So in this study, a line of maximum QI (QImax, cf. Sykes, 2001; Sykes and Snowdon, 2002; Pedersen et al., 2006; Petersen, 2006) can be outlined between Ro ~0.6% and ~1.0% or Tmax ~430°C and ~450°C. Petersen (2006) has postulated that the QImax line was defined between the apex of the upper limit (0.7%Ro or 435°C Tmax) of the QI-band defined by the majority of the humic coals and type III kerogen and the apex of the lower

limit (1.0%Ro or 455°C Tmax). Sykes and Snowdon (2002), Petersen (2006) have interpreted about the decline in QI that indicates the commencement of initial oil expulsion. Because type II kerogen possess a few bonds with low dissociated activation energy at the stage of low maturation, only few hydrocarbons of S1 and S2 can be produced and causing the decline of QI values. (Suggate et al., 1993; Hu, 2001; Sun et al., 2001; Chen, 2006). In other words, for the mixed samples of type II / III kerogen in the petroleum build-up stage, the QI value will be lower than while at the 0.6%Ro maturity, they accord with the results in this study. Therefore we can obtain the same results as HI value in the QImax .

3.5. BI-Band Defined by the Majority

During the evolution in BI with increasing thermal maturity (%Ro, Tmax), the bitumen Index (BI) (Sykes, 2001; Sykes and Snowdon, 2002; Petersen, 2006; Killops et al. 1998) corresponds to the maturity (%Ro or Tmax) at which the BI value begin to decline which represents the start of the efficient oil window, or indicates the efficient oil expulsion (Petersen, 2006). From the results of descriptive analysis for the mean value of respective parameter in the interval of %Ro or Tmax ($^\circ$C) span, the BI exhibits the highest value of mean at the ~0.75 %Ro (within 0.7-0.8%Ro) and ~1.0%Ro (within 0.9-1.1%Ro), ~440°C Tmax (within 438-450°C Tmax x) and ~452°C Tmax (within 442-460°C Tmax) (Table 9). The evolution in BI with increasing %Ro for 608 and 506 samples is shown in Figures 6a and 7a. and the evolution in BI with increasing Tmax for 608 and 506 samples in Figures 6b and 7b. The data are distributed in an area (BI band) can be outlined around the vast majority of the samples. Lewan (1994, 1997) suggest the early generated compounds are presumed to be bitumen or heavy crude oil which forms oil by partial decomposition at higher maturity. As pyrolysis hydrocarbon S2 is decomposed continuously, will cause the decrease of S2 and the increase of free hydrocarbon S1, further result in the increase of BI. The HI value is reduced and BI value will increase until the hydrocarbon reaches the equilibrium amount of efficient oil repulsion, then BI value starts to decline (Suggate et al., 1993; Hu, 2001; Sun et al., 2001; Chan, 2006). As for samples of type II kerogen, there are few bonds with low dissociated activation energy at the stage of low maturation, and the free hydrocarbon (S1) lacked more than type III kerogen which can produce free hydrocarbon (S1) from the decomposed hydrocarbon (S2).

Since the samples in this study include type II / III kerogen, so their BI values will decrease gradually in lower maturity (%Ro, Tmax) other than the samples of humic coals and type III kerogen. A cross-plot of BI vs. %Ro shows that the majority of the organic materials form a BI band (Figures 6a, 7a). Similarly a cross-plot of BI vs. Tmax shows that the majority of the organic materials form a BI band (Figures 6b, 7b). The upper limit of the BI band shows a rapid increase from ~14mg HC/g TOC at about a Tmax of ~430°C to ~40mg HC/g TOC at a Tmax of ~440°C (Figures 6b, 7b), which, according to the %Ro–Tmax relationship in Fig. 8, corresponds to ~0.73%Ro. Similarly BI vs. %Ro shows an increase in BI from ~0.55%Ro to above 32mg HC/g TOC at ~0.75%Ro (Figures 4a, 4c). Above ~0.75%Ro or ~440°C, the upper limit of the BI band decreases to very low yields at approximately ~1.95%Ro or ~525°C. The lower limit of the BI band marks low values up to a %Ro of ~0.85%Ro or Tmax of ~445°C. The maximum of the lower limit of the BI band may be set at a %Ro of ~1.05%Ro (corresponds to Tmax ~455.48°C cf. Figure 8) or Tmax of ~455°C beyond which the lower limit decreases to very low values around ~1.25%Ro or ~465°C. Tmax=~465°C corresponds to ~1.2%Ro according to the %Ro–Tmax relationship in Figure 8. The sharp increase in BI at a Tmax of 430-440°C (or corresponds to 0.54-0.73%Ro according to the %Ro–Tmax relationship in Figure 8) marks the onset of petroleum generation (Petersen, 2002; Sykes, 2001; Sykes and Snowdon, 2002).

3.6. Effective Oil Expulsion Window

Petersen (2002) defined an effective oil window for source rock between vitrinite reflectance of 0.85 to 1.8%. At a Ro of 1.8% S1 yields are very low and the remaining generation potential according to its HI was stabilized at low values, the threshold of petroleum generation was determined to start at a Ro of about 0.5–0.6% (Tissot and Welte, 1984) and build-up of liquid petroleum occurs from approximately 0.5–0.85%Ro. In addition, several studies have recognized the initial increase in HI up to a maximum value (HImax) with increasing maturity (Huc et al., 1986; Teichmüller and Durand, 1983; Sykes, 2001; Sykes and Snowdon, 2002). HImax corresponds to the effective HI, according to Sykes (2001) and Sykes and Snowdon (2002). The beginning of petroleum expulsion is come first by petroleum build-up to a maximum BI. Petersen (2002) proposed that the effective oil window for humic coals and TypeIIIkerogen starts at a Ro of 0.85%.

The studied worldwide organic material data set shows that ~0.75%Ro corresponds to the upper limit of the BI band, whereas the lower limit of the band reaches a maximum at approximately Ro ~1.05% or Tmax 455°C (Figures6 and 7). In accordance with the results of descriptive analysis for the mean value of respective parameter in the interval of %Ro span, BI has the highest mean value at the Ro ~0.75%, Tmax 440°C (Table 9a-b). As noted previously, we can also receive the same result from the Pearson correlation analysis (Table 8b-c, Figures 5-4). Ro from 0.17~0.6% up to 0.61~1.0%, the correlation of HI and QI with BI turned from medium to low, as BI was increased with thermal maturity Ro up to ~0.75% (Figures 5-7). On the other hand, during Ro=0.17~1.0% up to 1.0~3.43%, the correlation of HI and QI with %Ro increased, with correlative coefficient up to r=-0.72 (Table 8c). Matching with the principle of Sykes (2001) and Sykes and Snowdon (2002), the line between these two maxima (Ro 0.85–1.05%; Tmax 440–455°C) outlines the line for the efficient liquid petroleum expulsion. The maximum BI line thus corresponds to the maturity range within which the start of the oil expulsion window for worldwide data set of the studied.

3.7. Relationships among Parameters

From Figures 5, 9, we can examine the relationships among parameters (%Ro, Tmax, HI, QI, BI, PI, S2, TOC, S1+S2, S1) for dataset of 608 samples. Tables 7-8, show the Pearson correlation analysis and the mean value of respective parameter in the interval of %Ro and Tmax (°C) span for data-sets of 608, 506, and 1140 samples respectively. The results are follows: the couple of parameters (%Ro vs. Tmax, HI vs. QI, TOC vs. [S1+S2], S2 vs. [S1+S2], S2 vs. TOC) exhibit high correlation. The parameters have the highest value of mean with increasing thermal maturity at Ro= ~ 0.6% (within Ro 0.5~0.7%) for (HI, QI, S1, S2, S1+S2), at Ro= ~ 0.75% (within Ro 0.7~0.8%)for (BI, PI), and at Ro= ~ 0.9% (within Ro 0.8~1.0%) for (S1, S2, S1+S2). From dataset of 608 samples, the parameters have the highest value of mean with increasing thermal maturity at Tmax = ~ 430°C (within Tmax 425~435 °C) for (HI, QI), at Tmax = ~ 450°C (within Tmax 442~460 °C) for (HI, QI, BI). As for 506 samples, the parameters have the highest value of mean with increasing thermal maturity at Tmax = ~ 430°C for (HI, QI, S1, S2, S1+S2), at Tmax = ~ 450°C for (HI, QI, BI, PI, S1, S2, S1+S2). As in Table 9c (interval of %Ro span), the parameter (HI) has the highest value of mean with increasing thermal maturity at Ro = ~ 0.6% and ~1.0%. Furthermore, the parameter (HI) has the highest value of mean with increasing thermal maturity at the Tmax = ~ 430 °C and ~ 450 °C. Accordingly, we confirm the upper and lower limit of the HI and QI band, and to define to the lines of maximum HI and QI from Ro ~ 0.6 to ~1.0% or from Tmax ~ 430 to ~450°C. Similarly, the upper and lower limit of BI band, the line of maximum BI is defined from Ro ~ 0.75 to ~1.05% or from Tmax ~ 440 to ~455°C.

Table 8b-c, show the bivariate correlation analysis of the intervals of %Ro span for data-set of 608 samples. From Ro=0.17~0.6% to Ro=0.61~1.0% for 10 parameters, the changes of correlativity turned from medium to low for (BI vs. HI, BI vs. QI, S2 vs. QI). The correlativity rises from low to medium for PI vs. BI. From Ro=0.17~1.0% to Ro=1.0~5.6% for 10 parameters, the changes of correlativity turned from being independent to medium(negative) for (HI vs. %Ro, QI vs. %Ro), and from medium to high for Tmax vs. %Ro, and from high to medium for (TOC vs. S2, TOC vs. [S1+S2]). The multiple regression analysis of BI, S2, and %Ro with HI (S2/TOC) and QI ([S1+S2]/TOC) show the strong positive relationship. Similarly the multiple regression analysis of HI, QI, BI, PI, S2, TOC, Tmax, and S1 with %Ro also shows the strong positive correlation (Table 10). On the other hand, the multiple regression for 97 samples (Ro=1.0~5.6%) with R^2=0.94, reach significance level of regression coefficients of independent variable, and no multicollinearity according to eigenvalue, condition index, variance proportions (Table 10). In addition, the curve regression of HI and QI with %Ro based on 97 samples (Ro=1.0~5.6%), we can get the exponential equation of curve regression, $y = 994.8e^{-1.7x}$ (HI, R^2=0.62, R=0.79), and $y = 1646.2e^{-2.0x}$ (QI, R^2=0.72, R=0.87). Therefore, the oil generation potential rapidly exhausted with thermal maturity beyond 1.0%Ro and their Pearson correlation coefficient r=

-0.72 (Table 8c), in accordance with the result gained from the (Figures 1-4, 6-7).

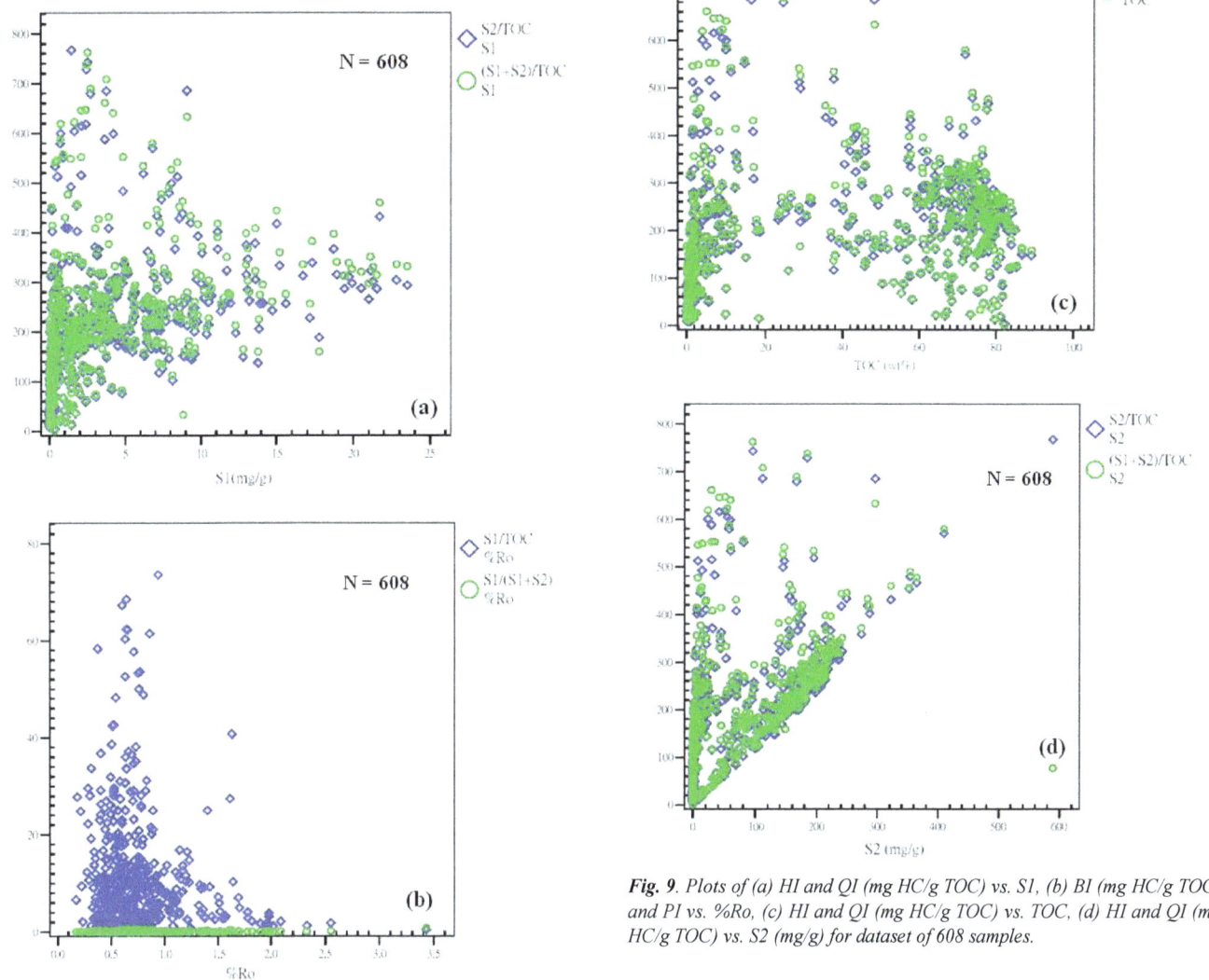

Fig. 9. Plots of (a) HI and QI (mg HC/g TOC) vs. S1, (b) BI (mg HC/g TOC) and PI vs. %Ro, (c) HI and QI (mg HC/g TOC) vs. TOC, (d) HI and QI (mg HC/g TOC) vs. S2 (mg/g) for dataset of 608 samples.

Table 10. Multiple regression analysis for different intervals of %Ro span in the datasets of 608 and 506 samples.

multiple regression analysis					
%Ro	0.17~0.60	0.61~1.00	0.17~1.00	1.00~5.60	0.17~5.60
Sample number	245	266	511	97	608
(A) adjusted R^2	0.01	0.02	0.07	0.75	0.36
(B) adjusted R^2	0.12	0.03	0.02	0.52	0.22
(C) adjusted R^2	0.19	0.25	0.46	0.94	0.81
(D) adjusted R^2	0.28	0.26	0.39	0.87	0.76
(E) adjusted R^2	0.48	0.44	0.46	0.75	0.51
(F) adjusted R^2	0.51	0.39	0.47	0.76	0.50

※   (A) Represent multiple regression analysis of HI, QI, BI, PI, TOC, S1, and S2 with %Ro

(B) Represent multiple regression analysis of HI, QI, BI, PI, TOC, S1, and S2 with Tmax

(C) Represent multiple regression analysis of HI, QI, BI, PI, TOC, S1, S2 and Tmax with %Ro

(D) Represent multiple regression analysis of HI, QI, BI, PI, TOC, S1, S2 and %Ro with Tmax

(E) Represent multiple regression analysis of BI, S2, and %Ro with HI

(F) Represent multiple regression analysis of BI, S2, and %Ro with QI

(a) For the 97 samples (%Ro=1.0~5.6), the t-values for regression coefficients of independent variables reach significance level ($p<0.05$), and no multicollinearity

(b) according to tolerance[>0], VIF[<10], eigenvalue[>0], condition index[<30], variance proportion[<1]

3.8. Comparison Analysis of Petroleum Potential

Synthesizing the above analysis, we can divide petroleum potential into four different regions, based on cross-plot of HI vs. %Ro (Figures 10a-b). The highest petroleum potential occurred in the second part (II), including 226 and 211 sample data (Ro=0.6-1.0%, HI>100) from the datasets of 608 and 506 samples (Figures 10a-b) respectively. The first part (I) represents the region of petroleum generation, and the forth part (IV) exhibits the region of lowest petroleum potential, and the oil generation potential was rapidly exhausted in the third part (III). The sample data of parameters for CJ1-38 samples, TW1-48 samples, ML1-59 samples, and AU1-13 samples were present in Table 11. The CJ1-38 samples from oilfield in eastern Junggar Basin, China (Chen, 2003), whereas the others samples were detected in this study. For CJ1-38, TW1-48, ML1-59, and AU1-13 samples, the evolution in the remaining petroleum potential (HI) with increasing thermal maturity is shown in Figure 10c. The majority of CJ1-38 samples are in the second part (II) (Figure 10c; the region of highest petroleum potential, cf. Figures 10a-b). The Pearson correlation analysis for datasets of 226 and CJ1-38 samples, the couple of parameters (BI vs. HI, BI vs. QI) have medium correlation (Table 12); as for the couple of the parameters ([S1+S2, S1] vs. [HI, QI, BI]) have no correlation for datasets of 226 and CJ1-38 samples in the second part (II). On the other hand, the couple of parameters (S2 vs. TOC, [S1+S2] vs. TOC, S1 vs. TOC) exhibit high correlation for datasets of 226 and CJ1-38 samples (Table 12, Figures 9), in accordance with the result that possess highest petroleum potential of CJ1-38 samples. The AU1-13 samples are in the third and forth parts (III/IV), with oil generation potential rapidly exhausted. The ML1-59 samples are in the first and second parts (I / II), the TW1-48 samples are in the first part (I), both represent the region of petroleum generation (Figure 10c).

They exhibit high correlation for the couple parameters (S2 vs. HI, [S1+S2] vs. HI, S1 vs. HI, S2 vs. QI, [S1+S2] vs. QI, S1 vs. QI, HI vs. BI, QI vs. BI) for AU1-13, ML1-59, and TW1-48 samples. Therefore, four datasets except CJ1-38 samples are similarly in the correlation for couple of parameters of respective datasets (Table 12). The non-parameteric tests (2-Independent Samples) for the datasets of TW1-48 and CJ1-38 samples, they exhibit significantly different ($p>\alpha$) for the distribution of data from respective parameters except S1 (Table 7d), also in accordance with CJ1-38 samples which possess the highest petroleum potential in the second part (II) (Figure 10c). According to Table 13, the CJ1-38 samples also have the highest mean value of HI and QI for the five data-sets (226, CJ1-38, TW1-48, ML1-59, AU1-13) although with the lowest mean value of TOC and S2. Because of the good correlation of S2, S1, and S1+S2 with TOC, they were believed to have few inert organic carbon (Dahl, 2004), that explained their highest petroleum potential (Table 13, Figures 9c-d).

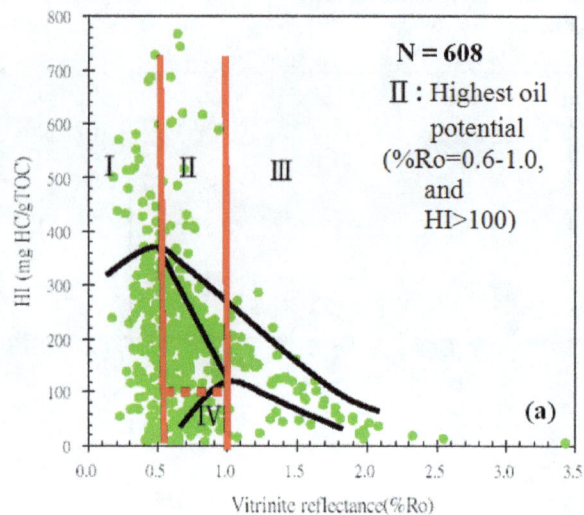

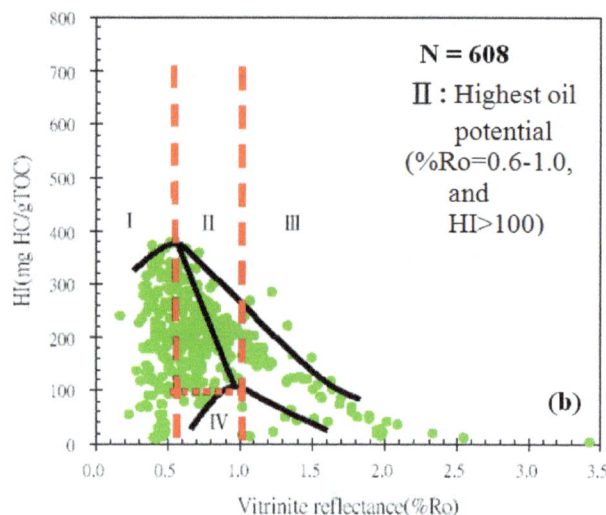

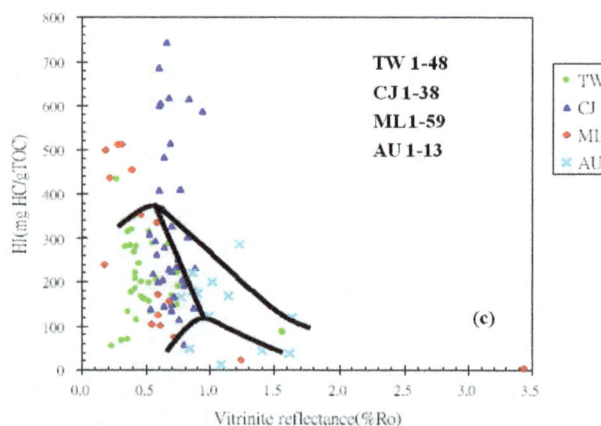

Fig. 10. The diagrams show the evolution in HI with increasing thermal maturity for datasets of 608 samples (a) and 506 samples (b). The plot (c) shows the evolution in the remaining petroleum potential (HI) with increasing thermal maturity for CJ1-38 samples, TW1-48 samples, ML1-59 samples, and AU1-13 samples.

Table 11. *This form shows the value of parameters for datasets of T1-48 samples, M1-59 samples, and AU1-13 samples. They were detected in this study.*

No.	Ro%	Tmax	S1	S2	TOC	HI	S1+S2	BI	QI	PI
ML001	0.68	438	0.41	12.02	5.26	229.00	12.43	7.79	236.31	0.03
ML002	0.73	433	9.43	183.88	78.40	235.00	193.31	12.03	246.57	0.05
ML003	0.59	429	3.75	133.26	67.01	199.00	137.01	5.60	204.46	0.03
ML004	0.82	440	21.26	220.09	73.13	301.00	241.35	29.07	330.03	0.09
ML005	0.52	438	0.57	15.05	4.88	308.00	15.62	11.68	320.08	0.04
ML006	0.53	436	0.26	3.27	2.37	138.00	3.53	10.97	148.95	0.07
ML007	0.57	437	0.61	16.92	5.76	294.00	17.53	10.59	304.34	0.03
ML008	0.63	435	0.79	17.15	8.38	205.00	17.94	9.43	214.08	0.04
ML009	0.88	443	1.06	12.17	5.26	231.00	13.23	20.15	251.52	0.08
ML010	0.70	447	0.50	6.06	1.85	328.00	6.59	27.03	356.22	0.08
ML011	0.69	446	2.09	29.43	5.71	515.00	31.52	36.60	552.01	0.07
ML012	0.70	438	0.34	5.44	3.67	148.00	5.78	9.26	157.49	0.06
ML013	0.64	443	2.19	9.73	3.51	277.00	11.92	62.39	339.60	0.18
ML014	0.63	445	3.22	19.58	5.34	367.00	22.80	60.30	426.97	0.14
ML015	0.63	445	6.47	44.53	12.31	362.00	51.00	52.56	414.30	0.13
ML016	0.72	447	0.11	2.01	1.22	165.00	2.12	9.02	173.77	0.05
ML017	0.94	441	3.64	29.09	4.95	588.00	32.73	73.54	661.21	0.11
ML018	0.78	435	0.77	6.10	2.96	206.00	6.87	26.01	232.09	0.11
ML019	0.66	449	2.51	95.43	12.85	743.00	97.94	19.53	762.18	0.03
ML020	0.79	444	0.38	4.07	2.12	192.00	4.45	17.92	209.91	0.09
ML021	0.83	448	2.10	41.43	6.74	615.00	43.53	31.16	645.85	0.05
ML022	0.64	444	4.86	34.25	7.09	483.00	39.11	68.55	551.62	0.12
ML023	0.70	440	0.05	1.13	0.83	136.00	1.18	6.02	142.17	0.04
ML024	0.75	448	0.29	2.17	1.88	115.00	2.46	15.43	130.85	0.12
ML025	0.60	441	1.21	7.37	1.80	409.00	8.58	67.22	476.67	0.14
ML026	0.60	447	3.75	111.71	16.31	685.00	115.46	22.99	707.91	0.03
ML027	0.61	447	1.63	54.31	8.99	604.00	55.94	18.13	622.25	0.03
ML028	0.60	446	0.74	23.64	3.94	600.00	24.38	18.78	618.78	0.03
ML029	0.76	441	1.03	20.23	4.94	410.00	21.26	20.85	430.36	0.05
ML030	0.68	446	2.38	50.84	8.23	618.00	53.22	28.92	646.66	0.04
ML031	0.87	441	0.34	2.23	1.58	141.00	2.57	21.52	162.66	0.13
ML032	0.79	444	0.11	0.64	1.14	56.00	0.75	9.65	65.79	0.15
ML033	0.64	439	0.39	5.22	3.58	146.00	5.61	10.89	156.70	0.07
ML034	0.55	434	0.20	3.47	1.59	218.00	3.76	12.58	236.48	0.05
ML035	0.58	437	0.20	4.20	1.60	263.00	4.40	12.50	275.00	0.05
ML036	0.61	436	1.30	43.32	12.59	344.00	44.62	10.33	354.41	0.03
ML037	0.70	440	2.19	51.91	23.28	223.00	54.10	9.41	232.39	0.04
ML038	0.74	431	6.45	124.79	49.29	253.00	131.24	13.09	266.26	0.05
M1	5.60	374	0.51	0.51	88.64	0.58	1.02	0.01	0.01	0.50

Table 11. (Continued)

No.	Ro%	Tmax	S1	S2	TOC	HI	S1+S2	BI	QI	PI
M2	0.69	424	0.97	160.54	73.80	217.54	161.51	0.01	2.19	0.01
M3	0.99	430	0.43	75.16	56.28	133.55	75.59	0.01	1.34	0.01
M4	0.59	421	1.74	176.40	66.50	265.25	178.14	0.03	2.68	0.01
M5	0.66	429	0.71	102.04	68.77	148.37	102.75	0.01	1.49	0.01
ML1	0.31	440	0.52	7.90	1.54	511.99	8.42	33.70	545.69	0.06
ML10	0.68	429	1.44	113.15	72.83	155.36	114.59	1.98	157.34	0.01
ML11	0.59	432	1.27	92.18	74.36	123.96	93.45	1.71	125.67	0.01
ML12	1.24	445	0.60	16.89	73.45	23.00	17.49	0.82	23.81	0.03
ML13	0.72	432	1.98	52.79	71.14	74.21	54.77	2.78	76.99	0.04
ML14	0.17	426	0.48	17.37	7.27	239.03	17.85	6.61	245.63	0.03
ML15	0.61	437	1.15	70.09	68.91	101.72	71.24	1.67	103.39	0.02
ML16	3.43	559	0.32	0.08	82.22	4.00	0.40	0.40	0.49	0.80
ML2	0.28	428	8.47	146.53	28.65	511.48	155.00	29.57	541.05	0.05
ML3	0.21	430	8.81	155.32	35.50	437.51	164.13	24.82	462.33	0.05
ML4	0.18	428	8.07	144.61	29.01	498.54	152.68	27.82	526.36	0.05
ML5	0.54	429	0.74	69.44	66.64	104.21	70.18	1.11	105.32	0.01
ML6	0.59	429	1.17	123.13	71.94	171.15	124.30	1.63	172.78	0.01
ML7	0.58	450	2.08	153.41	45.91	334.13	155.49	4.53	338.66	0.01
ML8	0.39	453	1.78	352.47	77.64	453.96	354.25	2.29	456.25	0.01
ML9	0.45	444	3.11	156.04	44.03	354.38	159.15	7.06	361.44	0.02
T1	0.44	429	1.39	160.64	75.30	213.34	162.03	0.02	2.15	0.01
T2	0.55	434	3.03	135.35	63.43	213.38	138.38	0.05	2.18	0.02
T3	0.53	425	5.82	211.86	70.46	300.68	217.68	0.08	3.09	0.03
T4	0.27	414	3.97	143.46	65.38	219.41	147.43	0.06	2.25	0.03
T5	0.29	419	3.26	194.45	72.43	268.45	197.71	0.05	2.73	0.02
T6	0.39	428	2.29	134.06	68.55	195.56	136.35	0.03	1.99	0.02
T7	0.77	437	8.31	216.84	78.62	275.80	225.15	0.11	2.86	0.04
T8	0.25	418	4.18	170.40	49.19	346.44	174.58	0.08	3.55	0.02
T9	0.34	413	3.14	157.22	50.09	313.89	160.36	0.06	3.20	0.02
T10	0.58	419	7.93	229.48	64.39	356.39	237.41	0.12	3.69	0.03
T11	0.26	421	4.63	217.11	61.73	351.73	221.74	0.08	3.59	0.02
T12	0.39	420	4.13	150.00	47.40	316.47	154.13	0.09	3.25	0.03
TW1	0.74	434	1.28	111.28	75.16	148.06	112.56	1.70	149.77	0.01
TW10	0.75	437	1.64	150.92	79.80	189.12	152.56	2.06	191.18	0.01
TW11	0.73	441	1.35	103.22	64.12	160.97	104.57	2.11	163.08	0.01
TW12	0.23	377	1.22	31.42	58.10	54.08	32.64	2.10	56.18	0.04
TW13	0.36	431	3.30	224.00	71.17	314.74	227.30	4.64	319.38	0.01
TW14	1.57	477	1.84	47.76	55.88	85.47	49.60	3.29	88.76	0.04
TW15	1.56	479	1.44	43.87	50.37	87.09	45.31	2.86	89.95	0.03
TW16	0.27	423	7.09	249.63	57.63	433.13	256.72	12.30	445.43	0.03
TW17	0.36	416	1.01	61.51	46.02	133.67	62.52	2.19	135.86	0.02

Table 11. (Continued)

No.	Ro%	Tmax	S1	S2	TOC	HI	S1+S2	BI	QI	PI
TW18	0.41	418	1.78	198.41	57.04	347.86	200.19	3.12	350.98	0.01
TW19	0.52	424	3.50	229.69	72.91	315.01	233.19	4.80	319.82	0.02
TW2	0.53	431	0.68	83.51	62.62	133.37	84.19	1.09	134.46	0.01
TW20	0.41	425	2.89	86.05	41.47	207.48	88.94	6.97	214.45	0.03
TW21	0.42	424	3.64	94.27	42.67	220.94	97.91	8.53	229.47	0.04
TW22	0.45	423	3.33	97.48	40.25	242.20	100.81	8.27	250.47	0.03
TW23	0.47	424	2.20	69.09	41.35	167.09	71.29	5.32	172.41	0.03
TW24	0.47	422	2.29	81.62	49.98	163.30	83.91	4.85	167.88	0.03
TW25	0.46	418	4.48	113.06	56.53	200.00	117.54	7.92	207.92	0.04
TW26	0.50	426	1.42	83.16	51.66	160.99	84.58	2.75	163.73	0.02
TW27	0.43	423	0.32	30.01	25.97	115.57	30.33	1.23	116.80	0.01
TW28	0.34	420	5.66	114.45	40.91	279.80	120.11	13.84	293.63	0.05
TW29	0.42	418	4.50	85.49	46.43	184.12	89.99	9.69	193.81	0.05
TW3	0.58	429	1.35	138.93	67.69	205.25	140.28	1.99	207.25	0.01
TW30	0.71	438	1.63	133.17	78.88	168.84	134.80	2.07	170.90	0.01
TW31	0.41	417	0.76	129.04	48.32	267.00	129.80	1.57	268.63	0.01
TW32	0.38	425	1.83	179.08	62.97	284.00	180.91	2.91	287.30	0.01
TW33	0.31	432	0.85	45.53	68.11	67.00	46.38	1.25	68.10	0.02
TW34	0.43	415	0.90	159.81	56.79	281.00	160.71	1.58	282.99	0.01
TW35	0.55	418	0.44	35.23	12.20	289.00	35.67	3.61	292.38	0.01
TW36	0.39	426	2.73	216.84	67.94	319.00	219.57	4.02	323.18	0.01
TW4	0.39	430	2.69	78.42	44.28	177.12	81.11	6.08	183.20	0.03
TW5	0.52	425	1.21	36.49	18.58	196.38	37.70	6.51	202.90	0.03
TW6	0.52	427	1.22	67.07	43.22	155.18	68.29	2.82	158.01	0.02
TW7	0.36	433	0.07	1.57	2.23	70.53	1.64	3.14	73.67	0.04
TW8	0.74	437	2.42	175.15	79.20	221.16	177.57	3.06	224.22	0.01
TW9	0.68	435	2.39	212.47	79.26	286.07	214.86	3.02	271.09	0.01
AU1	1.63	437	0.30	0.87	0.74	118.21	1.17	40.76	158.97	0.26
AU10	0.77	434	0.26	2.92	1.75	167.14	3.18	14.88	182.03	0.08
AU11	0.86	426	2.32	8.32	3.78	220.40	10.64	61.46	281.85	0.22
AU14	1.08	433	0.23	2.61	18.30	14.26	2.84	1.26	15.52	0.08
AU15	0.84	432	0.17	1.03	2.04	50.42	1.20	8.32	58.74	0.14
AU2	1.61	407	0.11	0.16	0.40	39.90	0.27	27.43	67.33	0.41
AU3	1.40	416	0.10	0.18	0.40	45.00	0.28	25.00	70.00	0.36
AU4	1.14	434	0.29	2.88	1.72	167.54	3.17	16.87	184.41	0.09
AU5	1.23	426	3.90	12.06	4.23	284.84	15.96	92.11	376.95	0.24
AU6	0.90	433	0.26	3.18	1.84	173.20	3.44	14.16	187.36	0.08
AU7	0.99	432	0.25	2.46	2.01	122.21	2.71	12.42	134.62	0.09
AU8	0.86	433	0.28	3.74	2.10	178.35	4.02	13.35	191.70	0.07
AU9	1.02	433	0.26	3.78	1.90	199.26	4.04	13.71	212.97	0.06

Table 12. *Distribution of Pearson's correlation coefficient of datasets of 226, CJ38, TW48, ML59, and AU13 samples*

Datasets		Pearson's correlation coefficient (r) r (couple of parameters)			
		G	M	L	EL
a	$0.6 \leqq \%Ro \leqq 1.0$ 226 samples	0.88(S vs C) 0.89(C vs A) 0.82(B vs P)	0.58(Q vs B) 0.54(C vs D)	0.44(H vs B)	(H vs S) (Q vs S) (B vs S) (H vs A) (Q vs A) (B vs A) (H vs D) (Q vs D) (B vs D)
b	$0.52 \leqq \%Ro \leqq 0.94$ CJ 38 38 samples	0.94(S vs C) 0.93(C vs A)	0.52(Q vs B) 0.78(C vs D) 0.67(B vs P)	0.43(H vs B)	(H vs S) (Q vs S) (B vs S) (H vs A) (Q vs A) (B vs A) (H vs D) (Q vs D) (B vs D)
c	$0.23 \leqq \%Ro \leqq 1.57$ TW 48 48 samples	0.83(H vs S) 0.81(Q vs S) 0.83(H vs A) 0.82(Q vs A) 0.88(B vs D)	0.66(S vs C) 0.65(C vs A) 0.55(H vs D) 0.58(Q vs D) 0.691(B vs P)	0.42(H vs B) 0.45(Q vs B)	0.16(C vs D)
d	$0.17 \leqq \%Ro \leqq 5.60$ ML 59 59 samples	0.81(H vs B) 0.83(Q vs B)	0.54(H vs S) 0.52(Q vs S) 0.65(H vs D) 0.66(Q vs D) -0.70(H vs C) -0.71(Q vs C)	-0.38(C vs D)	0.14(S vs C) 0.13(C vs A) -0.12(B vs P)
e	$0.77 \leqq \%Ro \leqq 1.63$ AU13 13 samples	0.82 (H vs S) 0.87 (Q vs S) 0.86(Q vs A) 0.82 (B vs A) 0.80(Q vs D) 0.92(B vs D)	0.61(H vs B) 0.75(Q vs B) 0.78(B vs S) 0.79 (H vs A) 0.68(H vs A)	0.48(B vs P)	0.15(S vs C) 0.14(C vs A) 0.08(C vs D)
※		1. The symbol of respective parameter ; %Ro(R), Tmax(T), HI(H), QI(Q), BI(B), PI(P), S2(S), TOC(C), [S1+S2](A), S1(D). 2. G represents high correlation ($|r| \geqq 0.8$). 3. M represents medium correlation($0.5 \leqq |r| < 0.8$). 4. L represents low correlation($0.3 \leqq |r| < 0.5$). 5. EL represents extremely low correlation ($|r| < 0.3$).			

Table 13. *Mean and 95% confidence interval of datasets of 226 samples, CJ1-38, TW1-48, ML1-59, and AU1-13 samples*

Parameters	Data sets				
	226 samples	38 samples	48 samples	59 samples	13 samples
	(Mean) 95% Confidence Interval for Mean				
%Ro	(0.73) 0.72~0.75	(0.68) 0.65~0.72	(0.58) 0.44~0.63	(0.71) 0.27~1.10	(1.15) 0.93~1.28
T max (℃)	(439.3) 437.2~439.2	(442.1) 439.3~442.7	(428.2) 422.2~433.2	(441.6) 426.1~460.3	(427.6) 423.8~434.1
HI	(237.2) 229.3~259.6	(326.6) 265.6~384.4	(208.6) 174.1~233.3	(254.3) 157.9~354.4	(137.2) 88.1~185.9
QI	(243.3) 229.3~259.6	(352.3) 286.7~411.5	(219.2) 177.5~237.4)	(266.5) 161.5~368.9)	(158.9) 103.9~222.6
BI	(11.8) 9.74~13.41	(24.1) 17.6~30.3	(4.28) 3.25~5.37	(9.46) 2.88~15.68	(27.1) 11.1~41.5
PI	(0.043) 0.038~0.050	(0.073) 0.058~0.086	(0.023) 0.018~0.027	(0.074) -0.028~0.179	(0.172) 0.096~0.239
S2	(111.2) 97.9~121.9	(40.6) 20.9~55.3	(116.9) 88.9~133.2	(106.2) 58.1~150.8	(3.12) 1.38~5.42
TOC	(47.2) 42.2~51.5	(13.4) 5.6~18.7	(54.2) 46.9~59.6	(52.7) 39.3~67.1	(3.27) 0.34~5.99
S1+S2	(113.6) 101.5~126.2	(42.3) 22.2~58.8	(117.8) 90.8~135.6	(110.2) 60.1~154.1	(3.98) 1.39~6.75
S1	(3.74) 3.29~4.49	(2.42) 1.11~3.60	(2.21) 1.63~2.67	(3.21) 1.04~4.21	(0.65) -0.01~1.35

(a) The 226 samples were obtained from the second part (the region of highest petroleum potential with Ro=0.6-1.0%,HI>100, cf. Figure 10) in the cross-plot of HI vs. %Ro for dataset of 608 samples.

(b) The 38 samples obtained from CJ1-38 (Chen, 2003, cf. table 6; TW1-48, ML1-21, AU1-13 cf. table 11).

4. Conclusion

The results from statistical analysis of 10 parameters data for all samples in this research work, not only be executed a linear regression, curve regression between any two parameters, and multivariate regression, but also be carried on the forecast of grey correlation grade of grey theory (include grey relational generating (Nominal-the-better-: Ro%; Larger-the-better-: T_{max}, HI, QI, BI, S2, S1+ S2, S1; smaller-the-better-: TOC, PI) and globalization grey relational grade). The highest petroleum potential is located in the range with %Ro=0.6-1.0%, and HI>100. The oil expulsion window is located in the range with %Ro=0.75-1.05% or T_{max}=440-455 °C. The range of by Ro%, T_{max}, cross-plots, Grey Model and statistical analysis are all in good agreement.

The hierarchial cluster analysis dendrogram (Q mode) based on data from 10 parameters of TW1-48 and CJ1-38 samples reveals a high similarity between the majority of CJ1-38 samples. In addition, the majority of CJ1-38 samples possess high petroleum potential in the second part (II), and the maturation of TW1-48 samples from immaturity to the oil window falls into first and second parts (I and II). Experimental and statistical analytical investigation reveals that the values of eight parameters (HI, QI, BI, PI, S2, S1+ S2, and S1) increase as the thermal maturity of organic materials increases during the initial stage of thermal maturation. At maturities greater than Ro=0.6-1.0% (Tmax = 430-450°C), the values of these parameters start to decrease gradually. At Ro>1.0%, the values of those parameters decrease rapidly, with a corresponding drop in petroleum potential.

Based on statistical analysis and cross-plots of HI, QI and BI versus the vitrinite reflectance (%Ro) and Tmax (°C), the HI_{max}, QI_{max} and BI_{max} lines are defined between the vertex of the upper and lower limits of the HI-band, QI-band and BI-band as defined by a majority of the samples. The constructed HI, QI and BI bands were broad at low maturities and gradually narrowed with increasing thermal maturity. The petroleum generation potential is completely exhausted at a vitrinite reflectance of 2.0-2.2% or a T_{max} of 510-520°C. An increase in HI and QI suggests extra petroleum potential related to changes in the structure of the organic material. A decline in BI signifies the start of the oil expulsion window and occurs within the vitrinite reflectance range 0.75-1.05% or a T_{max} of 440-455 °C. Furthermore, petroleum potential can be divided into four different parts based on the cross-plot of HI vs. %Ro. The area with the highest petroleum potential is located in section II with %Ro=0.6-1.0%, and HI>100. Oil generation potential is rapidly exhausted at section III with %Ro >1.0%. This result is in accordance with the regression curve of HI and QI with %Ro based on 97 samples with %Ro=1.0~5.6%. The exponential equation of regression can thus be achieved: curve, $HI = 994.8e^{-1.7Ro}$ and $QI = 1646.2e^{-2.0Ro}$ (R^2=0.72). The worldwide organic material dataset defines two oil expulsion windows represented by the upper and lower limits of the BI band: %Ro ~0.75 to ~1.95% or T_{max} ~440 to ~525°C, and %Ro ~1.05 to ~1.25% or T_{max} ~455 to ~465°C, respectively. The start of the oil expulsion window occurs within the %Ro range of ~ 0.75–1.05%Ro or the T_{max} range ~ 440-455°C and the total oil window extends to %Ro = ~ 1.25-1.95 or T_{max} = ~ 465-525°C.

As the results of this study, we expect to promote evaluation techniques for HC exploration and propose new guidelines for evaluating the petroleum potential of organic matter. A significant petroleum generation build-up occurs at an approximate Ro% interval of 0.6 to 0.75, as indicated by a worldwide sample dataset of Coal and Carbonaceous Materials. S_2 yields and HI values stabilize at low values –

approximately at a vitrinite reflectance of 2.2%, which indicates the exhaustion of the petroleum generative potential. For general purposes, the "oil window" for organic materials in this study is defined within a Ro ranging from 0.75 -1.95%. This re-considered "oil window" is very probably also valid for source rocks of Coal and Carbonaceous Materials. Moreover, samples with Ro%<0.6 or Ro%=0.6-1.0 and HI<100 are the least effective oil source rocks, and samples with Ro%=0.6-1.0 and HI>100 are the most effective oil source rocks.

Acknowledgements

Author would like to give thanks to anonymous reviewers of this manuscript for giving constructive suggestions and revision comments. Special thanks to all of the faculties and staff of the Geochemical Department of Exploration and Development Research Institute, CPC., for providing valuable samples and analyzing assistance. This research was financially supported by Grants NSC-102-2116-M-252-001- of National Science Council.

References

[1]　Akande SO, Ojo OJ, Erdtmann BD, Hetenyi M (1998) Paleoenvironments, source rock potential and thermal maturity of the Upper Benue rift basins, Nigeria : implications for hydrocarbon exploration. Organic Geochemistry29 (1-3): 531-542

[2]　Amijaya H, Littke R (2006) Properties of thermally metamorphosed coal from Tanjung Enim Area, South Sumatra Basin, Indonesia with special reference to the coalification path of macerals. International Journal of Coal Geology 66:271-295

[3]　Arfaoui A, Montacer M, Kamoun F, Rigane A (2007) Comparative study between Rock-Eval pyrolysis and biomarkers parameters: A case study of Ypresian source rocks in central-northern Tunisia. Marine and Petroleum Geology 24:566-578

[4]　ASTM (1975) Standard D-2797, ASTM Standard manual. Part 26:350-354

[5]　ASTM (1980) Standard D-2797, Microscopical determination of volume percent of physical components in a polished specimen of coal, ASTM. Philadelphia, Pa.

[6]　Banerjee A, Sinha AK, Jain AK, Thomas NJ, Misra KN, Chandra, K (1998) A mathematical representation of Rock-Eval hydrogen index vs. Tmax profiles. Organic Geochemistry28 (no.1/2):43-55

[7]　Bordenave ML, Espitalie' J, Leplat P, Oudin JL, Vandenbroucke M (1993) Screening techniques for source rock evaluation, In: Bordenave, M.L. (Ed.), Applied Petroleum Geochemistry. Editions Technip, Paris, pp 219–224

[8]　BostickNH, Daws TA (1994) Relationships between data from Rock-Eval pyrolysis and proximate, ultimate, petrographic, and physical analyses of 142 diverse U. S. coal samples. Organic Geochemistry21:35–49

[9] Canonico U, Tocco R, Ruggiero A, Suarez H (2004) Organic geochemistry and petrology of coals and carbonaceous shales from western Venezuela. International Journal of Coal Geology 57:151-165

[10] Chen J, Liang D, Wang X, Zhong N, Song F, Deng C, Shi X, Jin T, Xiang S (2003)Mixed oils derived from multiple source rocks in the Cainan oilfield, Junggar Basin, Northwest China. Part I : genetic potential of source rocks, features of biomarkers and oil sources of typical crude oil. Organic Geochemistry 34:889-909

[11] Chen J-P, Deng C-P, Wang H-T, Han D-X (2006) Genetic potential and geochemical features of pyrolysis oils of macerals from Jurassic coal measures, Northwest China. Geochimica 1:81-87

[12] Chiu J-H, Kuo C-L, Lin H-J, Chou T-H (1993) Geochemical modeling of source rock hydrocarbon generation potential. Exploration Development Research Report 16: 232-256

[13] Chiu J-H, Kuo C-L, Wu S-H, Lin L-H, Shen J-C, Chou T-H (1996) Simulation of the hydrocarbon generation potential of the coal samples. Exploration Development Research Report 19:420-453

[14] Cooles GP, Mackenzie AS, Quigley JM (1986) Calculation of petroleum masses generated and expelled from source rocks. Organic Geochemistry 10:235-245

[15] Dahl B, Bojesen-Koefoed J, Holm A, Justwan H, Rasmussen E, Thomsen E (2004) A new approach to interpreting Rock-Eval S2 and TOC data for kerogen quality assessment. Organic Geochemistry35:1461-1477

[16] Davis RC, Noon SW, Harrington J (2007) The petroleum potential of Tertiary coals from Western Indonesia: Relationship to mire type and sequence stratigraphic setting. International Journal of Coal Geology 70:35-52

[17] Deng J-L (1988) Essential topics on grey system: theory and application, China Ocean Press, pp 327

[18] Espitalie'J, Deroo G, Marquis F (1985) La pyrolyse Rock-Eval et ses applications. Revue Institut Franc- ais du Pe' trole Part I40:563–578, Part II40:755–784

[19] Espitalie' J, Laporte JL, Madec M, Marquis F, Leplat P, Paulet J, Boutefeu F (1977) Me' thode rapide de caracte' risation des roches me' res, de leur potentiel pe' trolier et de leur degre' d'e' volution. Revue Institut Franc- ais du Pe' trole 32:23–42

[20] Hu C-J (2001) The application of Rock-Eval 6 in geochemical exploration. Exploration Development Research Report 23:367-378

[21] Hunt JM (1996) Petroleum Geochemistry and Geology. W.H. Freeman and Company, New York, pp 743

[22] ISO 7404-5 (1994E)Methods for the petrographic analysis of bituminous coal and anthracite Part 5 Method of determining microscopically the reflectance of vitrinite, International Standard, 2nd ed., pp 1-13

[23] Karakitsios V, Rigakis N (2007) Evolution and petroleum potential of Western Greece, Journal of Petroleum Geology 30(3):197-218

[24] Katz BJ (1983) Limitation of 'Rock-Eval' pyrolysis for typing organic matter, Organic Geochemistry 4:195-199

[25] Keller G (2001) Applied statistics with Microsoft Excel, Thomson Learning Asia Pte Ltd, Beijing, pp 670

[26] Killops SD, Funnell RH, Suggate RP, Sykes R, Peters KE, Walters C, Woolhouse AD, Weston RJ(1998) Predicting generation and expulsion of paraffinic oil from vitrinite-rich coals. Org. Geochem 29(1-3):1-21

[27] Killops SD, Funnell RH, Suggate RP, Sykes R, Peters KE, Walters C, Woolhouse AD, Weston RJ, Boudou J-P (1998) Predicting generation and expulsion of paraffinic oil from vitrinite-rich coals. Organic Geochemistry 29:1–21

[28] Kotarba MJ, ClaytonJL, Rice DD, Wagner M (2002) Assessment of hydrocarbon source rock potential of Polish bituminous coals and carbonaceous shales, Chemical Geology184:11-35

[29] Kotarba MJ, Wiectaw D, Koltun YV, Marynowski L, Kusmierek J, Dudok IV (2007) Organic geochemistry study and genetic correlation of natural gas, oil and Menilite source rocks in the area between San and Stryi rivers (Polish and Ukrainian Carpathians). Organic Geochemistry 38(8): 1431-1456

[30] Kotarba M, Lewan MD (2004) Characterizing thermogenic coalbed gas from Polish coals of different ranks by hydrous pyrolysis. Organic Geochemistry35:615-646

[31] Lee H-T (2011) Analysis and characterization of samples from sedimentary strata with correlations to indicate the potential for hydrocarbons, Environmental Earth Sciences 64:1713-1728

[32] Lee, H-T, Sun, L-C (2013) The atomic H/C ratio of kerogen and its relation to organic geochemical parameters : implications for evaluating hydrocarbon generation of source rock, Carbonates and Evaporites 28(4):433-445

[33] Liu D-H, Zhang H-Z, Dai J-X, Sheng G-Y, Xiao X-M, Sun Y-H, Shen J-G (2000) The research and evaluation of forming hydrocarbon from micro-constituent of coal rock. Chinese Science Bull.45(4):346-352

[34] Magoon LB, Dow WG (1994) The petroleum system-from source to trap. AAPG Memoir 60, Tulsa, Oklahoma, U.S.A., pp 655

[35] Newman J, Boreham CJ, Ward SD, Murray AP, Bal AA (1999) Floral influences on the petroleum source potential of New Zealand coals. In: Masterlerz, M., Glikson, M., Golding, S.D. (Eds.), Coalbed Methane: Scientific, Environmental and Economic Evaluation. Kluwer Academic pp 461–492

[36] Norgate CM, Boreham CJ, WilkinsAJ (1999) Changes in hydrocarbon maturity indices with coal rank and type, Buller Coalfield, New Zealand. Organic Geochemistry30:985-1010

[37] Pedersen GK, Andersen LA, Lundsteen EB, Petersen HI, Bojesen-Koefoed JA, Nytoft HP (2006) Depositional environments, organic maturity and petroleum potential of the Cretaceous Coal-Bearing Atane Formation at Qullissat, Nuussuaq Basin, West Greenland. Journal of Petroleum Geology 29(1):3-26

[38] Pepper AS, Corvi PJ (1995) Simple kinetic models of petroleum formation. Part I: oil and gas generation from kerogen. Marine Petroleum Geology 12 (3):291–319

[39] Peters KE, Cassa MR (1994) Applied source-rock geochemistry, In: Magoon, L.B., Dow, W.G. (Eds.), The Petroleum System—From Source to Trap. American Association of Petroleum Geologists, Memoir 60:93–120

[40] Peters KE (1986) Guidelines for evaluating petroleum source rocks using programmed pyrolysis. American Association of Petroleum Geologists Bulletin 70:318–329

[41] Petersen HI (2002) A reconsideration of the "oil window" for humic coal and kerogen type III source rocks. Journal of Petroleum Geology 25(4):407-432

[42] Petersen HI (2006) The petroleum generation potential and effective oil window of humic coals related to coal composition and age. International Journal of Coal Geology 67:221-248

[43] Petersen HI, Nytoft HP, Nielsen LH (2004) Characterisation of oil and potential source rocks in the northeastern Song Hong Basin, Vietnam: indications of a lacustrine-coal sourced petroleum system. Organic Geochemistry 35:493-515

[44] Petersen HI, Rosenberg P, Andsbjerg J (1996) Organic geochemistry in relation to the depositional environments of Middle Jurassic coal seams, Danish Central Graben, and implications for hydrocarbon generative potential. AAPG Bulletin 80(1):47-62

[45] Petersen HI, Tru V, Nielsen LH, Due NA, Nytoft HP (2005) Source rock properties of lacustrine mudstones and coals (Oligocene Dong Ho formation), onshore Song Hong Basin, Northern Vietnam. Journal of Petroleum Geology28(1):19-38

[46] Powell TG, Boreham CJ, Smyth M, Russell N, Cook AC (1991) Petroleum source rock assessment in non-marine sequences : pyrolysis and petrographic analysis of Australian coals and carbonaceous shales. Organic Geochemistry17(3):375-394

[47] RabbaniAR, Kamali MR (2005) Source rock evaluation and petroleum geochemistry offshore SW Iran. Journal of Petroleum Geology 28(4):413-428

[48] Sachsenhofer RF, Privalov VA, Izart A, Elie M, Kortensky J, Panova E A, Sotirov A, Zhykalyak MV (2003) Petrography and geochemistry of Carboniferous coal seamsin the Donets Basin (Ukraine): implications for paleoecology. International Journal of Coal Geology55:225-259

[49] Suggate RP, Boudou JP (1993) Coal rank and type variation in Rock-Eval assessment of New Zealand coals, J. Pet. Geol. 16:73–88

[50] Sun X-G, Qin S-F, Luo J, Jin K-L (2001) A study of activation energy of coal macerals. Geochimica 30(6):559-604

[51] Sykes R (2001) Depositional and rank controls on the petroleum potential of coaly source rocks. In: Hill, K.C., Bernecker, T. (Eds.), Eastern Australasian Basins Symposium, a Refocused Energy Perspective for the Future. Petrol. Expl. Soc. Austral., Spec. Publ.pp 591–601

[52] Sykes R, Snowdon LR (2002) Guidelines for assessing the petroleum potential of coaly source rocks using Rock-Eval pyrolysis. Org. Geochem. 33:1441–1455

[53] Taylor GH, Teichmüller M, Davis A, Diessel CFK, Littke R, Robert P (1998) Organic Petrology. Gebrüder Borntraeger, Berlin, Stuttgart, pp 704

[54] Teichmüller M, Durand B (1983) Fluorescence microscopical rank studies on liptinites and vitrinites in peat and coals, and comparison with results of the Rock-Eval pyrolysis. Int. J. Coal Geol. 2:197–230

[55] Tissot BP, Welte DH (1984) Petroleum formation and occurrence ; a New approach to oil gas exploration. Springer-Verlag, Berlin, Heidelberg, New York, pp 699

[56] Tissot BP, Pelet R, Ungerer P (1987) Thermal history of sedimentary basins, maturation indices, and kinetics of oil and gas generation. AAPG Bull. 71:1445–1466

[57] Vassoevich NB, Akramkhodzhaev AM, Geodekyan AA (1974) Principal zone of oil formation. In: Tissot, B., Bienner, F. (Eds.), Advances in Organic Geochemistry 1973, Éditions Technip, Paris,pp309–314

[58] Veld H, Fermont WJJ, Jegers LF (1993) Organic petrological characterization of Westphalian coals from The Netherlands : correlation between Tmax, vitrinite reflectance and hydrogen index. Organic Geochemistry20(6):659-675

[59] Wang C-J (1998) A "folded-fan" method for assessment on the hydrocarbon generating potential of coals. Geochimica 27(5):4 83-492

[60] Wen K-L (2004) Grey Systems Modeling and Prediction, Yang's Scientific Research Institute, USA, pp 468

[61] Wu S-H, Weng R-N, Shen J-Q, Sun Z-X, Guo Z-L (2003) The study for constituent characteristics of the hydrocarbon compound of coal and coal shales in the NW Taiwan, Exploration Development Research Report 25:229-239

[62] Xiao XM, Hu ZL, Jin YB, Song ZG (2005) Hydrocarbon source rocks and generation history in The Lunnan Oilfield area, northern Tarim Basin (NW China). Journal of Petroleum Geology28(3):319-333

[63] Xiao X (1997) The organic petrological characteristics of Triassic source rocks and their hydrocarbon-generating potential in Tarim Basin. Geochimica 26(1):64-71

[64] Xiao X, Liu D, Fu J (1996) The evaluation of coal-measure source rocks of coal-bearing basins in China and their hydrocarbon-generating models. Acta Sedimentologica Sincia 14(supp.):10-17

[65] Zhang T-P, Zhang Y-C, Cai K-Z (2007) SPSS Statistic modeling and analytic procedure. Kings Information Co., Ltd., Taipei, pp 674

21

Energy Pricing for Households in Europe

Novosad Valentyna[1], Kolosova Viktoria[2]

[1]Science Company "MAE", Kyiv, Ukraine
[2]Department of Debt and International Financial Policy of the Ministry of Finance, Kyiv, Ukraine

Email address:

mae2010@meta.ua (N. Valentyna), centre_srr@hotmail.com (K. Viktoria)

Abstract: Every year the energy consumption of households increases. In this regard, the cost of the consumed energy becomes an essential part of the budget of each household. Because European countries import a significant part of required to them of energy resources, the prices of electricity and natural gas for households in these countries are higher than in other countries. However, the levels of abilities of households to pay the bills for the consumed energy in different countries are different. In this work we have tried to explore the relationships between the average salary and the prices for the most important energy resources for households in different European countries. These studies have shown that some relationships between salary and energy prices have impact on the political situation in some countries. Therefore it's necessary in the energy pricing for households taking into account the levels of the household budgets of their own country.

Keywords: Energy, Energy Pricing, Economy, Marketing

1. Introduction

With the development of modern society, improvement of the living standards, more energy is used at home. None of the modern homes today can do without energy resources. Energy charges are a significant part of the family budget expenses. Therefore energy pricing for households is a subject to review and adjustment by each country. This question is particularly important for European countries since these countries have the highest energy consumption per capita and the highest energy prices. High energy consumption indicators of the population in North America and in Europe are due to climatic conditions and a higher standard of living than in other continents.

Table 1. Electricity consumption per capita by population in the world.

	Country	Average electricity consumption by population in 2011, kWth per capita
1	Norway	7350
2	USA	4590
3	France	2390

	Country	Average electricity consumption by population in 2011, kWth per capita
4	UK	1840
5	Germany	1760
6	Spain	1700
7	Greek	1420
8	Ukraine	1384*
9	Czech Republic	1350
10	Bulgaria	1300
11	Italy	1160
12	Russia	915
13	Польша	760
14	Турция	610

Source of information [1]
*Source of information-[2].

2. Main Part

To understand what the needs are and what energy is consumed by the population in different countries, we look at the structure of their consumption according to the types of needs and energy types (see tables 2 and 3 below).

Table 2. Energy Consumption per Household by Final Energy Use (Gj/household/year).

Country	Common consumption	Space heating	Space cooling	Water heating	Cooking	Water heating and cooking	Lighting and other	Kerosene for lighting	Other
USA	101	46	6	17	4		25		3
Australia	53	21		15	3		14		

Country	Common consumption	Space heating	Space cooling	Water heating	Cooking	Water heating and cooking	Lighting and other	Kerosene for lighting	Other
UK	83	50		18	2		10		3
France	74	54		7	4		9		
Germany	74	58		7	2		7		
Japan	41	11	1	14	3		12		
South Korea	58	35		10	4		9		
China-urban	18	8	1			7	2		
China-rural	55	20				33	2		
India-urban	15					10	4	1	
India-rural	24					21	1	2	
Vietnam	6					1	3	2	

The source of Information: [3]

Table 3. Energy Consumption per Household by Energy Type (Gj/household/year).

Country	All sources	Kerosene	LPG	Natural gas	Goal	Electricity	Fuel wood	Agricultural waste	District heating	Other
USA	101	7	4	48	4	38				
Australia	53	1	1	17		24	10			
UK	83	5	1	56	3	17				1
France	74	15	3	24	1	19	12			
Germany	74	23	1	27	1	13	5		4	
Japan	41	10	6	10		15				
South Korea	58	18	6	21		10			3	
China- urban	18	3		2	7	4			2	
China –rural	55				8	2	18	26		1
India-urban	15	3		3	4	5				
India-rural	24	2				1	18			3
Vietnam	6						1	2		3

The source of information: [3]

The data in Table 2 demonstrates the various needs of the end consumer of energy resources. The basic needs of the population for energy is space heating, hot water, lighting and home appliances. The Table 3 shows the structure of the use of energy by type. These two tables reflect the level of development of the society in different countries, the conditions in which people live and the availability of certain natural resources. The population of the more developed countries and the more northern countries consume greater amounts of energy than the population of the less developed countries and countries with warm climate. Therefore the greatest energy consumption of one household per year is in the United States and the United Kingdom and the lowest is in Vietnam. The difference between the energy consumption in the United States and Vietnam is about 17 times.

There is a difference in the energy consumption of urban and rural populations. The urban population of China and India consume significantly less energy than rural. Natural gas and electricity are the main energy sources, which are used in everyday life European countries have.

These two tables show that the consumption of energy by households with the development of society will increase as natural gas and electricity in the near future will remain the main energy used by people in everyday life.

Taking into account the above factors we will consider the pricing of natural gas and electricity for the population as a major issue in the general policies of each country and as

such, the matter that can significantly affect the well-being of the population.

Since energy consumption in the European countries is a significant proportion of the budget of each European family we consider the current price of natural gas and electricity for the population through the prism of the average wage in European countries. To make it easier to compare countries with each other we divide them into several categories for research and take the most prominent representatives of these categories.

Category A – countries with the average salary of $2,000 to $3,000 per month.

Category B – countries with the average salary of $1,000 to $2,000 per month.

Category C – countries with the average salary of $500 to $1000 per month.

Category D – countries with the average salary up to $500.

Table 4. Country with the average salary of $2,000 to $3,000 per month.

	Country	Average salary, $/month (in 2014)
1	UK	3246
2	France	2660
3	Germany	2568
4	Italy	2404
5	Spain	2019

Source of information: [13]

Table 5. *Country with the average salary of $1000 to $2000 per month.*

	Country	Average salary $/month (in 2014)
1	Estonia	1052
2	Greece	1023

Source of information: [13]

Table 6. *Country with the average salary of $500 to $1000 per month.*

	Country	Average salary $/month (in 2014)
1	Czech Republic	877
2	Poland	843

Source of information: [13]

Table 7. *Country with an average salary up to $500.*

	Country	Average salary $/month (in 2014)
1	Romania	498
2	Bulgaria	416
3	Belarus	397
4	Moldova	243
5	Ukraine	206

Source of information [13], [8], [9], [11]

In order to understand the effect of the state on the level of average household income, we have indicated in the tables the average salary levels, and determined the relationship between the minimum established by the state and the average income of each country.

And now we will look at the price of natural gas and electricity in comparison to the medium and minimum levels of income in different categories of population and the possibility of improving the situation in some countries.

2.1. Natural Gas Prices

Natural gas prices for the population in Europe are very dependent on the imports of the natural gas, mainly from Russia. Therefore the price of natural gas for the population depends to some extent on the distance from country that is the main supplier of this type of fuel. The farther the selected country is from Russia the higher is the income and price for the natural gas. Countries located closer to Russia have lower level of development and consequently lower income, however, these countries have the ability to lower prices for natural gas as the cost of transportation to the borders of these countries is much cheaper. Below we will further consider the relationship between the income per capita and prices for 1000 m3 of natural gas. (Or, respectively, consumption).

Table 8. *Country of category A and their natural gas prices for households.*

	Country	Average salary $/month (in 2014)	Natural gas prices for households, $/1000m3	% prices in the salary
1	UK	3246	589	18%
2	France	2660	657	25%
3	Germany	2568	546	21%
4	Italy	2404	658	27%
5	Spain	2019	761	38%

Source of information: [13], [14]

Table 9. *Country of category B and their natural gas prices for households.*

	Country	Salary per month,$/month (2014)	Natural gas prices for households, $/1000m3	% prices in the salary
1	Estonia	1052	393	37%
2	Greece	1023	766	75%

Source of information: [13], [14]

Table 10. *Country of category C and their natural gas prices for households.*

	Country	Salary per month,$/month (2014)	Natural gas prices for households, $/1000m3	% prices in the salary
1	Czech Republic	877	500	57%
2	Poland	843	436	52%

Source of information:[13],[14]

Table 11. *Country of category D and their natural gas prices for households.*

	Country	Salary per month,$/month (2014-2015)	Natural gas prices for households, $/1000m3	% prices in the salary
1	Romania	498	169	34%
3	Belarus	397	168	42%
4	Moldova	243	355	146%
5	Ukraine	206	334	162%

Source of information:[13],[8],[9],[11],[14]

These tables indicate that the population in the majority of countries with lower income use all the features and benefits of a close geographical location to the country - main supplier of natural gas to Europe. However, ignoring the objective conditions by some countries leads to an unstable political situation in those countries. Adjusting the price of natural gas, it is necessary to consider not only the cost, the needs of the country in filling the budget, but also a prudent relationship between the ability of the population to pay bills without a sharp reduction of its well-being and needs of the gas industry and the state.

2.2. Electricity Prices

Electricity prices follow the similar pattern as many power plants use natural gas as fuel in the production of electrical and thermal energy. However, the overall situation in the electricity pricing is much better. With the development of market relations in the electricity industry, the opportunity to trade electricity between different regions of the country and even between countries themselves provides the great opportunity to align production schedules and reduce the price indices due to the timing differences and lower losses in the transmission systems.

Development of unconventional sources of electricity production provides additional opportunities for expanding of the frontiers of the domestic energy use and improving the ecological situation. The best ratio of the average wage and the cost of electricity for the population indicate the prospects for its further use.

Table 12. Country of category A and their electricity prices for households.

	Country	Salary per month, $/month (2014)	Electricity prices for households,$/100 kWh	% electricity prices in salary
1	UK	3246	25,7	0,8
2	France	2660	21,4	0,8
3	Germany	2568	40,3	1,6
4	Italy	2404	32,9	1,4
5	Spain	2019	30,3	1,5

Source of information: [13], [5,6]

Table 13. Country of category B and their electricity prices for households.

	Country	Salary per month,$/month (2014)	Electricity prices for households,$/100 kWh	% electricity prices in salary
1	Estonia	1052	17,6	1,7
2	Greece	1023	23,8	2,3

Source of information: [13], [5.6]

Table 14. Country of category C and their electricity prices for households.

	Country	Salary per month,$/month (2014)	Electricity prices for households,$/100 kWh	% electricity prices in salary
1	Czech Republic	877	17,4	2,0
2	Poland	843	19,1	2,3

Source of information: [13], [5.6]

Table 15. Country of category D and their electricity prices for households.

	Country	Salary per month,$/month (2014-2015)	Electricity prices for households,$/100 kWh	% electricity prices in salary
1	Romania	498	17,4	2,5
2	Bulgaria	416	11,2	2,7
3	Belarus	397	7,6	1,9
4	Moldova	243	12,2	5
5	Ukraine	206	3,7	1,8

Source of information: [13], [8], [9], [11], [5, 6]

3. Conclusions

The statistical data on the income levels of different countries and the prices for the main types of energy demonstrate the capabilities of various countries to provide a decent standard of living for its population.

While establishing the energy prices for the population, one should remember that these prices have a very significant impact on the budgets of each household. Therefore it is necessary to make use of all the opportunities available to decrease such price in order to avoid the negative impact on the overall political situation in the country due to worsening of the living standards of the population.

References

[1]   www.proatom.ru//files/doc/spravka.docxd

[2]   www.ukrstat.org/uk

[3]   Hidetoshi Nokagami, Chihanu Muracoshi and Yumiko Iwafune, Jyukankyo, Research Institute "International Comparison of Household Energy Consumption and its Indicator".

[4]   www.eea.europa.eu

[5]   www.m.ria.ru/infografica

[6]   www.oilexp.ru/new

[7]   www.cbr.ru

[8]   www.index.minfin.com.ua

[9]   www.myfin.by/info

[10]   www.usd.ru.fixexchangerate.com

[11]   www.ru.sputnik.md/society/20150923

[12]   www.calk.ru/kurs-MDL-USD

[13]   www.remisfisher.com.2014average-salary-european-union

[14]   Electricity and natural gas price statistics www.ec.europa.eu

22

Feasibility and Estimation of Technical Potential and Calculation of Payback Period of Roof-Top Solar PV System in the City of Majmaah, Province of Riyadh, K.S.A

Ahmed-Bilal Awan[1, *], Taher Shaftichi[1], Ahmed G. Abu-Khalil[1, 2]

[1]Electrical Engineering Department, College of Engineering, Majmaah University, Majmaah, Saudi Arabia
[2]Electrical Engineering Department, Faculty of Engineering, Assiut University, Assiut, Egypt

Email address:

a.awan@mu.edu.sa (Ahmed-Bilal A.)

Abstract: Depletion of fossil fuels in near future and accumulation of their emissions in the environment have attracted the world's attention to utilize renewable resources of energy. Oil and gas are the two main sources of power generation in Kingdom of Saudi Arabia. The problem of energy security, the aspect of environmental pollution and depletion of the known fuel reserves in future have created a scope for utilization of renewable resources. In this paper, the most feasible method of rooftop solar PV power generation will be evaluated. A techno-economic feasibility of rooftop PV solar power generation for the city of Al-Majmaah province of Riyadh, KSA is performed.

Keywords: Renewable Energy, Solar Energy, Greenhouse Gases, PV, Solar PV

1. Introduction

Depletion of fossil fuels in near future and accumulation of their emissions in the environment have attracted the world's attention to utilize renewable resources of energy. It is realized that a continuous reliance on fossil fuels will have catastrophic results because excessive carbon dioxide emission has dramatic global warming effects. Our earth could heat up by several degrees in future if we do not stop using non-renewable energy resources. Environmental pollution is supposed to be a serious threat to life on the planet. Therefore use of green energy sources is spreading day by day through out the world. Fossil fuel burning results in emission of hazardous gases. Accumulation of these gases in the environment is a serious threat [1]. Scientific confirmation that CO_2 emissions associated with fossil fuel energy combustion represent the largest source of greenhouse gas emissions from human activity [2, 3]. It is becoming increasingly evident that the level of CO_2 emissions associated with fossil fuel combustion is so voluminous that an effective technical fix to the problem is inevitable [4-7]. Concerning oil, the Japanese government which oversees an oil dependent economy estimates that commercially recoverable reserves of oil will be exhausted in 40 years [8]. In terms of natural gas, the global reserves-to-production ratio of natural gas has been estimated at 63–66 years [9]. In other words, many young people alive today might actually witness the depletion of these vital resources.

These concerns with fossil fuels have attracted the world's attention to utilize renewable resources of energy because of their environmental friendly features [10] and abundant availability. Renewable energy sources are inexhaustible, contrary to fossil fuels, and more widely spread over the Earth's surface [11, 12]. The Sun being exceptional energy source, produces plentiful energy for the world. Energy produced by the Sun is in the form of electromagnetic radiation. Solar energy reaching the earth's surface averages to 1353 W/m2 [13].

Commercial and residential buildings in Saudi Arabia consume about 50% of the total electrical energy consumed [14, 18]. Increasing demand of electrical energy is one of the main problems being faced by the power companies in KSA. In order to meet this increasing demand of electricity in the country, it is desirable to explore every possible option of generating electric power. Saudi Arabia has enormous oil resources. At the same time, the Kingdom is blessed with other resources, notably solar energy that may be the future electric supply in the Kingdom. Green energy presents many potential advantages to KSA. Firstly, due to its abundant resources, the Kingdom has a viable option for domestic consumption that would save a large amount of oil for export. Secondly, the green energy reduces atmospheric pollution and greenhouse gases emissions. The monthly average daily solar radiation of KSA varies from 3.03-7.51 kwh/m2, which is one of the highest in the world [15]. Most of the countries around the globe are interested to penetrate the RE in their power sectors to obtain economic and environmental benefits.

Research is being conducted throughout the world for the development of sustainable, renewable and new energy efficient systems. The increase in population and the development of the Industrial sector in the Kingdom of Saudi Arabia have resulted in an alarming increase in the use of fossil fuels. Though KSA is an oil rich country and the cost of fossil fuel is very low but the costs associated with emissions (in the result of fossil fuels) may affect the economy of the country. Fossil fuels, although produce useful energy, are responsible for production of harmful emissions like CO_2, SO_x, NO_x etc. The obvious choice available is to use renewable energy.

2. Global Solar Generation

The global share of renewable energy in the power sector was 20.3% at the end of year 2011 [16]. The hydroelectric generation being the oldest and most mature form of bulk power generation has a share of 15.3% whereas only 5% was contributed by other renewable generations [17].

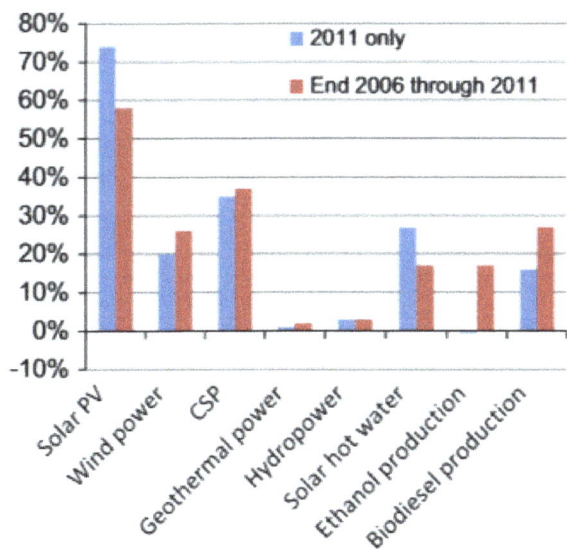

Fig. 1. Comparison of growth rates of renewable power generation.

The main methods of harnessing solar energy for power generation are solar thermal generation and Solar PV generation. Solar PV capacity has increased more rapidly during the last few years. The global installed capacity has increased from 2.2 GW in 2002 to 70GW up to the end of year 2011 [16]. The growth rates of different renewable energy technologies between 2006 and 2011 are shown in the Fig. 1 [17] It is evident from Fig. 1 that Solar PV has the

largest growth rate of 74% followed by CSP (35%), solar water heating (27%), wind power (20%), biodiesel (16%), hydropower (3%), geothermal power (1%) and ethanol production (-0.5%).

There are various methods to harness the power of the Sun and generate electricity. Fig. 2 shows the different Solar Power Generation options. Fixed PV system has the least operation and maintenance (O&M) costs. The tracking panel PV has intermediate output and O&M costs. In this study the fixed flat panel PV option has been taken into account because of its simplest design and lowest cost.

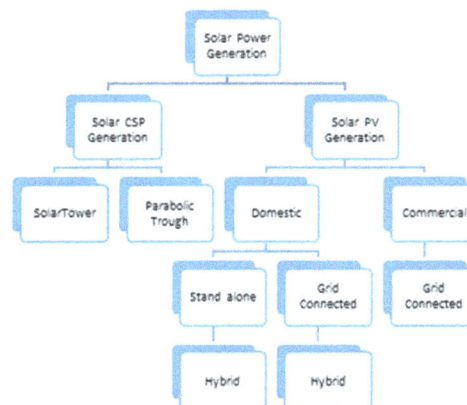

Fig. 2. Methods of Solar Power Generatins.

3. Methodology Adopted to Evaluate the Most Feasible Method of Solar Power Generation

The only viable method to evaluate the benefits of solar energy is the calculation and comparison of per unit cost of solar power generation. The crystalline silicon type are more efficient than the thin film type solar cells. Similarly the fixed flat panel PV has the lowest capital and maintenance costs. The tracking flat PV panel and concentrating PV(CPV) have more O&M costs and are technologically less matured in comparison. It has been proved that output of fixed plate system is maximum when inclined at an angle equivalent to latitude of the location. Therefore the PV panel system under study is fixed tilt poly crystalline PV system. The per unit installation cost of solar power genaration is given by the following equation:

$$Cost = \frac{Total\ Cost\ during\ life\ span\ in\ US\ \$}{Size\ of\ the\ installed\ system\ in\ KWp}$$

The most feasible method assesed by studying a 1 KW_p system which produces comparable results for other larger sizes of roof top PV systems and thus satisfy the objective of this research. The cost of system components was found through survey of the local market. The solar industry is progressing very fast and todays solar systems may become obsolete in 10 years, therefore the maufacturers are cautious when guarantees are concerned. The systems which have 25 years life time working warranties are German and Japanese.

Therefore the cost of solar PV system has been taken from distributors of German and Japanese manufacturers because they provide around 25 years of warranties on the solar system components. Such PV systems have efficiencies of around 15% and are warranted that the output will decline by no more than 80% of the rated output in 25 years. The calculation of the solar PV system comnponents and their costs are discussed in the following sections.

4. Size and Cost Calculations of PV Systems

Some solar projects are over 40 years of age and are still providing power. But solar projects are generally designed for an estimated life span of around 25 years. The Total Life Cycle Cost (TLCC) is given by:

$$TLCC = Fixed\ charges + variable\ charges$$

$$Fixed\ charges = Upfront\ cost$$

The upfront cost is the aggregate costs of system components required to initiate the project.

$$Cost_{upfront} = Cost_{panels} + Cost_{inverter} + Cost_{batteries}$$

Variable charges include operartion and maintenance (O&M) costs and cost of replacement of batteries, inverters, charge controllers etc. during the life cycle. For a fixed tilt system the O&M charges are negligible, and battery and other components replacement charges are given by:

Variable charges
= Cost of system component replacements during life period

4.1. Size of PV panels

Peak power delivered by PV panels is:

$$P_{peak,panels} = No.of\ Panels \times Power\ per\ Panel$$

It is not necessary for $P_{peak,\ usage}$ to be equal to $P_{peak,\ panels}$ since both the output and requirements change throughout the year. But inorder to maximize the benefits, the energy produced by solar panels should be equal to energy utilized. i.e $E_{panels} = E_{used}$

$$Peak\ power\ produced\ by\ panels = P_{peak,panels} = \frac{E_{used}}{T_{sun}}$$

$$T_{sun} = No.of\ hours\ of\ Sun\ shine$$

$$Cost\ of\ pannels = Cost_{panels} = No.of\ Panels\ \times SC_{panels}$$

$$SC_{panels} = panel\ specific\ cost\ in\ US\ dollars$$

A more specific way to ensure solar power availability is to calculate the minimum solar power generation and then design a solar PV system that can support the required load even during the minimum generation period. The solar PV is tested to produce rated output under Standard Testing Conditions (STC): 1000W/m² solar irradiance, 1.5 Air Mass and cell temperature of 25°C. From Table-I, it can be easily seen that the solar generation is minimum in June and can be upto 603.05 $Watt/m^2/hr$. There are many varieties of solar PV available in the market from 30W to 250W. After survey of market, Poly-crystalline PV with 250W, 12V ratings and dimensions of 1650 x 100 mm (1.65 m²) has been selected. Now the number of PV panels required for 1 KW$_p$ load can be calculated as in Table-II. Table-II depicts that surplus power will be available that could be used locally, stored or fed to utility for utilization during the months with irradiance higher than 603.05W/m². Selection of smaller PV size than required will result in decrease of battery life and shut down of PV system by the inverter when voltage reach cut-off value; with no backup for cloudy days.

Table 1. Detail of Solar power irradiance collected in Riyadh/Majmaah.

Month	G$_k$ Hor KWh/m²	Days	Sun Hours	KWh/m²/day	KW/m²/hr	W/m²/day	W/m²/hr
Jan	174.3	31	7.2	5.62	0.78	5622.58	780.55
Feb	177.3	28	8.0	6.33	0.79	6332.14	791.25
Mar	209.4	31	7.0	6.75	0.96	6754.84	964.28
Apr	186.6	30	8.2	6.22	0.76	6220.00	758.53
May	201.1	31	9.03	6.49	0.72	6487.09	721.11
Jun	201.9	30	11.16	6.73	0.60	6730.00	603.05
Jul	211.2	31	11.13	6.81	0.61	6812.90	611.85
Aug	217.0	31	10.16	7.00	0.69	7000.00	688.97
Sep	219.0	31	8.55	7.31	0.85	7064.52	854.97
Oct	222.2	30	10.33	7.16	0.69	7406.66	693.12
Nov	179.6	31	8.53	5.99	0.70	5793.55	702.23
Dec	168.6	30	7.03	5.43	0.77	5620.00	772.40
Total/Average	2368.2	365	8.86	6.49	0.74	6487.02	745.19

Table 2. Size Calculassions of PV panels.

Area of One PV Panel	$= 1.65$ m^2
Solar irradiance under STC to produce rated output of 250 W	$= 1650$ W
Thus the efficiency of the solar panel	$= 15.15\%$
Min.solar irradiance expected in June	$= 603.05$ W/m^2
Min. solar irradiance expected in June for panel with 1.65 m^2 area	$= 603.05 \times 1.65$ $= 995.03$ W
Min. generation of each solar panel during June	$= 995.03 \times 15.15\%$ $= 151$ W
No. of PV panels required to meet load demand of 1 KWp	$= 1000/151 = 7$ No.
Similarly the max. solar irradiance expected in the month of March	$= 964.28$ W/m^2
Max. solar irradiance expected in March for panel with 1.65 m^2 area	$= 964.28 \times 1.65$ $= 1591.06$ W
The max. generation of solar panels during March	$= 1591.06 \times 15.15\%$ $= 241$ W
Max generation from 07 No. Solar PV panels in March	$= 241 \times 7 = 1687$ W

4.2. Size of Charge Controller

The output of the PV panels vary due to change in weather conditions. The charge controller controls and maintains the current from the panels to the batteries. It is rated in amperes and is designed in accordance with the output voltage of the panels. They convert the dc output of the panel to ac and then back again to dc to feed regulated supply to the batteries. They are around 95%-97% efficient. The charge controllers protect the batteries from being overcharged. They also protect the PV panels from the back flow of charge from the batteries. For both Off-grid and on On-grid systems, the size of the controller depends on the short circuit current ratings (I_{sc}) of the panels, which is 8.5A for the selected panels. The controllers must have spare capacity since it is not a good idea to use them at full capacity. During cool sunny days, the output of the panels can increase the rated output by 25%. Since the panels are connected in parallel, therefore the size of charge controller can be calculated by:

Size of charge controller $= I_{sc}$ x No of panels x 1.25

$$= 8.5 \times 7 \times 1.25$$

$$= 74 \text{ A or } 100 \text{ A approx}$$

I_{sc} = Short circuit current of pannel

4.3. Size of Batteries

To keep the upfront cost to minimum, we are designing the PV system for a backup of one day only. The life of a battery is a function of amount of battery discharge in percentage. This is also known as depth of discharge (Dod). For instance, a battery allowed to discharge 20% only will have more life as compared to that allowed to discharge 40%. The dry batteries can be discharged upto 70% with the output voltage and temperature of the cells in permissible limits. The batteries themselves are around 85% efficient. First the voltage and size of batteries in apmere hours (Ah) is selected. For systems with longer cables and high load, 24V or 48 V system may be adopted to minimize the losses. But for smaller systems, the 12V system provides a simple solution. Therefore dry batteries of 200 Ah, 12V that could be discharged upto 70% without decline in their rated output voltage have been selected for the study. The number of batteries required can be calculated from the following formula:

$$N_{\text{B-req}} = \frac{Wh \ load}{(V_b \times \eta_b \times Dod \times Ah_b)} \times No. \ of \ days$$

For a load of 1 KW the energy required at 40% load factor $= 1000 \times 24 \times 40\% = 9600$ Wh/day

No of batteries required for an off-grid system with one day backup (incase of cloudy horizon) $= \frac{9600 \ Wh}{(12 \ V \times 0.85 \times 0.7 \times 200)} \times 1 = 6.7 = 7 \ batteries$

Similarly, for an on-grid system the backup required to cope with an occusional electrcity cut of 4 hours $= \frac{9600 \ Wh}{(12 \ V \times 0.85 \times 0.7 \times 200)} \times 0.167 = 2 \ batteries$

This concludes that for an off-grid PV system of 1 KW$_p$, around 7 batteries of 200 A will supply one day backup (in case of cloudy wheather), whereas 2 batteries are required to provide 4hours of continuous backup in an on-grid solar PV system. In an off grid system the batteries will be replaced after every five years. Whereas in an on-grid system, the battery sets will be replaced after every six years owing to increase in life due to less frequency of charging and discharging.

4.4. Size of Inverter

In an Off-gird solar PV system, the peak power delivered by the inverter should be equal to peak power usage i.e.

$$P_{peak,inverter} = P_{peak,usage}$$

Considering a load factor of 40%, the load is 400W for a 1KW system.

$$size \ of \ inverte = load(W) \times 1.3 = 400 \times 1.3 = 520W$$

In view of available solar power and possibility of future extension in load a 1 KW inverter is recommended for Off-grid system. In an On-Grid system, the inverters not only have to confront the above load but also has to redirect the additional solar generation to the grid. The charge controllers are around 97% efficient at the most, the batteries can produce 85% of their input and the inverters themselves are around 90% efficient. If 2% drop in the cables is taken into account, the efficiency of the system (excluding efficiency of the panels) calculates to around 0.727 % as follows:

$$\eta_{sys} = 0.98 \times 0.97 \times 0.85 \times 0.9 = 0.727 \%$$

The max. power (P_{max}) supplied by the solar system is therefore:

$$P_{max} = 250\,W \times 7\,(no.\,of\,panels) \times 0.727$$
$$= 1272.25\,Watts$$

Considering spare capacity for motive loads, the size of the inverter to be selected is around 2000 W.

Cost Comparison of Solar PV Systems

Saudi Arabia can save oil by generating electricity using Solar PV and as a result, can export more oil in international market. Saudi Arabia is an oil rich country and domestic oil prices are very low. Power generation in the country mainly comes from oil and gas power plants. A cost comparison is presented in this section to see the effectiveness of solar PV generation. According to U.S. DOE's EIA, 1.8barrels of oil is required to generate 1MWh of electricity [18]. The domestic price of oil charges to Saudi Electricity companty is $0.73/mmBTU or $4.1/barrel [18]. Therefore, cost of electicity for an oil based power plant in Saudi Arabia is $7.8 per MWh. There is some additional O&M cost which is assumed to be $5.39 per MWh. These figure are very difficult to beat but use of rooftop solar PV is still a very good alternative. Around 50% of the generated electricity in KSA is being used by residentiol consumer. Therefore use of rooftop solar PV will save huge amount of oil which could be exported in international market and in this way Kingdom

can earn revenue and sustain as a leading oil export country for longer time. A cost of crude oil in internation market is $61.0/barrel as of today. For each MWh of electricity, the country can save 1.8barrel of oil which is equivalent to $110 in internation market. According to these international rates the cost of electricity production is:

$$Cost\ per\ MWh = 61.0 \times 1.8 + O\&M\ cost$$
$$= 110 + 5.39 = \$115.19$$
$$Cost\ per\ KWh = \$0.115$$

For the cost comparison, the existing solar solution provider companies were surveyed in this matter and the cost of equipment with extended warranties during the entire term of the project was found. The component wise cost of a 1 KW$_p$ system is detailed in Table-III. The only difference between the off-grid and on-grid system is the net metering cost and hence increase in labor installation. Addition of batteries further increase the labor cost. The output of solar PV system in converted into energy value in US dollars according to oil prices in the international market. It had been observed that oil prices are showing an annual rise of 3%-8% per year in the last two decades. We assumed the minimum rise of 3% in our calculations over the life span of the project. Table IV indicates the energy value per year and the accumulative returns or payback. The payback period of each method is as under:

- Payback period of Off-grid rooftop solar PV system without batteries is 12 years
- Payback period of On-grid rooftop solar PV system without batteries is 13 years
- Payback period of On-grid rooftop solar PV system with batteries is 17 years
- Payback period of Off-grid rooftop solar PV system with batteries is 25 years

Table 3. Cost of Solar PV Power System.

Cost Type	System Components / Equipment	Quantity required for 1 KW$_p$roof-top solar PV system	Cost of Solar System in US$				
			Per Unit Estimated Cost (US $)	1KWp Off-grid without batteries (US $)	1KWp Off-grid with batteries (US $)	1KWp On-grid without batteries (US $)	1KWp On-grid with batteries (US $)
	Solar Panel	07 No (250 W$_p$ PV panels)	2.13/W	3733	3733	3733	3733
	Charge controller	01 No	2.7/Amp	-	270	-	270
	Invertor	01 No (1 KW for Off-Grid) (2 KW for On-Grid)	0.135/W	135	135	270	270
Installation Cost / Upfront Cost or Fixed Cost	Batteries	07 No Off-Grid 02 No On-Grid	0.93/Ah	-	1302	-	375
	BOS (Cost of frames, cables, installation cost, transport cost, net metering cost etc.)	Once	500 to 800	500	600	700	800
Variable Cost	Batteries	one set after every 5 years	0.93/Ah	-	1302x4 = 5208	-	375x4 = 1500
Total cost	-	-	-	4,338	11,248	4,703	6,948

The cheapest solution for roof-top solar power generation at domestic level is Off- grid without batteries. But there is no other backup than solar power and the reliability of this system is least. To create a back-up source, batteries are required in an off-grid system for use in the evening and night and backup for cloudy days. But this makes the Off-grid system with batteries, the most expensive solar PV system. Similarly when the solar power generation is more than demand or someone has the capability to install a solar system of higher capacity than his load requirements, the surplus generation will go waste. Therefore to maximize the benefits of solar power generation, the most feasible solar power method for solar power generation would be On-Grid without batteries. The per unit installation costs are comparable but the benefits expected with On-Grid solar PV system without batteries are far more. The payback period of this system is 13 years.

Table 4. Payback period of Solar PV system.

Year	Solar Irradiation on tilted plane (KWh/m2)	Total area of solar cells (m^2)	Efficiency of Solar module	Efficiency of Solar system	Solar Output (KWh)	Expected Utility KWH Rate (US $)	Energy Value (US $)	Cumulative Returns (US $)
Year 1	2368.2	11.55	15.15%	72.7%	3013	0.115	346	346
Year 2	2368.2	11.55	15.03%	72.7%	2989	0.118	353	699
Year 3	2368.2	11.55	14.91%	72.7%	2965	0.122	362	1061
Year 4	2368.2	11.55	14.79%	72.7%	2941	0.126	371	1432
Year 5	2368.2	11.55	14.67%	72.7%	2917	0.130	379	1811
Year 6	2368.2	11.55	14.55%	72.7%	2893	0.134	388	2199
Year 7	2368.2	11.55	14.43%	72.7%	2870	0.138	396	2595
Year 8	2368.2	11.55	14.31%	72.7%	2846	0.142	404	2999
Year 9	2368.2	11.55	14.20%	72.7%	2824	0.146	412	3411
Year 10	2368.2	11.55	14.09%	72.7%	2802	0.150	420	3831
Year 11	2368.2	11.55	13.96%	72.7%	2776	0.154	427	4258
Year 12	2368.2	11.55	13.85%	72.7%	2754	0.159	438	4696
Year 13	2368.2	11.55	13.74%	72.7%	2732	0.164	448	5144
Year 14	2368.2	11.55	13.63%	72.7%	2710	0.169	458	5602
Year 15	2368.2	11.55	13.52%	72.7%	2688	0.174	468	6070
Year 16	2368.2	11.55	13.40%	72.7%	2665	0.179	477	6547
Year 17	2368.2	11.55	13.29%	72.7%	2643	0.184	486	7033
Year 18	2368.2	11.55	13.18%	72.7%	2621	0.189	495	7528
Year 19	2368.2	11.55	13.07%	72.7%	2599	0.195	507	8035
Year 20	2368.2	11.55	12.97%	72.7%	2579	0.200	516	8551
Year 21	2368.2	11.55	12.87%	72.7%	2559	0.206	527	9078
Year 22	2368.2	11.55	12.76%	72.7%	2537	0.212	538	9616
Year 23	2368.2	11.55	12.65%	72.7%	2516	0.218	548	10164
Year 24	2368.2	11.55	12.55%	72.7%	2496	0.224	559	10723
Year 25	2368.2	11.55	12.45%	72.7%	2476	0.231	572	11295
Total	-	-	-	-	68,411	-	11,295	-

5. Conclusion

The purpose of this research is to optimize the power generation cost while reducing hazardous emissions. Feasibility of rooftop solar PV system in Riyadh region in Kingdom of Saudi Arabia is evaluated in this paper and the most feasible method of rooftop solar PV power generation is evaluated. A comparison of various solar PV generation options is conducted by studying their design in detail and finally the payback periods is calculated in light of costs and benefits of each project. It is found that On-grid rooftop solar PV system without batteries is the most feasible system from both reliability and econimical point of view.

References

[1]   G. W. Crabtree and N. S. Lewis, "Solar Energy Conversion," Physics Today, vol. 60, pp. 37-42, 2007.

[2]   Chalvatzis KJ, Hooper E. Energy security vs. climate change: theoretical framework development and experience in selected EU electricity markets. Renew Sust Energy Rev 2009; 13: 2703–9.

[3]   IPCC. Climate change 2007: synthesis report. Geneva, Switzerland: Intergovernmental Panel on Climate Change (IPCC); 2007.

[4]   Bruggink JJC, Van Der Zwaan BCC. The role of nuclear energy in establishing sustainable energy paths. Int J Global Energy Issues 2002; 18: 151–80.

[5]   Trainer T. Can renewables etc. solve the greenhouse problem? The negative case. Energy Policy 2010; 38: 4107–14.

[6]   Stern N. The global climate change imperative. 2007: 90.

[7]   Zidansek A, Blinc R, Jeglic A, Kabashi S, Bekteshi S, Slaus I. Climate changes, biofuels and the sustainable future. Int J Hydrogen Energy 2009; 34: 6980–3.

[8]   ANRE. Energy in Japan 2006: Status and Policies. Tokyo, Japan: Agency for Natural Resources and Energy, Japan Ministry of Economy, Trade and Industry (ANRE); 2006. pp. 1–28.

[9] EIA. International Energy Outlook 2008. Washington, USA: Energy Information Administration, U.S. Department of Energy; 2008, p. 1–260.

[10] G. W. Crabtree and N. S. Lewis, "Basic research needs for solar energy utilization," 2005.

[11] Braun FG, Hooper L, Wand R, Zloczysti P. Innovation in concentrating solar power technologies: a study drawing on patent data. Discussion papers. Berlin: DIW Berlin, German Institute for Economic Research; 2010.

[12] Amita Ummadisingu, M.S. Soni, Concentrating solar power – Technology, potential and policy in India, Renewable and Sustainable Energy Reviews, Volume 15, Issue 9, December 2011, Pages 5169-5175, ISSN 1364-0321.

[13] Goswami, D. Yogi, Kreith, Frank, and Kreider, Jan F., Principles of Solar Engineering, 2nd edition. Taylor and Francis, Philadelphia, PA, 2000.

[14] Shaahid SM, Elhadidy MA. Economic analysis of hybrid photovoltaic–diesel–battery power systems for residential loads in hot regions—a step to clean future. Renewable and Sustainable Energy Reviews 2008; 12: 488–503.

[15] S.M. Shaahid, L.M. Al-Hadhrami, M.K. Rahman, Review of economic assessment of hybrid photovoltaic-diesel-battery power systems for residential loads for different provinces of Saudi Arabia, Renewable and Sustainable Energy Reviews, Volume 31, March 2014, Pages 174-181, ISSN 1364-0321.

[16] Renewables 2012 Global Status Report. (http://www.map.ren21.net/GSR/GSR2012_low.pdf).

[17] Ahmad Bilal Awan, Zeeshan Ali Khan, Recent progress in renewable energy – Remedy of energy crisis in Pakistan, Renewable and Sustainable Energy Reviews, Volume 33, May 2014, Pages 236-253, ISSN 1364-0321.

[18] Kaffine Daniel T, McBee, Brannin J, Lieskovsky, Jozef, Emissions Savings from Wind Power Generation in Texas, The Energy Journal, Volume 34, Issue 1, January 2013, Page 155.

[19] Ali Ahmad, M.V. Ramana, Too costly to matter: Economics of nuclear power for Saudi Arabia, Energy, Volume 69, 1 May 2014, Pages 682-694, ISSN 0360-5442.

23

Photoelectrochemical Cell Based on Natural Pigments and ZnO Nanoparticles

Getachew Yirga[1], Sisay Tadesse[3], Teketel Yohannes[2]

[1]Department of Physics, Haramaya University, Dire Dawa, Ethiopia
[2]Materials Science Program, Addis Ababa University, Addis Ababa, Ethiopia
[3]Department of Chemistry, Hawassa University, Hawasa, Ethiopia

Email address:
gyirga10@gmail.com (G. Yirga), teketely@yahoo.com (T. Yohannes), sisayt@hu.edu.et (S. Tadesse)

Abstract: Natural pigments extracts from *Bougainvillea spectabilis, Carissa Ovata, Hibiscus sabdariffa, Amarathus iresine herbisti, Beta vulgaris*, are used as natural sensitizers for a dye sensitized solar cell (DSSC). ZnO nanoparticles were synthesized using sol-gel method and the size of the nanoparticle was determined using effective mass approximation model. Devices were Assembled using ZnO nanoparticles and natural sensitizers. DSSCs based on ZnO nanoparticles and ethanol extract of *Amarathus iresine* sensitizers have shown relatively better conversion efficiency of 0.039. Incident photon to current conversion efficiency (IPCE), short circuit current density (J*sc*) and open circuit voltage (V*oc*) were measured for all the sensitizers.

Keywords: Natural Dyes, Electrolytes, Solar Cells, Titanium Dioxide (TiO_2)

1. Introduction

The development of dye sensitized solar cells (DSSCs), which have derived inspiration from photosynthesis, has opened up exciting new possibilities and prototypes for producing solar photovoltaics possibly at lower cost. Early DSSC designs involved transition metal coordinated compounds (e.g., ruthenium polypyridyl complexes) as sensitizers because of their strong visible absorption, long excitation lifetime, and efficient metal-to-ligand charge transfer. Because of these, such type of DSSC's are highly effective with maximum efficiency of 11% [1, 2]. The costly synthesis and undesired environmental impact of those prototypes call for cheaper, simpler, and safer dyes as alternatives. Natural pigments, including chlorophyll, carotene, and anthocyanin, are freely available in plant leaves, flowers, and fruits and fulfill these requirements. Experimentally, natural dye sensitized TiO_2 solar cells have reached an efficiency of 7.1% and high stability [3]. Higher efficiency over 8.0% has been obtained using similar synthetic organic dyes [2].

Natural dyes have become a viable alternative to expensive and rare organic sensitizers because of its low cost, easy attainability, abundance in supply of raw materials and no environment threat. Various components of a plant such as the flower petals, leaves and bark have been tested as sensitizers. The nature of these pigments together with other parameters has resulted in varying performance. In this study pigments *Bougainvillea spectabilis, Carissa Ovata, Hibiscus sabdariffa, Amarathus iresine herbisti*, and *Beta vulgaris* were extracted as sensitizers and as a semiconductor ZnO and TiO_2 were used. All the photoelectrochemical performance was measured and characterized for all the above sensitizers and semiconductors.

The solar cell efficiency was determined by its current-voltage (J-V) characteristics under standard illumination conditions. A standard solar spectrum of air mass 1.5 (AM 1.5) with an intensity of 1000 W/m^2 is used for characterization. The V_{oc} is the difference in potential between the two terminals in the cell under light illumination when the circuit is open. J_{sc} is the photocurrent per unit area (mA/cm^2). The degree of the squared shape of the J-V curve is given by the fill factor (FF), which measures the ideality of the device and is defined as the ratio of the maximum power output per unit area to the product of V_{oc} and J_{sc}.

$$FF = \frac{J_{max.power} \, V_{max.power}}{J_{sc} V_{oc}} \quad (1)$$

The solar to electric power conversion efficiency η is given by the ratio of the maximum extractable power to the incident solar power (P_{in}) given by Equation 2.

$$\eta = \frac{P_{max}}{P_{in}} = \frac{J_{max}.V_{max}}{P_{in}} = \frac{J_{sc}.V_{oc}.FF}{P_{in}} \qquad (2)$$

Where P_{in} is the incident power, Pout is the output power, FF is the fill factor, η is the efficiency, J_{sc} is the short circuit current density, and V_{oc} is the short circuit voltage. IPCE is one of the fundamental measurements of the performance of the solar. It is also known as the external quantum efficiency and describes how efficiently the light of a specific wavelength is converted to current that is (electrons out) / (photons in). The IPCE can be calculated according to the following equation.

$$IPCE(\%) = \frac{1240.J_{sc}}{\lambda.P_{in}}.100 \qquad (3)$$

J_{sc} is the short circuit current density, λ is the wavelength of the incident light and P_{in} is the intensity of the incident light. The factors determining the IPCE can be expressed as:

$$IPCE(\lambda) = \eta_{lh}(\lambda).\eta_{inj}(\lambda).\eta_{col}(\lambda) \qquad (4)$$

Where $\eta_{lh}(\lambda)$ the light harvesting efficiency of the sensitized oxide layer, $\eta_{inj}(\lambda)$ is the electron injection efficiency from the sensitizer into the oxide, and $\eta_{col}(\lambda)$ is the electron collection efficiency.

2. Materials and Experimental Methodology

2.1. Synthesis of ZnO Nanoparticles

Zinc acetate (BDH), sodium hydroxide pellets (Scharlau), and polyethylene glycol (Applied Science) were used as received. All the materials were first cleaned and rinse with distilled water and dried. All the chemicals were weighed with analytical balance and mixed in cleaned round bottom flask. For the sol-gel method 2.7 g of zinc acetate dehydrate and 0.5 g of polyethylene glycol (PEG) were taken and dissolved in in 250 ml of distilled water separately. 2 g of sodium hydroxide was dissolved in 500 ml of distilled water with vigorous stirring. The sodium hydroxide solution was added to the zinc acetate solution drop wise and the mixture was refluxed for 8 hrs. at 120°C using the setup shown in Fig 1.

Fig. 1. Experimental set-up for refluxing the solution.

The obtained ZnO solution was centrifuged to solid matter and solution. The solid matter was washed first by distilled water repeatedly, finally dried in Furnace to obtain ZnO. Diagrammatically the whole experiment can be summarized below. After preparing the colloidal nanoparticles optical absorption was done using Gensys-2 PC spectrometer to determine the size of colloidal nanoparticles. About 5 ml of the colloid were taken for measurement. Here for measurement the colloid nanoparticles were taken as soon as they reach the final temperature. We record the UV-visible absorption spectra between 275 and 500 nm. The solvent was used as a blank solution.

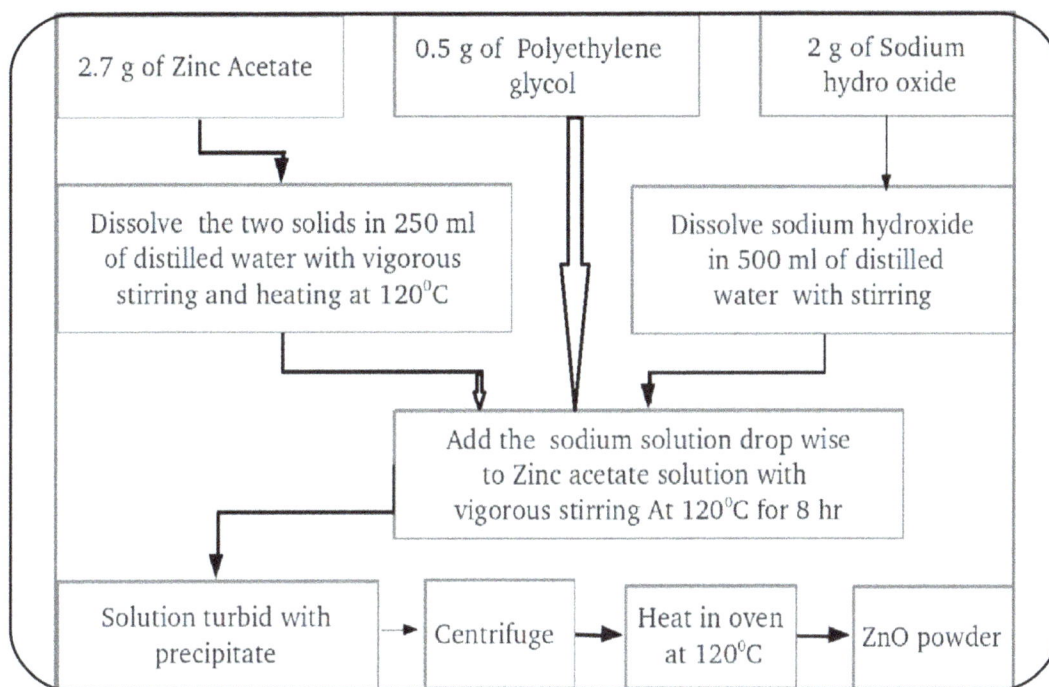

Fig. 2. *Steps of ZnO synthesis.*

2.2. Natural Dyes Extraction

To select the dye that has good absorption in the visible region, fresh fruits, leaves, and flowers of plants including, *Beta vulgaris, Bougamvillea, Amarathus iresine herbisiti, Hibiscus sabdariffa,* and *Carissa ovata* collected. The collected plant was dried at room temperature in a shade to prevent pigment degradation (see Fig.3). After drying for about 2 months the samples are completely dried in an oven at 70^0C to avoid some moisture from it. Then after, the samples were crushed with Micro Plant Grinding machine to produce the powder of the respective plant materials. The dye extraction from the powder was done as follows; 2 g of each powder sample was taken and soaked in 50 ml of ethanol for extracting using ethanol and in 50 ml water for extracting with water in separate bottle.

The solution is stored at room temperature for about 6 hours to dissolve the powder completely. Then the solution was filtered with glass filter to separate the solid from the pure liquid. After filtration the extracted pigment in different solvent are shown in Fig. 4.

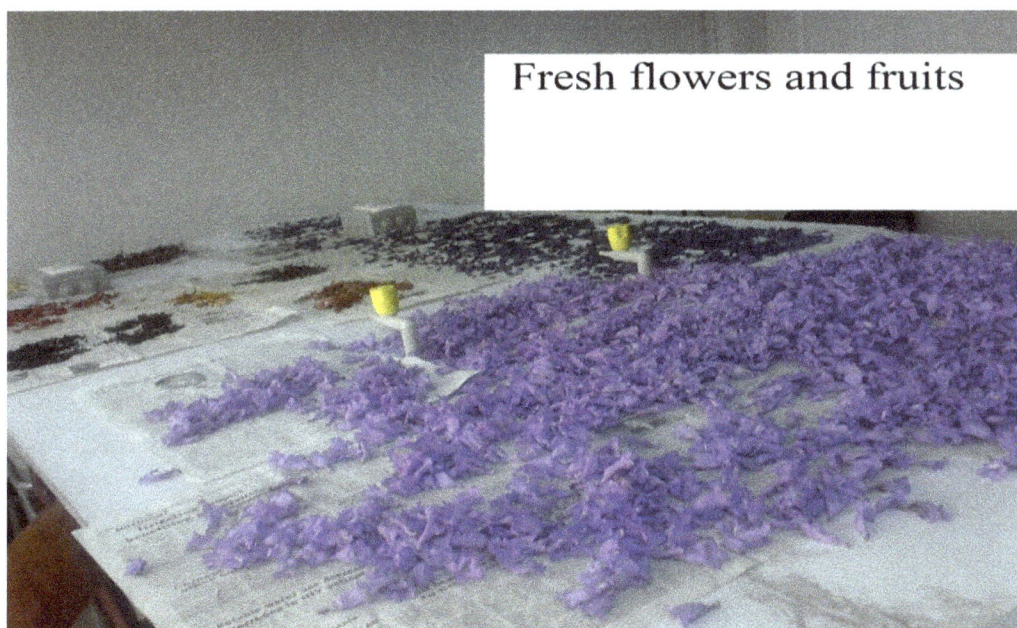

Fig. 3. *Fresh flowers and fruits under room temperature.*

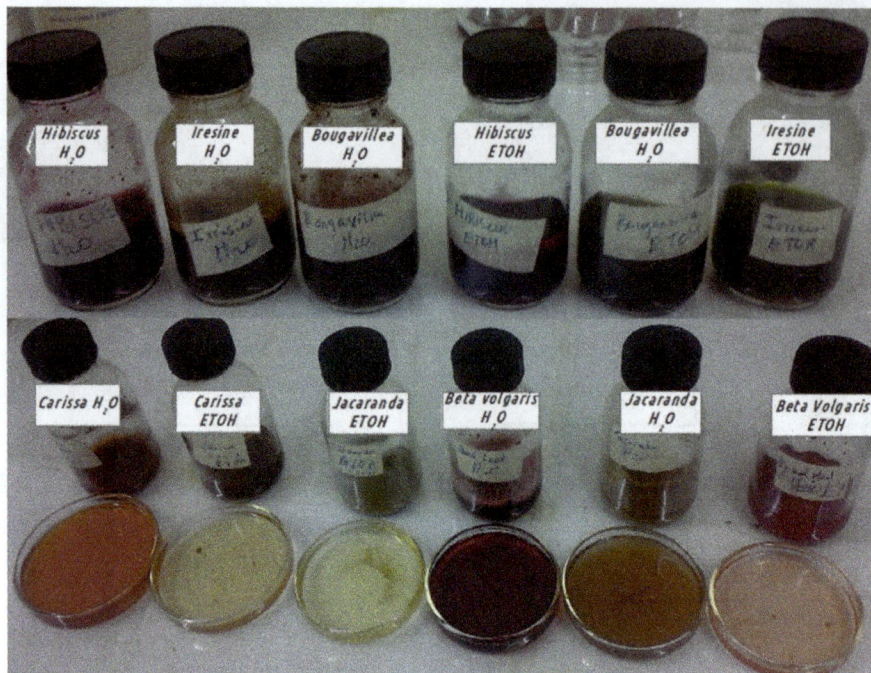

Fig. 4. *Samples of extracted pigment.*

The pigment shown in Fig.4 are ready for soaking the electrode inside it, and also used for measuring the UV-vis absorption spectra. The absorption measurement was carried using the Gensys-2 PC spectrometer for each extract.

2.3. Fabrication of Dye-sensitized Solar Cells

The materials and the fabrication method employed for the fabrication of dye-sensitized Photo electrochemical cells based on ZnO nanoparticles are described below.

2.3.1. Substrate Cleaning

The substrate cleaning is believed to be a key process that influences the final performance of the devices. A significant effect on the photovoltage behavior can be observed experimentally depending on the extent of cleaning [4]. For

this reason; TCO glass substrates have been thoroughly cleaned before film deposition. The glass cleaning protocol was, first cleaned with Acetone then with isopropanol finally with Ethanol. In all case the cleaning process was done in ultrasonic bath for about 20 minute in each solvent.

2.3.2. ZnO Paste Formation

30 mg of polyethylene glycol (dispersing agent) and 10 ml of distilled water was mixed. 1.8 g of ZnO powder continuously grinded down by using procelain mortar to break down the aggregated particle. Then 2.5 ml of the PEG solution was slowly added to the powder and completely mixed with each other by using the mortar. Finally the paste was ready for deposition.

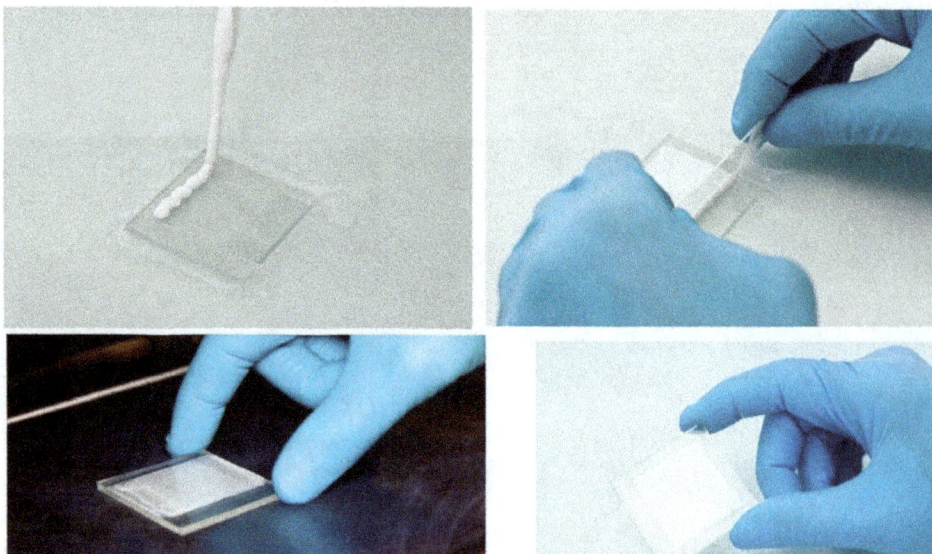

Fig. 5. *Deposition of ZnO oxide paste.*

2.3.3. Deposition of ZnO Films

The simplest and most widely used method for depositing ZnO paste on a substrate is the so called doctor blade method. The technique is also known as slot coating in its mechanized version. It uses a hard squeegee, or doctor blade, to spread a portion of ZnO paste onto the glass (see Fig.5).

With the conductive side facing up, we apply four parallel strips of tape on the edges of the glass plate, covering about small portion of glass. After making the paste ready for deposition we apply the portion of paste near the top edge of the TCO glass between the two pieces of tape with the help of a glass rod, then, the paste spread across the plate with the support of the tape on all sides. The preparation of electrode was completed by firing the deposited layer. The organic solvent burns away, leaving the ZnO nanoparticles sintered together. This process ensures electrical contact between particles and good adhesion to the TCO glass substrate.

2.3.4. Electrode Sensitization

Natural dyes were used for electrode sensitization in the course of this thesis. Before immersion in the dye solution, films were warmed up to higher temperature (80^0C) to minimize the water vapor content inside the porous of the semiconductor electrode. The sensitization was always performed at room temperature. The best method of adsorbing a natural sensitizer to the oxide layer is by dipping the electrode in a solution of the dye already prepared. After sensitization, the films were rinsed in the same solvent (ethanol, water) as employed in the dye solution.

2.3.5. Quasi-Solid State Electrolyte

The polymer gel electrolyte was prepared according to the method developed by L. Fan et al as described below [5]. 0.9 M of 1-ethylene-3-methyl immidazolium iodide (EMIM-I) was added into acetonitrile (Aldrich) under stirring to form a homogeneous liquid electrolyte. In order to obtain a better conductivity, 0.5 M of sodium iodide (BDH) was dissolved in the above homogeneous liquid electrolyte, and then 0.12M iodine and 35 %(w/w) of PVP (Aldrich) were added. Then, the resulting mixture was heated at 70 -80^{0}C under vigorous stirring to dissolve the PVP polymer, followed by cooling down to room temperature to form a gel electrolyte.

2.3.6. Coating Counter Electrodes

The poly (3, 4-ethylenedioxythiophene) (PEDOT) film for the counter electrode was formed by electrochemical polymerization of 3, 4-ethylenedioxy-thiophene (EDOT) (Aldrich), in a three electrode one-compartment electrochemical cell. The electro- chemical cell consisted of a pre cleaned ITO-coated glass working electrode, platinum foil counter electrode and quasi Ag/AgCl reference electrode dipped in LiClO4 (Aldrich) acetonitrile (sigma-Aldrich)

solution. The solution used for the polymerization contained 0.1 M EDOT and 0.1 M LiClO4 in acetonitrile (Sigma-Aldrich). The monomer was used as received. The polymerization was carried out potentiostatically at +1.8 V. At this potential, the electrode surface becomes covered with blue-doped PEDOT film. The film was then rinsed with acetonitrile and dried in air.

2.3.7. Assembly of DSSCs for Characterization

Here the sensitized electrode was washed by the solvent of the dye then by ethanol and dried using hair dryer to dry the electrode, and then the non-covered part of the film by the paste was covered by a tape spacer in all side by leaving some place for electrical contact. By facing the active sides of the photoanode and the cathode, the two electrodes are pressed together after putting the quasi electrolyte on the photoanode. Then the devices are ready for characterization.

3. Results and Discussion

3.1. ZnO Nanoparticles Size Determination

Fig 6 is the plot of absorbance versus wavelength of the colloidal ZnO nanoparticles. From the absorption maxima the corresponding wavelength λ = 366 nm was obtained which is used for calculation of energy bandgap of the nanoparticle.

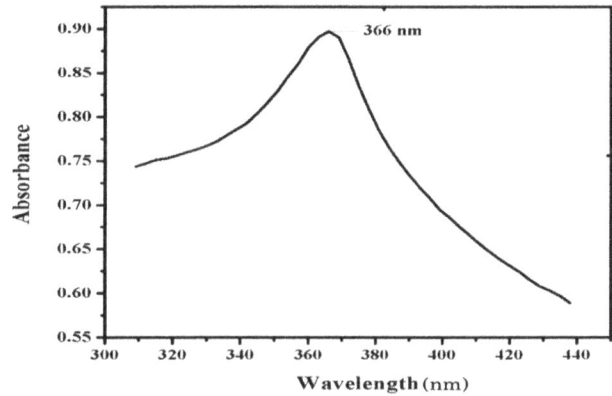

Fig. 6. Plot of absorbance versus wavelength for colloidal ZnO nanoparticle.

Using this wavelength the energy bandgap of the nanoparticle E_g was calculated using Equation 1:

$$E_g^{(nano)}(R) = E_g^{(bulk)} + \frac{h^2}{8m_0R^2}\left(\frac{1}{m_e^*}+\frac{1}{m_h^*}\right) - \frac{1.8e^2}{\varepsilon_r\varepsilon_0 R} \quad (5)$$

Where m0, m_e^*, and m_h^* are the rest mass of electron, effective masses of electron, and effective mass of hole respectively [6,7,8]. Rearranging Equation 5 and solving for R gives:

$$R = \frac{\left(-\dfrac{1.8e^2}{\varepsilon_r\varepsilon_0}\right) \pm \sqrt{\left(\dfrac{1.8e^2}{\varepsilon_r\varepsilon_0}\right)^2 \left(E_g^{(nano)} - E_g^{(bulk)}\right)\dfrac{h^2}{8m_0R^2}\left(\dfrac{1}{m_e^*}+\dfrac{1}{m_h^*}\right)}}{E_g^{(nano)} - E_g^{(bulk)}} \quad (6)$$

For $E_g^{(nano)}$ of 3.39eV, the radius of nanoparticle, R was obtained to be 5.7 nm. Finally it is possible to say that the nanoparticle prepared by the method which was mentioned in the experimental part has an approximate average particle size of 11.4 nm.

3.2. Optical Absorption Measurements

The absorption spectrum of the prepared dye was measured by using the Genesys-2 PC spectrometer. Here first, some amount of the final dye solution was put in the quartz cuvette, and the absorbance *versus* wavelength measurement was taken for each sample of ethanol and H$_2$O extract. The results of absorbance measurement of each sample was shown in Fig. 7.

In water extract of Bougainvillea spectabilis [Fig. 7 (a)] two peaks were found: the first one around 485 nm which can be associated to the presence of indicaxanthin (which is a type of betaxanthin found in plant pigment), while the second one at 531 nm is attributable to the betacyanin pigment. The pigment extracted from *Carissa Ovata* using water extracted shows an absorption peak at 553 nm [Fig. 7 (b)]. In water extract, the spectra show an absorption peak in the region of 520 - 550 nm which is the peak of anthocyanin containing dyes.

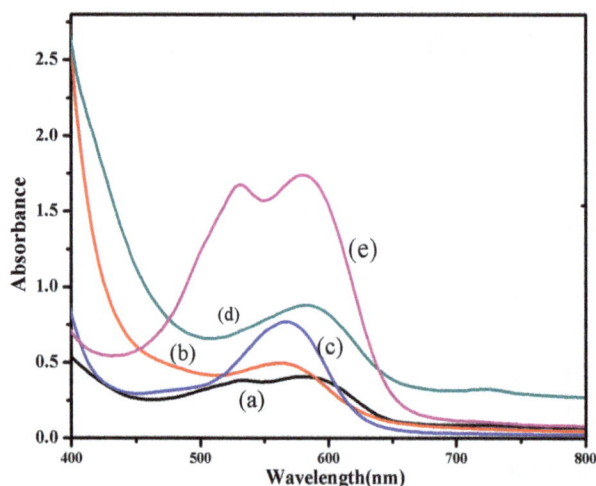

Fig. 7. Light Absorption spectra of dye solutions extracted with water of (a) Bougainvillea spectabilis, (b) Carissa Ovata, (c) Hibiscus sabdariffa, (d) Amarathus iresine herbisti, (e) Beta vulgaris.

This is because of the diverse pigmentation from orange to red, purple, and blue pigment which are found in anthocyanin containing pigment and shows an absorption in the visible region (approximately 490 - 550 nm) [9].

Water extract of *Hibiscus sabdariffa* shows absorption peak at 519 nm. The peak ascertains the presence of anthocyanin pigment [Fig. 7 (c)]. In water extract of *Amarathus iresine herbisti* [Fig. 7 (d)] (λ_{max}= 532 nm) anthocyanin containing pigment were observed in the absorption spectra. *Beta vulgaris* showed an intense absorption peaks in the region 400 - 600 nm [Fig. 7 (e)]. Here also water extract shows strong absorption peak of betalians,

which are at 470 nm and 533 nm due to the mixed contributions of the yellow-orange betaxanthins, and of the red-purple betacyanines at around 480 nm and 540 nm respectively [10].

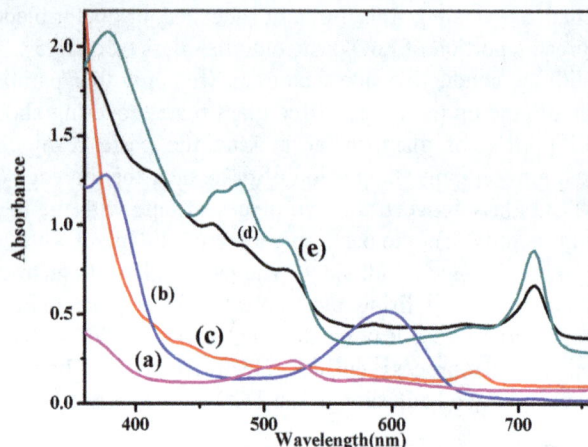

Fig. 8. Light Absorption spectra of dye solutions extracted with ethanol of (a) Beta vulgaris, (b) Hibiscus sabdariffa, (c) Carissa Ovata, (d) Bougainvillea spectabilis, (e) Amarathus iresine herbisti.

Ethanol extract of *Bougainvillea spectabilis* shows an absorption peak at 664 nm which has also an absorption peak below 500 nm [Fig. 8 (d)]. From these peak it is clearly shown that ethanol extract contain both chlorophyll a and b which have an absorption peak in between 400 - 500 nm and 600 - 700 nm [11]. The pigment extracted from *Carissa Ovata* [Fig. 8 (c)] using ethanol shows an absorption peak at 664 nm. In ethanol extract of *Carissa ovata* the extracted pigment was contain chlorophyll pigment which shows the characteristics absorption peak of chlorophyll.

Both water and ethanol extract of *Hibiscus sabdariffa* [Figs. 7(c) and 8(b))] shows absorption peaks at 549 nm and 519 nm respectively. The peaks are associated with presence of anthocyanin pigment. *Beta vulgaris* [Fig. 8 (a)] showed an intense absorption peaks in the region 400 - 600 nm. Here also both water and ethanol extract shows strong absorption peak of betalians, at 470 nm and 533 nm because of the mixed contributions of the yellow-orange betaxanthins and the red-purple betacyanines (480, and 540 nm respectively) [5] Ethanol extract of *Amarathus iresine herbisti* (λ_{max}= 433, 464, 664 nm) indicates the presence of chlorophyll which absorb most of the blue and red light [Fig. 8 (e)].

3.3. Current Density Versus Voltage Characteristics of ZnO Based DSSCs

The J-V characteristic of all sensitizers were measured and plotted for analysis and made comparison as shown below. Figure 9 (b) is ethanol extract of Beta vulgaris, the current density decreases as compared to the water extract [Fig. 10 (d)], but the reverse is true for the open circuit voltage. This is due to the low injection rate of electron into the semiconductor conduction band which decreases the short circuit current.

Water extract of Hibiscus sabdariffa [Fig. 10 (e)] has showed better current than the corresponding ethanol extract [Fig. 9 (c)] with some increment in short circuit current density.

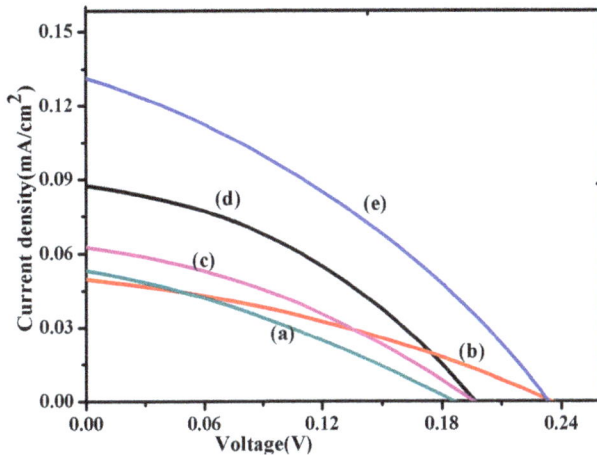

Fig. 9. *J-V of ZnO based DSSCs extracted with ethanol, (a) Carissa ovate, (b) Beta vulgaris, (c) Hibiscus sabdariffa, (d) Bougainvillea, (e) Amarathus iresine herbisti.*

In Fig. 10 (c), (d) and Fig. 10 (b), (e) both water and ethanol extract almost they have the same open circuit voltage but the current density relatively higher for ethanol than the water extract which means that there is only a difference in electron injection efficiency. Most of the natural dye which have a good and a broader absorption in the visible spectrum are expected to show a good rectification of the J-V curve that is responsible for good current density and power conversion efficiency. In these studies ethanol extract of *Carissa ovata, Bougainvillea spectabilis,* and *Amarathus iresine herbisti* shows a better rectification which results relatively good photoelectrochemical performance for ethanol extract than the water extract. In ethanol and water extract of Hibiscus sabdariffa, Bougainvillea spectabilis, and Carissa ovata, during socking, it was clearly an observed that, the film starts to dissolve.

This is because of protons derived from sensitizers make the dye solution relatively acidic which leads to the dissolution of the ZnO colloid.

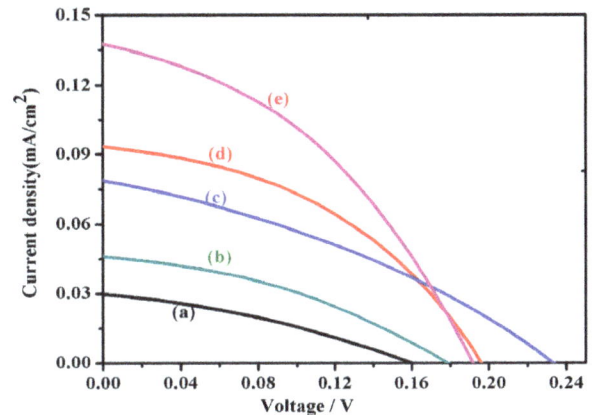

Fig. 10. *J-V of ZnO based DSSCs extracted with water, (a) Bougamvillia spectabilis, (b) Carissa ovata, (c) Amarathus iresine herbisti, (d) Beta vulgaris, (e) Hibiscus sabdariffa.*

The low local pH at the surface of ZnO during dye sensitization leads to the dissolution of Zn^{2+} ions from the ZnO surface [Fig. 11 (2)]. These Zn^{2+} ions form complexes with the dye [Fig. 11 (3)], which accumulate in the pores of the semiconductor film. It is assumed that only dye molecules directly attached to the ZnO surface can inject efficiently electrons and contribute to the photocurrent [12]. Therefore, Zn^{2+}/dye complexes, in spite of absorbing light, do not inject electrons. So that the low current as well as low power conversion efficiency may arises due to the above reasons. The instability of ZnO in acidic dyes results due to its amphoteric nature. In general, in a solution, the surface of the oxide is predominantly positively charged at a pH below the point of zero charge and negatively charged above this value, while the point of zero charge of metal oxides is defined as the pH at which the concentrations of protonated and deprotonated surface groups are equal.

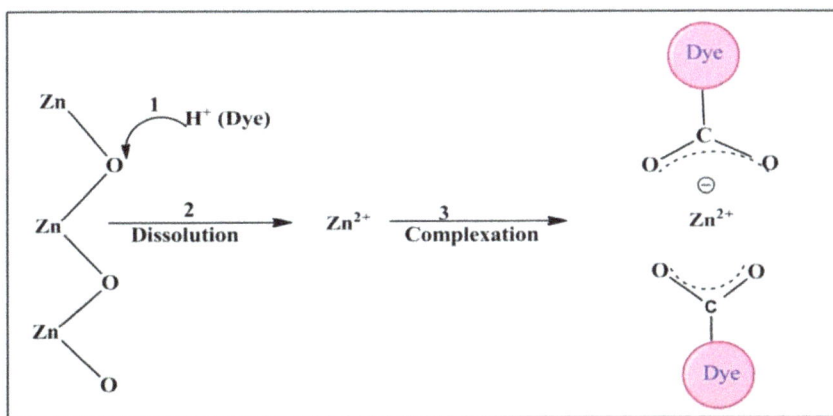

Fig. 11. *Zn^{2+}/dye complexes formation.*

For example the ZnO sensitization process with Ru-complex dye, the (pH = 5) is much lower than the point of zero charge of ZnO (≈ 9). That means that the ZnO surface is positively charged. Thus, the protons adsorbed on the ZnO surface will dissolve the ZnO. Table 1 Summarized the performance of the DSSCs in terms of short-circuit photo-current (J_{sc}), open-circuit voltage (V_{oc}), fill factor (FF), and energy conversion efficiency (η) compared to those of other extracts.

The efficiency of cell sensitized by the *Amarathus iresine herbistii* extracted with ethanol was the best among the others. This is due to broader absorption range of the sensitizers, higher interaction between ZnO nanocrystaline film and the pigment extracted from *Amarathus iresine herbistii* which leads to a better charge transfer. The current density and open circuit voltage it has higher value than the others which are 0.29 mAcm-2 and 0.29 V respectively. The fill factor also shows best value of 48.5% for ethanol extract of *Amarathus iresine herbisti*.

Table 1. Photovoltaic performance of ZnO based DSSCs with different sensitizers.

Sensitizers	Solvent	J_{SC}(mAc m^{-2})	Voc	FF (%)	η (%)
Carissa Ovata	Water	0.041	0.18	36.8	0.0027
	Ethanol	0.053	0.19	30.7	0.0031
Beta vulgaris	Water	0.094	0.20	41.3	0.0077
	Ethanol	0.050	0.24	33.0	0.0039
Bougainvillea spectabilis	Water	0.030	0.17	31.5	0.0016
	Ethanol	0.088	0.20	37.4	0.0066
Hibiscus sabdariffa	Water	0.137	0.20	38.1	0.0104
	Ethanol	0.063	0.20	34.3	0.0043
Amarathus iresine	Water	0.079	0.24	32.6	0.0062
	Ethanol	0.290	0.29	48.5	0.0390

As reported in literature, [13] that the extracting solvent can affect the DSSCs performance. The efficiency of the DSSCs was found to increase immensely when ethanol was used for extracting pigments [13]. In this study, similar finding was also obtained. This might be due to the fact that our extracted pigments are more soluble in ethanol as a result; the aggregation of dye molecules decreases as expected. Better dispersion of dye molecules on the oxide surface could exactly improve the efficiency of the system.

3.4. Photocurrent Action Spectra of ZnO Based DSSCs

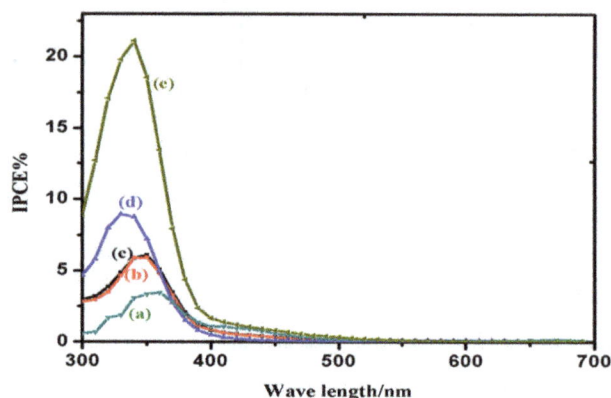

Fig. 12. IPCE of ZnO based DSSC sensitizers extracted with ethanol, (a) Amarathus iresine herbisti, (b) Carissa ovata, (c) Bougainvillea spectabilis, (d) Hibiscus sabdariffa, (e) Beta vulgaris.

The Photo action spectra [Figs. 12 and 13] provided further insights on the photoelectrochemical behavior of natural dyes. All the following figures shows the incident photon to current conversion efficiency (IPCE) spectra of ZnO electrodes sensitized with all of our extracted natural dyes as a function of wavelength. Some of the IPCE spectra of the organic dyes adsorbed on the ZnO electrodes are broader than the absorption curves of the dyes in solution.

Ethanol extract of *Bougainvillea spectabilis* [Fig. 12(c)], *Carissa ovate* [Fig. 12(b)], *Hibiscus sabdariffa* [Fig. 12(d)], *Beta vulgaris* [Fig. 12(e)], *and Amarathus iresine herbisti* [Fig. 12(a)], shows broad IPCE spectrum up to 610,720, 430, 460, 550, 710 nm, respectively [Fig. 12]. Water extract of *Bougainvillea spectabilis, Carissa ovata, Hibiscus sabdariffa, Jacaranda mimosifolia, Beta vulgaris,* and *Amarathus iresine herbisti* from 300 up to 550, 730, 700, 460, 550, 650 nm, respectively [Fig. 13], which showed some sensitization effect, but there spectrum is dominated by the spectrum of the ZnO nanoparticles. The IPCE of ZnO sensitized by *carisa ovate* and *Hibiscus sabdariffa* shows a red shift on the absorption onset for water extract than ethanol extract.

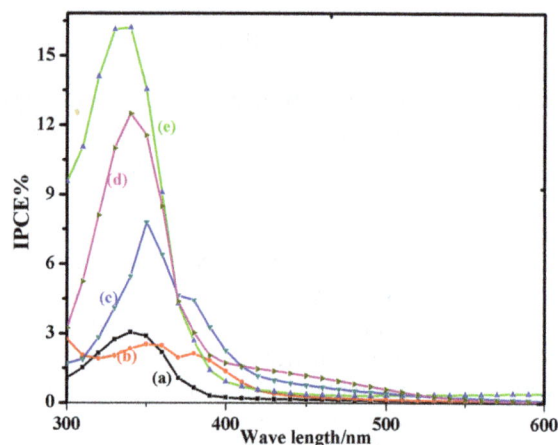

Fig. 13. J-V of ZnO based DSSCs sensitizers extracted with water, (a) Bougamvillia spectabilis, (b) Carissa ovate, (c) Amarathus iresine herbistii, (d) Beta vulgaris, (e) Hibiscus sabdariffa.

Low IPCE results in either inefficient light harvesting efficiency (LHE) by the dye, or inefficient charge injection into ZnO, or inefficient collection of injected electrons. A low LHE can be due to a low dye absorption coefficient over the solar spectrum, a low dye concentration, a thin ZnO film to support large concentration of adsorbed dye which absorbs a significant fraction of the incident light, insufficient light scattering within the film, absorption of light by ZnO or the redox electrolyte, and dye degradation [14].

Low η_{inj} can be due to dye desorption, dye aggregation or the excited state levels of the dye lying below the conduction band edge of ZnO, the presence of the surrounding electrolyte, the wavelength of the exciting photon. Low η_{cc} is due to competition between fast recombination of photo-injected electrons with the redox electrolyte or oxidized dye and electron collection. The formation of Zn^{2+}/dye complex can agglomerate which comes from dissolution of the nanostructured film form a thick covering layer instead of a monolayer, and is therefore inactive for electron injection which also limit the incident photon to current conversion efficiency. In IPCE of H_2O extract of Carissa ovata, both H_2O, and ethanol extract of *Amarathus iresine* shows additional spectral peaks which are a quite strong evidence of aggregate formation [15]. The presence of aggregated dyes or non-injecting dyes at the surface affects all parameters. In all DSSCs device studied a small value of the incident to photon conversion efficiency, short circuit current, and fill factor were obtained that leads to small solar to electricity conversion efficiency [16].

In view of the instability of ZnO in acidic dyes, the development of new types of photosensitizers for use in ZnO DSSCs is a subject that has already been widely investigated. These photosensitizers are expected to be chemically bonded to the ZnO semiconductor, be charge transferable with high injection efficiency, and be effective for light absorption in a broad wavelength range. New types of dyes should be developed with the aim of fulfilling these criteria. For our study some of the chlorophyll extract shows relatively good efficiency for ethanol extract than water extract.

In water extract of natural dye, increased socking time of the film will leads to the dissolution of ZnO nanoparticle on the ITO, this is because of the extraction solvent (water) possesses an acidic nature, protons derived from dissolution of the dye makes the dye solution relatively acidic and this may be one of the reason for decrease in the short-circuit photocurrent. Also upon long socking time, formation of aggregate are formed which can be center of charge recombination and decrease the photovoltaic effects.

Generally, natural dyes suffer from low V_{oc}, and J_{sc} which leads lower power conversion efficiency than an equivalent commercial N719 sensitized cell. This is because of most of natural dyes didn't completely attached to the ZnO, and TiO_2 nanoparticles, even upon washing by solvent to avoid aggregation from the socked film, the sensitizers leave the surface of the film, this is due to, more of the natural dye attached with film of the nanoparticles by physical attaching or by a weak bonding force.

4. Conclusion

In this work, ZnO nanoparticle was successfully synthesized which is one of the major semiconductors for DSSCs, using polyethylene glycol as dispersing agent. The average particle size of these semiconductor nanoparticles was determined using an approximation model called the effective mass model. From optical absorption spectra the maximum absorption λ_{max} was determined which was used for finding the energy band gap of the nanoparticles. From the effective mass approximation which relates the radius of nanoparticle with its energy bandgap the size of ZnO nanoparticle was determined to be 1.14 nm.

DSSCs device are fabricated using ZnO nanoparticles for the six sensitizers and their photovoltaic performance were determined. For ethanol extract of *Bougainvillea spectabilis,*

Carissa ovata, Amarathus iresine herbistii, Beta vulgaris, and *Hibiscus sabdariffa*, efficiency of 0.0066%, 0.0031%, 0.039%, 0.0039%, and 0.0043% respectively were obtained. The efficiency of water extract of *Bougainvillea spectabilis, Carissa ovata, Amarathus iresine herbistii, Beta vulgaris,* and *Hibiscus sabdariffa* were found 0.0016%, 0.003%, 0.0062%, 0.0077%, and 0.0104% respectively.

In general, ZnO nanoparticles based DSSCs shows low photoelectrochemical performance as compared to commercial TiO_2 based DSSCs. Some of the limiting factor for this was insufficient attachment of natural dyes with the nanoparticles, formation of aggregation between the nanoparticles up on film formation, low injection rate, low regeneration of electron, and formation of Zn^{2+} /dye complex. Therefore, modification of ZnO aggregates with a very effective shell material on ZnO, preventing the generation of Zn^{2+}/dye agglomerates will be a good method for improving the stability of ZnO nanoparticles.

References

[1] B. O'Regan and M. Gratzel, "A low-cost, high-efficiency solar cell based on dye-sensitized colloidal TiO_2 films", Nature, vol. 353, pp. 737-740, 1991.

[2] Y. Yen, W. Chen, C. Hsu, H. Chou, J. Lin, M. Yeh, "Arylamine-Based Dyes for p-Type Dye-Sensitized Solar Cells". Org. Lett., vol. 13, pp. 4930, 2011.

[3] W. Campbell, K. Jolley, P. Wagner, K. Wagner, P. Walsh, K. Gordon, Schmidt-Mende, M. Nazeeruddin, Q.Wang, M. Grätzel, "Highly Efficient Porphyrin Sensitizers for Dye-Sensitized Solar Cells" J. Phys. Chem. C, vol. 111, pp. 11760, 2007.

[4] P. J. Cameron, L.M. Peter, and S. Hore. "How important is the back reaction of electrons via the substrate in dye-sensitized nanocrystalline solar cells", J. Phys. Chem., vol. 109, no.2, pp. 930-4936, 2005.

[5] L. Fan, S. Kang, J. Wu, S. Hao, Z. Lan, J. Lin, "Quasi-Solid State Dye-sensitized Solar Cells Based on Polyvinylpyrrolidone With Ionic Liquid", J. Energy Sources A. vol. 32, pp. 1559. 2010.

[6] J. Singh, "Effective mass of charge carriers in amorphous semiconductors and its applications, Journal of Non-Crystalline Solids", vol. 299, pp. 444-448, 2002.

[7] J. Zhang, P. Zhou, J. Liu, J. Yu, "New understanding of the difference of photocatalytic activity among anatase, rutile and brookite TiO_2", Phys. Chem. Chem. Phys., vol. 16, pp. 20382-20386, 2014.

[8] N. H. Quang, N. T. Truc, Y. M. Niquet, "Computational Materials Science", vol. 44, pp. 21-25, 2008.

[9] K. Wongcharee, V. Meeyoo and S. Chavadej, "Dye-sensitized solar cell using natural dyes extracted from rosella and blue pea flowers", Sol. Energy Mater. Sol. Cells, Vol. 91, pp. 566-571, 2007.

[10] H. Horiuchi et al., "Electron injection efficiency from excited N3 into nanocrystalline ZnO films: Effect of (N3-Zn2+) aggregate formation" J. Phys. Chem. B, vol.107, pp. 2570-257, 2003.

[11] H. Chang and Y.-J. Lo, "Pomegranate leaves and mulberry fruit as natural sensitizers for dye-sensitized solar cells," Solar Energy vol. 84 no.10, pp. 1833-1837, 2010.

[12] S. Hao, J. Wu, Y. Huang, J. Lin, "Natural dyes as photosensitizers for dye-sensitized solar cell," Sol. Energy vol. 80, no.9, pp. 209–214 2006.

[13] Z.-S. Wang and F. Liu, "Structure-property relationships of organic dyes with D-π-A structure in dye-sensitized solar cells," Front. Chem. China, vol. 5, pp. 150-161, 2010.

[14] H. Zhou, L. Wu, Y. Gao, T. Ma, "Dye-sensitized solar cells using 20 natural dyes as sensitizers," J. Photochem. Photobiol. A: Chemistry, vol. 219, pp. 188-194, 2011.

[15] H. Hubert B. Michael, M. Peter, G. Thilo, "Natural pigments in dye-sensitized solar cells", Applied Energy vol. 115, pp. 216–225, 2014.

[16] Z. Qifeng, S. Christopher, Z. Xiaoyuan, "ZnO Nanostructures for Dye-Sensitized Solar Cells" Adv. Mater. Vol. 21 no. 41, pp. 4087–4108, 2009.

24

Biofuel Energy for Mitigation of Climate Change in Ethiopia

Abreham Berta, Belay Zerga

Department of Natural Resource Management, Wolkite University College of Agriculture and Natural Resource, Wolkite, Ethiopia

Email address:

abresh1240@gmail.com (A. Berta), belayzerga@Gmail.com (B. Zerga)

Abstract: There is a large interest in biofuels in Ethiopia as a substitute to petroleum-based fuels, with a purpose of enhancing energy security and promoting rural development. Ethiopia has announced a national biofuel production in the GTP in order to secure energy in the rural part and urban of Ethiopia. Its implications need to be studied intensively considering the fact that Ethiopia is a developing country with high population density and large rural population depending upon land for their livelihood. Ethiopia plan to reduce importing oil from foreign since the oil imported has huge potential of polluting the environment. Therefore, biofuel is free from pollutants; it is the main energy source for Ethiopia in the near future since Ethiopia economy is experiencing high growth rate, which may lead to enhanced demand for food, livestock products, timber, paper, etc., with implications for land use. The assessment is largely focused on first generation biofuel crops, since the Ethiopia program is currently dominated by many biofuel crops/trees, the process of biofuel and the relation environment and climate mitigation potential of biofuel. Technological,policy options, awareness creation required for promoting sustainable biofuel production in Ethiopia as recommendation from the review study.

Keywords: Biofuel, Climate Change, Mitigation, Oil, Petroleum

1. Introduction

Currently, worldwide annual energy use exceeds 9 billion tons of oil (*e.g.* fossil fuel). Of this, transport energy is responsible for 1.8 billion tons (20%) and results in over 6.3 billion tons of Carbon Dioxide (CO_2) emissions (WBCSD and EEA, 2004). There is an appetite for action on climate change. But global warming has not been reduced to safe levels(UNEP, November 2010) The world is currently on track for warming of at least 3 to 4oC by 2100,14 which would have far reaching consequences for food security, fresh water availability, and the frequency and intensity of storms(Intergovernmental Panel on Climate Change, 2014). At this level of climate change, the efficacy of adaptation strategies would be severely limited.

That is why the Paris summit 2015 is important because a strong international agreement will: allow countries to push ahead with strategies and policies for carbon emissions reduction, knowing that others are doing likewise; provide a predictable framework for a global low carbon economy; allow developing countries to pursue low carbon development strategies and adapt to the effects of climate change; improve international efforts to protect the natural environment(Rebecca Willis, 2014).

In response to the need for more sustainable forms of energy many new technologies have been introduced. In many countries the use of biofuel focuses on ethanol and biodiesel (*i.e. Jatropha* and Castor been *etc.*) has increased significantly in recent years (Tobin, 2005).

International Energy Agency research warns that delayed action would result in substantial additional costs, as high carbon investments made now would quickly lose their utility and value(OECD, 2014)

The last few years have witnessed both a dramatic increase in the price of oil and an increase in the production of biofuels like ethanol and biodiesel (Martinot, 2005).In many African countries there is a numerous political drive to establish domestic fuel industries. Biofuels are being up as the solution to rise fossil fuel import, to rural development, to the climate change challenge and all manner of other national development priorities (UNCT Ethiopia, 2008). Biofuel has several advantages over fossil fuels especially in landlocked countries likeEthiopia.

Recent trends also show that interest in biofuels is expanding towards developing countries where production costs are relatively inexpensive and gives possibility for biofuel to contend with fossil fuel prices. Currently, Ethiopia's economic and social development efforts are geared towards raising the country to a middle income level within the coming 20 years. To this effect every sector pursues its own targets since the vision is of national interest. Energy development, if designed in line with the needs of agriculture, industry, transport and other related sectors, would highly accelerate the achievement of this goal. (FFE, 2008).

Ethiopia's energy consumption is predominately based on biomass energy sources (94%).Traditional Energy Sources are Fuel wood, Charcoal, dung. Modern Energy is mainly products of petroleum and Electricity and Petroleum product is the major part from modern energy and it is mainly used for transport sector. The major source of electricity is hydro power plant. Ethiopia has potential of more than 45,000 MW from hydropower, Geothermal resource potential is estimated 1070 MW, Coal reserve is 70 Million Tons and Natural Gas Reserve 4 TCF (Terra Cubic Feet). Estimated potential of biogas cogeneration is greater than 263 MW, Woody biomass 787 Million tones Solar, wind and animate energy potential exist at considerable s size. There are huge potential on biofuel such as ethanol and biodiesel, currently 8 million liters of ethanol and huge potential of biodiesel in Ethiopia (Melis Teka, 2006).

The initiative for biofuels development in Ethiopia originally came from the private sector, though it did not take too long to get the government to buy-in. Mitigation of climate change is often presented by governments as a key policy goal for biomass fuel developments, but in the case of Ethiopia, the government is explicit about its reasons for promotion of biofuels. The reasons, among others, are energy security through the use of biofuels and to improve the balance of trade by import substitution and new export market development (Ministry of Mines and Energy, 2008).

Ethiopia has suitable environment for the growth of biofuel plants. It mean that Ethiopia has apple and fertile land, water resource and man power but have no technology, skill and experience of biofuel extraction.

The overall Objective of the study is assessment or review of biofuel production for mitigation of climate change in Ethiopia condition.

Significance of the study

There is an opportunities for the growth of biofuel crop like jatrohpa, caster bean, sugarcane, maize to produce ethanol and biodiesel but the imported energy like petroleum and coal require high price that affect economic development of Ethiopia and the energy produce from biofuel is free from emission of gases to the atmosphere. There is also 85 % of the Ethiopia communities are living in rural area and their lively hood is dependent on agriculture then this new getting form of energy is not understand by the communities of Ethiopia. Even the process of biofuel to produce biodiesel or ethanol is lack of knowledge in Ethiopia. What types of crops are used for ethanol or biodiesel, how propagate or multiplication the species for the sustainability of biofuel and the relation of biofuel with environment especially climate change mitigation is not clearly understand by Ethiopian society. This assessment research is avoiding the gap between all the above mentioned problems.

2. Data Requirements and Sources

This study has used data from primary and secondary sources. The primary sources of data were of *Jatropha* in benishangulgumze region of Ethiopia the two companies, Amabasel Jatroph Project and Sun Biofuels Eth/NBC the land coverage and their objectives as well as the production systems. All the necessary data regarding *Jatropha* cultivation and processing that are used to assess the energy balance, GHG effect and financial and economic feasibility of *Jatropha* production interviewed the companies.

The secondary data was collected from the following relevant sources: Agriculture and rural development offices at local and regional level, Ethiopia, minster of Environmental and forestry at deferent levels of Ethiopia, Swiss ministry of energy, irrigation and water of Ethiopia, Published and unpublished materials and internet. Personal observation and informal discussions with out-growers and biofuel developers' were used to generate primary data. After getting the data, there could be compiling in coherent form as reviewer of bio-fuel energy for climate change mitigation.

3. Literature Review

3.1. Biofuel Crops and Technologies

Traditionally solid-biomass (fuel wood and agro-residue) has been used as a source of energy for cooking, heating and even for power generation. Biofuel is another application of biomass. There are many sources of biomass feedstock for biofuel production, which include sugar crops, starchy crops, cellulosic material and oil crops, which are presented in Fig. 1 (FAO, 2008). However, the expansion of biofuel crops for the production of liquid biofuels for transport, in the recent years, is largely based on agricultural crops as feedstock such as sugarcane and maize for ethanol production and oil palm, soybean and rapeseed for biodiesel production. For example 85% of the global production of biofuels is in the form of ethanol from sugarcane and maize, where Brazil and USA accounts for 87% of global ethanol production (FAO, 2008).

3.2. Policy and Strategy of Ethiopia Government on Biofuel

It was planned during the PASDEP period to distribute 3 million improved energy saving biomass ovens these have been distributed as planned. As result it is estimated that about 26176 ha of forest have been conserved from deforestation and carbon dioxide emissions have been mitigated by about 36575 tones. (EFDRE, MFED, 2010)

The aim of the Ethiopia government in undertaking biofuel development include: To substitute fossil fuels used in transport, to leverage biofuel's suitability to assist with rural development, both for local energy consumption purposes and to establish a new production industry, to promote a climate friend energy source and for maximize the financial benefits to farmers (UNCT Ethiopia, 2008).

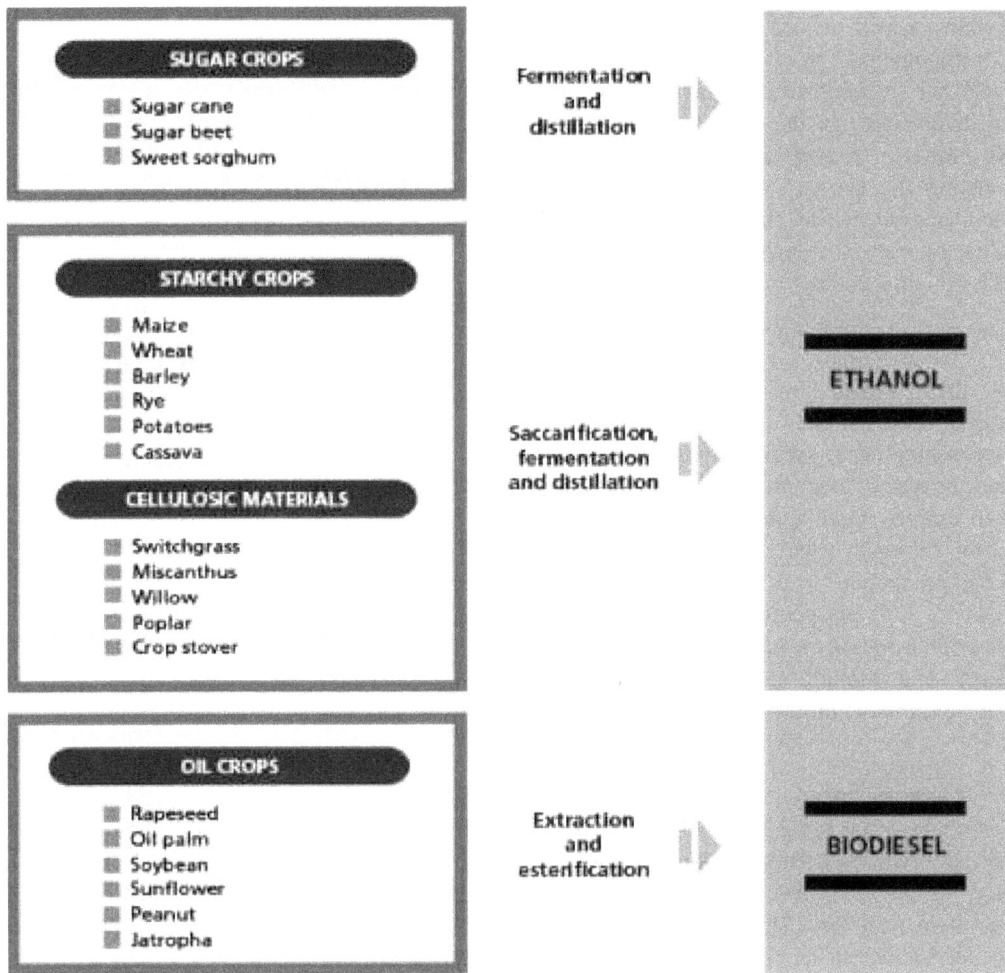

Source: FAO (2008)

Fig. 1. Biofuel crops, feed-stocks and fuels.

The strategic directions during GTP period are development of renewable energy, expansion of energy infrastructure, and creation of an institutional capacity that can effectively and efficiently manage such energy sources and infrastructure. During GTP period, the gap between energy demand for and supply of electricity will be minimized. Per capita consumption of electric city of households is expected to increase during the GTP period. In order to promote and realize the country's green development strategy, ongoing initiative to generate electricity from hydropower and renewable energy source like biofuel, solar and wind will be remain the strategic direction of GTP period. In addition new technological innovations will be utilized to insure that the energy sub-sector's doesn't emit additional carbon dioxide. The distribution of wood saving materials and technologies throughout the county will be continued (GTP Ethiopia, 2000).

Ethiopia emmision of CO_2 are very low compare to developed countries. Nevertheless , access to a range of reliable , affordable and clean energy sources is criteria for susutanable growth. One of the potential means to realize the

shift to more sustainable fuel is the production of biofuel and other renwable energy source , as part of ethiopia's green development strategy.as stated in the biofuel development strategy , ethiopia has sutaible land for biofuel(bio ethanol and biodesel) developpmet. Implementing a biofuel development initiative , in line with country's green development , will enable bio fuel to substitute for imported petroluem fuel even significant for export.more over job will be created locally at different stage of biofuel production process and thus contrubute to raizing living standards. Action will be taken to expand bio-fuel plantations,intruducing use of bio-fuel in house holds for cooking and lightining, standardize and refine information relating to land, technologies and markets for bio-fuel development, and promote knowlegde about biofuel use in each region. Benefite of biofuel development include inprove carbon sink of ethiopia, reduction deforestation and reduction in time spent on and burden to wemen and children in searching firewood.

The bio-fuel development program will implement strategies during the paln period to creat a network of implementing abencies, research institues and unversities to adopt and promote sustaianble development of biofuel tecknology. The network will collect and organize data on bio-fuel land, technologies and markets, coordinate involvement of relevant stockholders, support and motivate private investors' involvement in the bio-fuel development activities, involve farmer in production and supply of bio-diesel by coordinating the network of relevant extension services, and facilitate experience sharing with countries where bio-fuel development is at more advanced stage of development (GTP Ethiopia, 2000).

3.3. Biofuel Development in Ethiopia

Biofuel is not directly related to electricity issues but considering the current situation that

90% of energy consumption is supplied by wood or dung; biofuel can be an alternative to improve energy efficiency. Also, the export of biofuel could be an important source of foreign currency for Ethiopia, where there is currently no significant fossil fuel production.

This is important for Ethiopia particularly at the moment when the oil price surge is hitting the country's balance sheet. Ministry of energy and mining and other government agencies are starting to promote biofuel development now.

3.3.1. Policy

The Ethiopian Government plans to allocate 2,300ha for biofuel production counting on a big EU market as well as domestic demand. However, even Brazil, which is the most advanced country in biofuel development and which possesses 8 times more land than Ethiopia, has developed less than 2,300ha for biofuel so far. Therefore, the number might not be realistic but it shows the enthusiasm of the government.

3.3.2. Biodiesel

Biodiesel is made of Jetropha and Castor Oil. Since harvesting of these fruits cannot be mechanized, it requires a lot of manpower. The production cost can be minimized in Ethiopia, which possesses a plentiful workforce at a reasonable price. In addition to the labor cost, the land is leased by the government for a very low price, which also reduces the cost.

On the other hand, the transportation cost is an issue. Harar, which is the highest potential area for biodiesel production, and the second candidate site, Welo, is relatively near to the

Djibouti port, but the rest of the candidate sites are more than 1,000km from the port and the transportation cost would be substantial. One of the solutions is to sell the products on the domestic market near the production sites.

Castor Oil is in higher demand not only for biodiesel but also for medicines, lubricants and cosmetics. Therefore, it would be relatively expensive for biodiesel (1,500 US $/ t),while Jetropha costs only 1,000 US $/t, which gives it a higher potential as material for biofuel.

50% of the weight of a Castor fruit can be processed to crude oil, compared to only33-38% in case of the Jetropha.

The productivity improvement of Jetropha is ongoing.

3.3.3. Bio-ethanol

Bio-ethanol can be made of sugar cane. 8mln Litter of ethanol is currently produced annually from sugar cane in Ethiopia and bio-ethanol has been starting to be mixed with petrol at the rate of 5% since October 2008. According to ministry of , the usage area and the ratio will be gradually expanded.

Ethanol is more efficient than wood and dung, so it is good to promote ethanol usage at home as a short term measure. Some are already using bio-ethanol for cooking at home.

Also, a joint venture of American and Ethiopian companies is producing a cooking stove using ethanol, which may gradually increase the domestic demands (Embassy of Japan in Ethiopia, 2008).

3.4. Current Biofuels Development Status in Ethiopia

In order to ensure the country's continued development program, the national fuel security and environmental sustainability, it is important to increase fuel utilization and satisfy the demand partly by locally produced fuels such as biofuel. "The Ethiopian Biofuel Development and Utilization Strategy" is targeted to supply fuels from locally produced biofuel (*Jatropha*, Castor bean, Palm oil and sugar cane) and the objective of the strategy is to ensure the production of biofuel without affecting food self sufficiency, to improve balance of payment and to reduce environmental impacts. The strategy sets out that

106,997 million tons of ethanol from sugar cane will be produced by 2016 (nearly 7 times of the annual gasoline consumption of the country). The strategy also stipulates that with the participation of 1.25 million farmers in *Jatropha* biodiesel production the imported 1 billion liters of petrol diesel per annum can be substituted. By involving biodiesel producing companies the production will be in excess of local consumption and it will be a good commodity for export (MME, 2007).

Biofuels development as a primary product was first initiated by the private sector when Sun Biofuels Ethiopia (National Biodiesel Corporation), a subsidiary of a UK based private limited company, was allocated the first land for cultivation of jatropha for production of biodiesel in Benshangul Gumuz regional state in 2006. The coming of Sun biofuels awakened other players in the sector, including the government, the private sector, NGOs and civil society organizations. As the first project in the country, there were several drawbacks in the legal process of business formation and actual implementation of the project in the field. Since then several private companies have come to the scene. Fincha Sugar Factory, however, has been producing bioethanol as a by-product MELCA.

Several local and international private and non-private biofuels developers have registered in Ethiopia. Most of these companies have the intention of going for large-scale commercial development. Currently there are over 50 developers registered for the cultivation of energy crops for biodiesel production. For bioethanol, however, there are only six developers in the country of which four of them are government owned sugar estates. At present only one of the sugar estates, Fincha, is producing ethanol. The rest are at the pre-implementation stage either retrofitting existing factories for ethanol development (MELCA Mahiber, 2008).

According to MELCA only five regions were identified to conduct a brief assessment about the current development of biofuels in Ethiopia. These regions are identified based on the current trend of biofuels development expansion in the country.

So far, as shown above table 1,327,094 ha of land have already been allocated for investors. Over 80% of these developments are happening in arable land, forest land and woodlands. Many of these companies are still requesting for more lands for further expansion of biofuels production, in and around their current production sites.

Table 1. Regional distribution of biofuels energy company name and land hold within region in the country.

NO.	Company Name	Region	Land acquired (ha)	Out Growers Land(Ha)	crop type
1	Sun Biofuels Eth/NBC	Benshangul	80,000		Jatropha
2	Amabasel Jatroph Project	Benshangul	20,000		Jatropha
3	Jatropha Biofuels Agro Industry	Benshangul	100,000		Jatropha
4	IDC Investment	Benshangul	15,000		Jatropha
5	ORDA	Amahara	884		Jatropha
6	BDFC Ethiopia Industry	Amahara	18,000	30,000	Sugarcane/sugar beat
7	Jemal Ibrahim	Amahara	7.8		Castor bean
8	A Belgium Company	Amahara	2.5		Castor bean
9	Flora Eco Power Ethiopia	Oromia	10,000	5,000	Castor bean
10	Petro Palm Corporation Ethiopia	Oromia	50,000		Castor/Jatropha
11	VATIC International Business	Oromia	20,000		NA
12	Global Energy Ethiopia	SNNPR	2,700	7,500	Castor bean
13	OmoSheleko Agro Industry	SNNPR	5,500		palm
14	Sun Biofuels Eth/NBC	SNNPR	5,000		Jatropha
Total			327,094	42,500	

3.5. Socio Economic Importance of Biofuel

3.5.1. Land Inventory

There is a pressing need to clearly identify the availability of land for biofuel development, subjected to ownership, eco-geographical and food supply constraints. There is, in principle, considerable land available for biofuel production in Ethiopia but it is not clear that the actual status of the areas considered having potential. This need to be investigates as priority as it will be more difficult to obtain certification for Ethiopia biofuel if the land use impact associated with their production.

3.5.2. Maintenance and Sustaniability

Biodiversity –as with new agro-economic system, growing a biofuel crop will alter local habitates and resources that will affect speciess' distrubution and aboundance.

3.5.3. The Food Verse Fuel Dilemma

Biofuel production can increase the demand for agricultural input such as land and water, and this can jeopardize the production of food. In principle the scaling up of agricultural crops for biofuels can be done without competing with food production and without resulting forest clearance.

Biofuels will increase the price food because either food crops are converted to fuel or energy crops displace food crops on agricultural lands. The ultimate impact on a region will depend on several factors including the intensity of cultivation of biofuel crops and the extent of trade in food-related commodities. One can envision several scenarios. Developed regions such as the EU and the United States will experience price increase but may be able to absorb the price rise more easily than developed countries. One reason for this could be that since food-processing costs comprise a large share of the total cost, there will be a lesser impact on the final consumer price. The food processing industry will, however, be negatively affected due to higher input costs and lower demand for food. Developing countries that are net importers of food will be negatively affected due to higher food prices irrespective of whether they adopt biofuels or not. A region that is autarkic and does not adopt biofuel crops should, however, be isolated from such developments. If biofuel crops are cultivated exclusively on set-aside lands or marginal lands, with little competition with food crops, the impacts on food prices can be theoretically minimal. But in reality biofuels may still compete for other resources like water or labor and thus impact food production (UNCT Ethiopia, 2008).

3.6. Drivers for Biofuels

Increasing consensus about the end of cheap oil, the risks to supply due to political instability in major oil-producing regions, and the consequences of carbon emissions from fossil fuels have caused a spurt in the search for alternative sources of oil (Runge, 2007). But today, plant-based fuels like ethanol and biodiesel seem to be emerging as a serious alternative fuel ahead of technologies. According to runge, there are several reasons for the excitement surrounding biofuels.

1. Biofuels are replenishing: Biofuels are an inexhaustible resource since the stock can be replenished through agriculture.

2. Biofuels can reduce carbon emissions: Biofuels are sometimes considered as a solution to climate change. While this may be too optimistic, it is true that direct carbon emissions from combustion of biofuels are insignificant compared to fossil fuels. That said, it is hard to generalize about indirect carbon emissions (from agriculture and processing) and emissions of other harmful pollutants, which can be significant.

3. Biofuels can increase farm income: Today decline in farm income is a problem the world over (Gardner, 2003). With biofuels, most countries will be able to grow one or more types of crops in which they possess a comparative advantage and use them to meet either domestic or foreign demand or both. This increased demand for agriculture is expected to increase farm income.

4. Biofuels can improve energy security: The fact also means that countries can produce their own fuel, and reduce their dependence on foreign sources for energy (Hazell and Pachauri 2006).

5. Biofuels can create new jobs: Biofuels are more labor intensive than other energy technologies on per unit of energy delivered basis (Kammen, Kapadia, and Fripp 2004). The production of the feedstock and the conversion require greater quantities of labor compared to that required for extraction and processing of fossil fuels or other industrially based technologies like hydrogen and electric vehicles.

6. Biofuels have physical and chemical properties similar to oil: Several physical and chemical characteristics of biofuels such as their liquid state, specific energy density, viscosity, and combustion characteristics are more similar to gasoline or diesel than for alternatives. They are combustible in existing internal combustion engines with minor modifications. As a result, adapting to biofuel-based infrastructure (at least at low levels of blending like 10% or 20%) can be achieved more cost effectively than adapting to hydrogen, battery, or natural gas-based automobiles (Ugarte, 2006).

Biofuels are simple and familiar: biofuels have an aura of being simple and familiar to consumers, producers, and policymakers alike. Ethanol has been used as an additive or as a blend with gasoline in several countries for over two decades

3.7. Biofuel Sources and Conversion Technologies

Like any production system, inputs like fuel, capital, and labor are combined to produce the energy using a chemical conversion process. In the process pollution and other useful crop products are also produced. Table 3 shows the key differences between traditional and modern bioenergy systems in terms of these inputs, conversion technology, and the outputs. Traditional forms of biomass use are characterized by low capital, low conversion efficiency, poor utilization of fuel, and poor emission controls whereas modern forms of biomass use are characterized by higher

capital, higher conversion efficiency, better utilization of fuel, and better emission controls (WorldBank,2007).

3.7.1. Traditional Biomass

Traditional biomass implies the use of sources like wood, crop residues, animal dung, and charcoal for cooking and heating at the household level. This is often done using three stone stoves or in some cases using improved cook stoves or biogas stoves. Traditional use of biomass has the following characteristics. Firstly, traditional biomass is usually gathered or collected (often by women and children) from common lands or privately owned lands and are therefore largely an informal activity. The only cost to users is the opportunity cost of time invested in collecting fuel wood.

Second, combustion of biomass is characterized by low efficiency due to poor design of stoves. As a result, biomass is overused and is associated with deforestation, fodder scarcity, and depletion of soil quality (due to non availability of animal manure and other residues for soil). Third, uncontrolled and open burning of biomass in traditional stoves, in poorly ventilated chambers has serious health implications for women and children (Kammen 2005).But such attributes are not inherent to bioenergy and are the consequence of socio-economic and political factors, which can be addressed with the aid of appropriate policies. For example, dissemination of improved cook stoves and biogas systems, better ventilation in the kitchen area, sustainable harvesting of wood, etc., can make traditional biomass more sustainable (Kammen, 2006). Investments in improving the efficiency and reducing emissions from traditional biomass use will have impacts as wide ranging as improving gender equity and halting environmental degradation given its high use of child and women labor and the high fuel use per unit of delivered energy.

3.7.2. Modern Biofuels

Liquid biofuels for transportation like ethanol and biodiesel are one of the fastest-growing sources of alternative energy in the world today and are poised to reverse the historical trend of decline in the share of biomass in the global primary energy supply. Like traditional biomass, modern biofuel systems also encompass a variety of feedstock, conversion technologies. They are used mostly for generation of electricity or transportation as opposed to cooking and heating. The technological and commercial maturity and scalability of the various biofuel pathways are also diverse. Sugar and starch-based crops and the associated conversion technologies are the most mature for ethanol production today, while oilseed crops are the most mature sources of biodiesel. But since they have low yield per hectare and are also used for food, they are not well suited for large-scale expansion. Cellulose-based fuels are considered the most promising for the future but are not commercially and technically mature today. The production of electricity from biomass, using wood and agricultural and municipal wastes while technologically mature, is not commercially widespread. The reasons for low commercial maturity are several including high cost, under compensation for environmental benefits, etc. (Roos, 1999).

3.8. Major Types of Biofuels

A variety of biofuels can be derived from biomass. Ethanol and biodiesel are the most widely used biofuels for transportation today. In the future, butanol and Fischer-Tropsch fuels have the potential to become competitive as liquid fuels. Synthesis gas produced by gasification of wood is used mainly for electricity generation. Fuelwood and biogas produced by anaerobic digestion of plant and animal wastes are used for cooking and heating at the household level.

The term feedstock refers to the raw material used in the conversion process, which can be a crop, crop residue, or agricultural and municipal waste. The main types of feedstock are described in detail below

1. Sugar and starch-based crops: Crops rich in sugar and starch like sugarcane and corn (maize), respectively, supply almost all the ethanol that is produced today. Other major crops being used include wheat, sorghum, sugar beet, and cassava.

Technologies for conversion of sugar and starch are also the most technologically and commercially mature today. The major drawback of such crops is that they are important food crops and their use for fuel can have adverse impacts on food supply. Another drawback is these crops are intensive in the use of one or more among inputs like land, water, fertilizer, pesticides, etc., which have other environmental implications (Pimentel and Patzek 2005; Ulgiati 2001; Giamipietro, Ulgiati, and Pimentel 1997; Farrell 2006). In the future cellulosic sources are expected to displace such crops as the major source of ethanol.

2. Oilseed crops: In contrast to ethanol, biodiesel is produced from oilseed crops like soybean, rapeseed, and oil palm (Demirbas 2001, Sheehan 2000). But like sugar and starch crops, oilseed crops are also characterized by low yield and high use of inputs. In the future non edible crops like Jatropha curcas and Pongamia pinnata, which are considered to be low-input and suited to marginal lands, may become major sources of biodiesel especially in the dry and semi-arid regions of Asia and Africa. But the economic viability of crops these crops under conditions of low inputs and poor land quality are considered highly uncertain (Prayas, 2007).

3. Wood: Wood is predominantly used for cooking and heating at the household level and to a lesser extent for producing electricity at a small scale. When used directly at the household level, it is often collected from forests or other lands. Commercial plantations of woody trees like poplar and willow in temperate zones and eucalyptus and acacia exist today albeit on a small scale. The predominant use of commercial plantations today is for the supply of wood to paper and pulp industries (Ravindranath and Hall, 1995). Future cellulosic technologies, which permit the conversion of wood to ethanol, may compete with current uses of wood.

4. Wastes and residues: According to Kim and Dale (2004), there are about 73.9 Million tonnes of dry wasted crops and about 1.5 billion tonnes of dry ligno-cellulosic biomass from seven crops namely, maize, oats, barley, rice, sorghum, wheat,

and sugarcane (Kim and Dale, 2004). These could potentially yield about 490 billion liters of ethanol or about 30% of global gasoline use today. Furthermore, lignin-rich fermentation residue, which is the crop product of ethanol made from crop residues and sugarcane bagasse, can potentially generate both 458 TWh5 of electricity (about 3.6% of world electricity production) and. The utilization of this feedstock is contingent upon the successful commercialization of cellulosic technologies. The economics of collection and processing of residues is also not clear. The low specific energy density of residues can imply high transportation costs that might render a large fraction of this resource uneconomical (OECD, 2006).

5. Dedicated cellulosic crops: Cellulose is the substance that makes up the cell walls of plant matter along with hemicellulose and lignin. It is the primary structural component of green plants comprising more than 50% of the phyto-matter incorporated annually in plants (FAO, 2007). It is much more abundant than starch, sugar, and oil, which are concentrated only in seeds and fruits. Perennial grasses like switchgrass and Miscanthus are two crops considered to hold enormous potential for ethanol production. Perennial crops also confer other advantages like lower rates of soil erosion and higher soil carbon sequestration. However, technologies for conversion of cellulose to ethanol are just emerging and not yet technically or commercially mature. Cellulose conversion technologies will allow the utilization of non grain parts of crops like corn stover, rice husk, sorghum stalk, bagasse from sugarcane, and the woody parts (Wyman, 1999 and Lynd, 1996)

3.9. Conversion Technologies

A number of conversion technologies are available depending on the types of feedstock, fuel, and end use that are desired (Faaij 2006). We will provide a brief review of each of these conversion technologies.

Direct combustion: This is the most common and oldest form of conversion that involves burning organic matter in an oxygen-rich environment mainly for the production of heat. The most common use of this heat is in the production of steam for industrial use or for electricity generation. In some cases, the goal of burning might simply be reduction in the volume of waste without energy recovery as is the case with disposal of agricultural or medical waste. Examples of applications of direct combustion include burning of biomass like wood, dung, and agricultural wastes in homes for cooking and heating, co-firing of biomass with coal in electricity production, the burning of wood for process heat in chemical industries, etc. Typical flame temperatures for combustion and incineration range between 1,500° F and 3,000° F (Demirbas 2001).

Thermo-chemical conversion: In contrast to direct combustion, thermo-chemical conversion utilizes heat and pressure in an oxygen-deficient environment to produce "synthesis gas". Syn-gas is composed mainly of carbon monoxide and hydrogen, and can either be combusted to produce heat or converted to other fuels like ethanol and hydrogen. Thermo-chemical conversion is cleaner compared to other conversion pathways. Thermo-chemical conversion pathways include processes such as gasification, pyrolysis, plasma arc, and catalytic cracking. A detailed description of these technologies can be found in a report on conversion technologies by the California Integrated Waste Management Board (CIWMB). While gasification processes vary considerably, typical gasifiers operate from 1,300° F and higher and from atmospheric pressure up to five atmospheres or higher (CIWMB, 2005).

Biochemical conversion: Unlike thermal and thermo-chemical processes, biochemical conversion processes occur at lower temperatures and have lower reaction rates. Higher moisture feedstock is more easily converted through biochemical processes. Fermentation and anaerobic digestion are two common types of biochemical conversion processes. The main use of fermentation is in conversion of sugar and starch, found in crops like sugarcane, corn, wheat, etc., to ethanol. The fermentation of alcohol yields coproducts like distiller dried grains, which can be used as feed for livestock. Anaerobic digestion involves the bacterial breakdown of biodegradable organic material in the absence of oxygen over a temperature range from about 50° to 160° F. The main end product of these processes is called biogas, which is mainly methane (CH_4) and carbon dioxide (CO_2) with some impurities such as hydrogen sulfide (H_2S). Biogas can be used as fuel for engines, gas turbines, fuel cells, boilers, and industrial heaters, and as a feedstock for chemicals (with emissions and impacts commensurate with those from natural gas feedstock) (Demirbas, 2001 and CIWMB, 2005). Conversion of cellulosic feedstock using acid or enzymatic hydrolysis is another type of biochemical process, which is expected to become commercially very important in the future.

Transesterification: This is the most common method of producing biodiesel today.

Transesterification is a chemical process by which vegetable oils (like soy, canola, palm, etc.) can be converted to methyl or ethyl esters of fatty acids also called biodiesel. Biodiesel is physically and chemically similar to petro-diesel and hence substitutable in diesel engines. Transesterification also results in the production of glycerin, a chemical compound with diverse commercial uses. This process is carried out at a temperature of 60° C to 80° C (Demirabas 2001, 2003; Sheehan 2000; Crabbe2001).

Emerging technologies

A variety of other technologies for conversion of biomass to fuels, or substitutes for fossil fuel-derived products like plastics, is being researched and developed.

Cellulosic ethanol: Cellulosic conversion implies the transformation of nongrain or nonfruit parts of phytomatter, which are mostly comprised of cellulose such as the stem, wood, grass, leaves, etc., into ethanol. Switchgrass and Miscanthus are two perennial grasses that are undergoing trials as feedstock while a variety of chemical and biochemical processes including acid-based and enzymatic processes, are being developed simultaneously for breaking down cellulose into ethanol. Similar to sugar refineries that utilize bagasse for cogeneration of electricity, cellulosic conversion can also be accompanied by the combustion of lignin to supply heat and steam for conversion. This will have the added benefit of offsetting electricity produced from fossil fuels (Lynd, 1996).

Fischer-Tropsch fuels: These are synthetic substitutes to gasoline and diesel, which are produced by a process in which carbon monoxide and hydrogen are catalytically transformed into liquid hydrocarbons (HC). Although coal and natural gas are considered as the main sources for carbon monoxide and hydrogen, gasification of biomass feedstock is considered a more environmentally benign conversion pathway for Fischer-Tropsch fuels (Hamelinck 2004). Another line of research involves production of "biocrude" through high-temperature/ pressure and chemical breakdown of biomass into liquids, using hydrothermal upgrading (HTU) or pyrolysis. All these pathways are currently expensive and technically immature (Fulton 2005).

Biobutanol: Biobutanol is butanol (i.e., butyl alcohol), which is produced biologically from biomass through a process called acetone butanol ethanol (ABE) fermentation. As a result of low butanol yield, ABE fermentation was considered uneconomical. However, it is expected to be viable at a gasoline price of $3.00 per gallon or greater (Ramey 2004).

Algae biodiesel: Biodiesel production from algal oil is another technology, which is considered to have significant potential to replace diesel use. However, the major difficulties are in finding an algal strain with a high lipid content and fast growth rate that isn't too difficult to harvest, and has a cost-effective cultivation system (Sheehan 1998).

Biobased products and bioplastics: Agricultural feedstock can also be used to produce other industrial products called bioproducts and bioplastics, which are substitutes to chemicals, plastics, hydraulic fluids, and pharmaceuticals produced from fossil fuels. Agricultural feedstock which are considered as candidates for making such products, include a variety of crops, wood and plant oils, and agricultural and forestry residues. Bioproducts are considered to require less energy to produce than the fossil and inorganic products they replace (USDA, 2007).

3.10. Estimates of Future Potentials for Bioenergy

There are several studies that estimate the global potential of biofuels in absolute units of energy and as percentages of global energy that they can supply. Estimates of such potential can be classified into three categories, namely, biophysical, technical, and economic. Each category in the list comprises the ones following it, so that the three categories are of decreasing magnitude. Biofuels can in principle supply a large fraction of global energy need, and this is called the theoretical potential. The biophysical potential is determined primarily by natural conditions and describes the amount of biomatter that could be harvested at a given time. The technical potential depends on the available technologies and therefore evolves as technology progresses. Estimates of biophysical and technical potential vary depending on assumptions about land availability, yield levels in energy crop production, future availability of forest wood and of residues from agriculture and forestry, etc. The economic potential depends on at least two additional factors, namely, energy prices and policies toward renewable and clean technologies. However, oil prices are uncertain with respect to time, while policies vary both with time and also from region to region (Fischer and Schrattenholzer 2001). As a result, economic potential is hard to predict. For example, Brazilian ethanol is economically viable when oil sells at $35 per barrel whereas U.S. ethanol is viable only at around $50 per barrel (Ugarte 2006, OECD 2006).

Most studies report an increase in the supply of bioenergy over time. A review of 17 earlier studies on this subject by Berndes, Hoogwijk, and van den Broek (2003) reveals that estimates for potential contribution of biomass in the year 2050 range from below 100 EJ/yr to over 400 EJ/yr (Berndes, Hoogwijk, and van den Broek 2003). In comparison to the current level of bioenergy of 45 EJ/yr, this represents a doubling to a tenfold increase. A study by the International Institute of Applied Systems Analysis and the World Energy Council predicts that bioenergy would supply 15% of global primary energy by 2050 (Fischer and Schrattenholzer 2001). A study by the Natural Resources.

Defense Council predicts that an aggressive plan to develop cellulosic biofuels between now and 2015 could help the United States produce the equivalent of nearly 7.9 million barrels of oil per day by 2050. This is equal to more than 50% of the current total oil use in the transportation sector (Greene 2004). A majority of the increase is accounted by cellulosic biomass like switchgrass. However, it is also possible to envision scenarios that involve reduction in cropland while meeting the future food needs for a larger and wealthier population. One of the drawbacks of the above assessment is that it is static and does not take into account future changes in technologies and the demand for food. An analysis of the demand for cropland based on fundamental forces responsible for expansion of cropland by Waggoner and Ausubel (2001) suggests that sustained technological progress in crop production could meet the recommended nutritional requirements for a population of 9 billion and simultaneously reduce cropland by 200 million hectares by the year 2050. It is even claimed that under the best-case scenario the land withdrawn from agriculture could be as high as 400 million hectares. At the same time, they warn that such improvements would come about only through sustained investments in productivity, experimentation, and deployment of better technologies (Waggoner 1996, Waggoner and Ausubel 2001).

3.11. Diverse Solutions for a Diverse World

Biofuels have played a vital role in meeting the energy needs of human beings. There is reason to believe they will continue to do so in the future albeit in a different manner.

Traditional forms of biomass energy are still prevalent among the rural poor in developing countries that use it for cooking and heating. Modern forms of bioenergy are expanding in the developed countries largely for use in automobiles and electricity generation. With economic growth, the share of traditional biomass will decline while that of modern energy sources will increase so that transportation and electricity production may be the dominant end uses one day as opposed to cooking and heating.

However, given the slow pace of expansion of rural electrification and access to clean cooking fuels in developing countries, such a change may be a long while coming.

Traditional or modern, biofuels can make a positive contribution to all three pillars of sustainable development— economic, social, and environmental. But the diversity in the social, economic, and environmental impacts proscribes a "one size fits all" approach.

Most people contend that no single source of biomass or conversion technology or type of biofuel will suffice because of the disparate agro-climatic, ecological, technological, and socioeconomic and political economic factors that need consideration. Modern biofuels can in some cases be more detrimental to the poor than traditional biofuels. The appropriation of food crops for ethanol production may have adverse impacts of food prices (Runge 2007, Msangi 2006, OECD 2006, FAPRI 2005). The commercialization of cellulosic technologies may result in conversion of fodder resources for livestock or conversion of wood used by household into fuel for automobiles. The use of marginal lands for biofuel plantations can also worsen the energy poverty of the landless poor who would lose access to fuelwood and fodder from such lands (Gundimeda 2004, Rajagopal2007, Karekezi and Kithyoma 2006). In the case of poor rural households in developing countries, the use of biomass for providing cleaner energy for cooking and providing electricity may be more beneficial overall rather than using them to produce transportation fuels8. Along with technological progress, innovative policies will be necessary to ensure a smooth transition to a future where modern biofuels can be a significant supplier of energy. This chapter has provided a historical and technological perspective. In the following chapters, we will discuss the environmental, economic, and political aspects of biofuels.

3.12. Biofuels Climate Mitigation Potential

One of the major arguments behind support for biofuels is the perception that they are more climate friendly than oil. Biofuels are sometimes even claimed as being carbon neutral and fossil free. (Pimentel and Patzek 2005, Farrell 2006). In reality, biofuels consume a significant amount of energy that is derived from fossil fuels. Inputs to production include tillage, fertilizers, pesticides, irrigation, operation of machinery for harvesting and transport, steam and electricity for processing, etc., all of which embody fossil energy, leading to a significant net carbon addition to the atmosphere by the time the biofuel is ultimately consumed (Giampietro 1997, Lal 2004, Pimentel and Patzek 2005, Farrell 2006). Equally important is the fact that production of biofuels has other.

Nonclimate-related environmental impacts such as soil erosion due to tilling, eutrophication due to fertilizer runoffs, impacts of exposure to pesticides, habitat, and biodiversity loss due to land-use change, etc., which have not received the same attention as GHG emissions (Fearnside 2002, Curran 2004). In fact, already the fear of rainforest destruction due to the EU's biofuel mandate led commissioner Peter Mandelson to recently declare, "Europeans won't pay a premium for biofuels if the ethanol in their car is produced unsustainably by systematically burning fields after harvests, or if it comes at the expense of rainforests. We can't allow the switch to biofuels to become an environmentally unsustainable stampede in the developing world." Given these trade-offs, characterizing the overall environmental impact of biofuels is complex and challenging. (Nan Shi, Doris Chen Yu, Hui (Becky) Li 2009).

Biofuels benefit environmental by reducing GHGs and reducing local pollution, Bioethanolis water soluble, non toxic and biodegradable, Ethanol can replace 10% of the world's gasoline without clearing more rainforests and by doing less harm to the environment than current agriculture, Biofuels offer low levels of carbon dioxide emission, Hydrocarbon - based fuels produce greater noxious by productions like carbon monoxide compared to biofuel which produces compounds like nitrogen oxides; but the biofuel by product is less dangerous to human health and the environment, Fossil fuels: derived from *long dead* biological materials. Therefore, biofuel has huge potential of mitigation of climate change by substitute the oil and petroleum that import from foreign and maximize Ethiopia economy.

4. Conclusion

Concerns about climate change, security, and reliability of energy supply and the growing demand for oil are likely to make biofuel sever more attractive. As far as energy is concerned, the main contribution of biofuel will be in augmenting the supply of fuels for transportation, for electricity and heat. For the most part, the future of biofuels will depend on energy policies and technologies that will affect demand for liquid fuels. Increase in fuel efficiency, hybrid and fuel cell vehicles, and carbon taxes will reduce the demand for fuel, while increase in income and highways will increase the demand for fuel. Such increases are more certain in developing countries like Ethiopia.

A major motivation for biofuel is that they will raise farm income, which will have attendant political and economic benefits. But such gains may not be realized when domestic production competes with imports that are cheaper. This is the reason biofuel crops like other agricultural goods are also subject to barriers in the form of duties, quotas, and bans on imports. The rationale for such protection can be several such as the need to support domestic farmers, enable the development of a domestic infant industry, keeping food prices low, and environmental regulation.

Biofuels will have both environmental benefits and costs. They may reduce carbon emissions, but there will be negative impacts on agricultural activity.

The use of marginal lands or the utilization of crop residues for biofuel production will deny access to fuel wood and fodder, which can hurt the poor. In such cases the production of biomass suited to local needs may be socially more optimal than the production of biofuels for cars in cities. Demand-side policies that discourage the consumption of fossil fuels or carbon emissions as a way of improving energy security or protecting the environment seem should also be given greater attention by researchers and policymakers despite the political economic barriers to such policies.

Reduce fossil fuel imports–The use of ethanol can reduce the use of gasoline. A reduction in the use of gasoline reduces some of the dependence on unstable foreign sources of oil. Because the biofuel feed stock can easily obtain from Ethiopia then the oil from biofuel can process in Ethiopia. Ethiopia lost a lot of money in each year due to importing oil and pollute the environment from the emission of the oil but biofuel energy is make reverse.

There is Health benefits from reduced global warming since biofuel energy is free from emitting gas to the atmosphere as comparing oil that formerly used in Ethiopia. There has to be some kind of regulation to make sure there is no greater deforestation, land degradation of low - fertility tropical soils, and concomitant rise in CO_2 output.

Recommendation

- More research is required on crops, grass, woods etc that are not used or competition for food.
- Determination of the amount of carbon sequesters or sinks by the biofuel comparing petroleum and coal that currently Ethiopia used are essential.
- Policy and Strategies will be required to tackle the competition between the food dilemma and biofuel energy production especially Energy and Environment Policies of Ethiopia should look at carefully to overcome the problem.
- Most of biofuel producers have lack of knowledge on processing, cultivation and value chain therefore Technical Knowledge and awareness is required.
- Jatropha can be easily grown in Ethiopia since it is used as a fence at this moment, it need further replication and propagation in large amount and should start production of ethanol to overcome the problems of energy fuel.
- There is cost and energy when biofuel feed stocks is producing. Therefore, the amount of emission gas should calculate.

References

[1] Embassy of Japan in Ethiopia, 2008. Study on the Energy Sector in Ethiopia. Addis Ababa, Ethiopia.

[2] Gardner, B., 2003. "Fuel Ethanol Subsidies and Farm Price Support: Boon or Boondoggle?" working paper, Department of Agricultural and Resource Economics, University of Maryland.

[3]   Green Alliance, 2013, The global green race: a business review of UK competitiveness in low carbon markets.

[4]   Hazell, P., and R. K. Pachauri (eds.) 2006. *Bioenergy and Agriculture: Promises and Challenges*. International Food Policy Research Institute 2020.

[5]   Intergovernmental Panel on Climate Change, 2014, Climate Change 2014: impacts, adaptation and vulnerability.

[6]   International Renewable Energy Agency, November 2012, Renewable power generation costs: summary for policymakers.

[7]   Kammen, D., K. Kapadia, and M. Fripp 2004. "Putting Renewables to Work: How Many Jobs Can the Clean Energy Industry Generate," report of the Renewable and Appropriate Energy Laboratory, Energy and Resources Group/Goldman School of Public Policy at University of California, Berkeley.

[8]   Martinot, E. Renewables 2005: *Global Status Report*. World watch Institute, 2005.

[9]   MELCA Mahiber 2008. Rapid Assessment of Biofuels Development Status in Ethiopia and Proceedings of the National Workshop on Environmental Impact Assessment and Biofuels, Addis Ababa, Ethiopia.

[10] MelisTeka, 2006 .Energy Policy of Ethiopia. Addis Ababa, Ethiopia.

[11] Ministry of Mines and Energy Ethiopia, 2008. Ethiopian Biofuels Development and Utilization Strategym, Addis Ababa, Ethiopia.

[12] MME (Ministry of Mines and Energy), 2007. *The Biofuel Development and Utilization Strategy of Ethiopia*. Addis Ababa, Ethiopia.

[13] Nan Shi, Doris Chen Yu, Hui (Becky) Li, 2009. Introduction to Biofuel.

[14] OECD, 2014, The cost of air pollution.

[15] Rebecca Willis, 2014. Paris 2015: getting a global agreement on climate change. Published by Green Alliance August 2014 Green Alliance, Buckingham Palace Road London. www.green-alliance.org.uk (Access October 2, 2015).

[16] Rebecca Willis, 2014. Paris 2015 Getting a global agreement on climate change. Christian Aid, Green Alliance, Greenpeace, RSPB, and WWF

[17] Review of Environmental, Economic and Policy Aspects of Biofuels Deepak Rajagopal and David Zilberman, the World Bank, 2007.

[18] Roos, A., R. L. Graham, B. Hektor, and C. Rakos.1999. "Critical Factors to Bioenergy Implementation," *Biomass and Bioenergy* 17,: 113-126.

[19] Runge, C., and B. Senauer, (2007). "How Biofuels Could Starve the Poor," *Foreign Affairs*.

[20] Tobin, J., 2005. *Lifecycle Assessment of the production of Biodiesel from Jatropha*. An Msc Thesis presented to the School of Construction Management and Engineering of Reading University.

[21] Ugarte, D. G. de la Torre. 2006. "Developing Bioenergy: Economic and Social Issues," *Bioenergy and Agriculture: Promises and Challenges*, International Food Policy Research Institute 2020 Focus No. 14.

[22] UNCT ethiopia, 2008. Biofuel a vaibe alternative source of energy? knowledge sharing forum UN ethiopia.

[23] UNEP, November 2010, The emissions gap report: are the Copenhagen Accord pledges sufficient to limit global warming to 2 degrees C or 1.5 degrees C?

[24] WBCSD and EEA (The World Business Council for Sustainable Development & the International Energy Agency), 2004. *The IEA/SMP Transportation Model*.

Permissions

All chapters in this book were first published in JENR, by Science Publishing Group; hereby published with permission under the Creative Commons Attribution License or equivalent. Every chapter published in this book has been scrutinized by our experts. Their significance has been extensively debated. The topics covered herein carry significant findings which will fuel the growth of the discipline. They may even be implemented as practical applications or may be referred to as a beginning point for another development.

The contributors of this book come from diverse backgrounds, making this book a truly international effort. This book will bring forth new frontiers with its revolutionizing research information and detailed analysis of the nascent developments around the world.

We would like to thank all the contributing authors for lending their expertise to make the book truly unique. They have played a crucial role in the development of this book. Without their invaluable contributions this book wouldn't have been possible. They have made vital efforts to compile up to date information on the varied aspects of this subject to make this book a valuable addition to the collection of many professionals and students.

This book was conceptualized with the vision of imparting up-to-date information and advanced data in this field. To ensure the same, a matchless editorial board was set up. Every individual on the board went through rigorous rounds of assessment to prove their worth. After which they invested a large part of their time researching and compiling the most relevant data for our readers.

The editorial board has been involved in producing this book since its inception. They have spent rigorous hours researching and exploring the diverse topics which have resulted in the successful publishing of this book. They have passed on their knowledge of decades through this book. To expedite this challenging task, the publisher supported the team at every step. A small team of assistant editors was also appointed to further simplify the editing procedure and attain best results for the readers.

Apart from the editorial board, the designing team has also invested a significant amount of their time in understanding the subject and creating the most relevant covers. They scrutinized every image to scout for the most suitable representation of the subject and create an appropriate cover for the book.

The publishing team has been an ardent support to the editorial, designing and production team. Their endless efforts to recruit the best for this project, has resulted in the accomplishment of this book. They are a veteran in the field of academics and their pool of knowledge is as vast as their experience in printing. Their expertise and guidance has proved useful at every step. Their uncompromising quality standards have made this book an exceptional effort. Their encouragement from time to time has been an inspiration for everyone.

The publisher and the editorial board hope that this book will prove to be a valuable piece of knowledge for researchers, students, practitioners and scholars across the globe.

List of Contributors

Kenechukwu Emmanuel Ugwu, Anthony Chibuzo Ofomatah and Samson Ifeanyi Eze
National Centre for Energy Research and Development, University of Nigeria, Nsukka, Nigeria

Hsien-Tsung Lee and Li-Chung Sun
Department of Electrical and Information Technology, Nan Kai University of Technology, Nan Tou County, Taiwan

Anekwe Ozioma Juliana and Ajiwe Vincent Ishmael Egbulefu
Department of Pure and Industrial Chemistry, Nnamdi Azikiwe University, Awka, Anambra State, Nigeria

El-Naggar M. M. and Ibrahim H. A. H.
Microbiology Lab., Environ. Div., National Institute of Oceanography and Fisheries (NIOF), Alexandria, Egypt

Abdul-Raouf U. M.
Botany and Microbiology Department, Faculty of Science, Al-Azhar University-Assuit Branch. Egypt

El-Sayed W. M. M.
Microbiology Lab., Environ. Div., National Institute of Oceanography and Fisheries (NIOF), Hurghada, Egypt

Tahar Tafticht, Yamina Azzoug
Electrical Engineering Department, College of Engineering, Majmaah University, Majmaah, KSA

Danladi Eli, Idodo Maxwell and Aungwa Francis
Department of Physics, Nigerian Defence Academy, Kaduna, Nigeria

Muhammad Ahmad and Sunday Sarki
Department of Physics, Kaduna State University, Kaduna, Nigeria

Danladi Ezra
Department of Agricultural Science, Kaduna State University, Kaduna, Nigeria

Hani Muhaisen Alnawafleh
Faculty of Engineering, Al-Hussein Bin Talal University, Ma'an, Jordan
Faculty of Engineering, Tafila Technical University, Tafila, Jordan

Feras Younis Fraige
Faculty of Engineering, Al-Hussein Bin Talal University, Ma'an, Jordan

Faculty of Engineering, King Saud University, Al-Muzahmiayah Branch, Riyadh, Kingdom of Saudi Arabia

Laila Abdullah Al-khatib
Faculty of Engineering, Al-Hussein Bin Talal University, Ma'an, Jordan

Mohammad Khaleel Dweirj
Faculty of Engineering, Al-Hussein Bin Talal University, Ma'an, Jordan

Shaila Siddiqua, Abdullah Al Mamun and Sheikh Md. Enayetul Babar
Biotechnology and Genetic Engineering Discipline, Khulna University, Khulna, Bangladesh

Ahmed Bilal Awan
Electrical Engineering Department, College of Engineering, Majmaah University, Majmaah, KSA

Nirmala Thivyanathan
Principal, Research Centre of Zoology, Jayaraj Annapackiam College for Women, Periyakulam, Theni District, Tamilnadu, India

Alejandro Amaya, Mariana Corengia, Andrés Cuña, Jorge De Vivo, Andrés Sarachik and Nestor Tancredi
DETEMA, Facultad de Quimica, Universidad de la Republica, Montevideo, Uruguay

Valentyna Novosad
Scientific Company "MAE", 33 Horyva st., Kyiv, Ukraine

Omer Khalil Ahmad, Ahmed Hassan Ahmed and Obiad Majeed Ali
Technical institute of Hawija, Foundation of technical education, Baghdad, Iraq

Valentyna Novosad
Scientific company "MAE", Horyva st, Kyiv, Ukraine

Ahmed G. Abo-Khalil
Electrical Engineering Department, Assiut University, Assiut, Egypt

Sameh S. Ahmed
Mining and Metallurgical Engineering Department, Assiut University, Assiut, Egypt

Abel Malyango Masota
Tanzania Forest Services (TFS), P.O. Box 40832, Dar es Salaam, Tanzania

Eliakimu Zahabu and Rogers Ernest Malimbwi
Sokoine University of Agriculture, Department of Forest Mensuration and Management, P.O. Box 3013, Morogoro, Tanzania

Ole Martin Bollandsås and Tron Haakon Eid
Norwegian University of Life Sciences, Department of Ecology and Natural Resource Management, P.O. Box 5003, 1432 As, Norway

T. A. Babalola, A. O. Boyo and R. O. Kesinro
Department of Physics, Lagos State University, Ojo, Lagos State, Nigeria

Dagninet Amare, Asmamaw Endeblhatu and Awole Muhabaw
Bahir Dar Agricultural Mechanization and Food Science Research Centre, Bahir Dar, Ethiopia

Mohammad Ismail Hossain and Wayesh Qarony
American International University-Bangladesh, Kemal Ataturk Avenue Banani, Dhaka, Bangladesh

Hsien-Tsung Lee
Department of Electrical and Information Technology, NanKaiUniversity of Technology, Nan Tou County, Taiwan

Novosad Valentyna
Science Company "MAE", Kyiv, Ukraine

Kolosova Viktoria
Department of Debt and International Financial Policy of the Ministry of Finance, Kyiv, Ukraine

Ahmed-Bilal Awan and Taher Shaftichi
Electrical Engineering Department, College of Engineering, Majmaah University, Majmaah, Saudi Arabia

Ahmed G. Abu-Khalil
Electrical Engineering Department, College of Engineering, Majmaah University, Majmaah, Saudi Arabia
Electrical Engineering Department, Faculty of Engineering, Assuit University, Assiut, Egypt

Getachew Yirga
Department of Physics, Haramaya University, Dire Dawa, Ethiopia

Sisay Tadesse
Department of Chemistry, Hawassa University, Hawasa, Ethiopia

Teketel Yohannes
Materials Science Program, Addis Ababa University, Addis Ababa, Ethiopia

Abreham Berta and Belay Zerga
Department of Natural Resource Management, Wolkite University College of Agriculture and Natural Resource, Wolkite, Ethiopia

Index

A

Absorber Plate, 85-87, 92
Albizia Lebbeck, 62
Anionic Liquid Soap, 1-2
Astm, 2, 7, 19, 22, 24, 76, 133, 151
Atomic H/c Ratio, 5-10, 14-15, 17-19, 152

B

Battery Energy Storage, 34-35
Binders, 75, 77, 79-80
Biofuel, 24, 26, 79, 175-180, 183-185
Brackish Water, 115-118
Branches and Stem Volume, 104
Burial Metamorphism, 5

C

Calorific Value, 48, 51-53
Carbon Dioxide Emissions, 56-57, 176
Carica Papaya, 39-43
Charcoal, 4, 75-80, 119, 121-124, 176, 180
Coal, 1-10, 14-20, 56, 58-60, 133, 151-153, 176, 181-182, 184
Coal Briquettes, 7
Coal Liquefaction, 1
Coal Slurries Preparation, 2
Coal Water Fuel (CWF), 2
Coal-water Slurries, 1-4
Colocynthus Vulgaris, 21-24
Comparative Analysis, 21
Cyperus Esculentus, 21-24

D

Destructive Sampling, 104-107, 110, 112, 114
Dinitrosalicylic Acid (DNSA), 27
Distillation, 76, 115-116
Dsscs, 39-40, 42, 165, 169-173
Dye Extracts, 39

E

Ethylbiodiesels, 21
Eucalyptus Wood, 75, 80

F

Fischer-tropsch Process, 1
Fluorescence Spectroscopy, 6
Fluorine Doped Tin Oxide (FTO), 40

Foraging Canopy, 61, 67-69
Foraging Method, 61-63, 68-70, 72-73
Foraging Substrate, 61, 64, 67-70
Form Factor, 104-106, 108, 111-112

G

Gc-ms, 21-22
Greenhouse Gases, 54-55, 59, 158-159
Guild, 61-62, 64, 66, 69-74

H

Hibiscus Sabdariffa, 39-40, 42-43, 165, 167, 170-173
Hplc, 28, 31
Hydrocarbon Atomic Ratio (H/C), 11-13
Hydrogen Index, 11, 13-14, 17-18, 132, 134, 151, 153

I

Inca Edx Analyzer, 41
Inorganic Constituents, 1, 44
Interface Height, 3

L

Lateral Variability, 44-45, 47
Leitz Mpv Compact Microscope, 7

M

Maceral Analysis, 5, 7, 9, 14-15, 133
Maximum Power Point Tracking (MPPT), 34, 97
Mercury Emissions, 55-56, 60
Mosfet, 36
Mppt Method, 34
Mushan Formation, 132

N

Nanchuang Formation, 7, 15, 20, 132
Natural Dyes, 39-40, 43, 165, 167, 169, 172-174
Natural Gas Market, 81-84, 93
Natural Gas Pricing, 81-82, 93, 95
Natural Resources Authority (NRA), 44
Ncerd, 4
Niche Overlap, 61, 63, 67, 69, 73
Nigerian Coal, 1

O

Oil Shale (OS), 44
Optical Microscope, 7
Optimization, 26-29, 31-33, 48-49, 52, 54, 98, 127, 130

P

Photovoltaic, 34-35, 38-39, 59-60, 97-100, 102-103, 164, 172-173

Photovoltaic (PV), 34

Plackett-burman Design, 28-29

Plackett-burman Experiment, 26, 28-30

Pseudoalteromonas Piscicida, 26, 28, 31-33

Pv Optimal Control, 36

R

Renewable Energy, 32, 54-55, 57-60, 75, 79, 115, 118, 158-159, 164, 177, 185

Resource Partitioning, 61, 74

Rheological Behaviour, 2

Rock-eval Pyrolysis Analysis, 6, 8, 16, 19

Rock-eval Pyrolysis Tmax, 5

S

Saccharification Process, 26-29, 32

Scanning Electron Microscopy (SEM), 27

Sedimentary Material, 6, 16, 19

Sem Image, 41

Sensitization, 39, 42-43, 169, 171-172

Sesamum Indicum, 21-24

Settling Rate, 1, 3

Shiti Formation, 7, 15

Solar Energy, 24, 39-40, 43, 55, 57, 59, 85-88, 92, 98, 100, 117-118, 131, 158-159, 163-164

Storage Solar Collector, 85-88, 92

T

Tmax, 5-10, 12, 14-19, 132-145, 147-151, 153

Toc, 5-10, 12, 14-15, 17-19, 132-137, 140-150, 152

Toshka Project, 97-99, 103

Total Organic Carbon (TOC), 14

Transesterification, 21, 48-49, 182

Triglyceride, 49

Tropical Mixed Dry Deciduous Forest, 61-70, 72

U

Ulva Lactuca, 26-27, 31, 33

V

Variability, 44-47

Viscosity, 1-4, 21-22, 24-25, 180

Vitrinite Reflectance, 5-8, 10-11, 15-20, 132-134, 137, 144, 151, 153

W

Water Pumping, 97-99, 103

www.ingramcontent.com/pod-product-compliance
Lightning Source LLC
Chambersburg PA
CBHW050456200326
41458CB00014B/5196